Accuracy and Quality Control of Remote Sensing Data

Accuracy and Quality Control of Remote Sensing Data

Guest Editors

Mohammadmehdi Saberioon

Daniel Spengler

MDPI

Basel • Beijing • Wuhan • Barcelona • Belgrade • Novi Sad • Cluj • Manchester

Guest Editors
Mohammadmehdi Saberioon
GFZ Helmholtz Center
for Geosciences
Potsdam
Germany

Daniel Spengler
Tech Department
constellr GmbH
Potsdam
Germany

Editorial Office
MDPI AG
Grosspeteranlage 5
4052 Basel, Switzerland

This is a reprint of the Special Issue, published open access by the journal *Remote Sensing* (ISSN 2072-4292), freely accessible at: https://www.mdpi.com/journal/remotesensing/special_issues/remote_sensing_data.

For citation purposes, cite each article independently as indicated on the article page online and as indicated below:

Lastname, A.A.; Lastname, B.B. Article Title. *Journal Name* **Year**, *Volume Number*, Page Range.

ISBN 978-3-7258-6860-5 (Hbk)
ISBN 978-3-7258-6861-2 (PDF)
https://doi.org/10.3390/books978-3-7258-6861-2

Contents

About the Editors

Mohammadmehdi Saberioon

Mohammadmehdi Saberioon is a researcher at the GFZ Helmholtz Centre for Geosciences. His research focuses on remote sensing, soil spectroscopy, water quality monitoring, and AI-driven environmental assessment. He specializes in multi-sensor data integration, combining in situ measurements, UAV observations, satellite hyperspectral imagery, and ground-based radiometric systems to improve soil health evaluation, biodiversity monitoring, and sustainable agricultural management. He also has extensive experience in calibration and validation (Cal/Val) of hyperspectral and multi-sensor Earth observation data, ensuring data consistency, traceability, and cross-platform comparability. His work integrates machine learning and environmental modelling approaches to enhance soil, crop, and water-quality assessment across spatial scales.

Daniel Spengler

Daniel Spengler is CTO of the constellr GmbH, operating a constellation of high-resolution thermal satellites. His scientific interest is in integrating multi-source earth observation data into operation use leveraging advanced methods of machine learning and AI. Dr. Spengler oversees the full end-to-end responsibility from sensors design of next-generation constellr satellites to satellite operations, data processing, pipeline integration, and cloud environment.

Strong focus is given to Cal/Val to ensure data quality. Dr. Spengler received his PhD at the TU Berlin in the field of hyperspectral remote sensing and was senior scientist at GFZ Helmholtz Centre for Geosciences for more than 15 years before joining constellr.

Preface

Remote sensing research focuses heavily on calibration and validation (CAL/VAL) to ensure instrument reliability. Studies include absolute radiometric calibration of GF-7 and SDGSAT-1, and SkySat data quality assessment (SNR/MTF). The methodologies introduced are the RCRII model for vicarious calibration that corrects the adjacency effect, a Quadrifocal Tensor SFM technique for HOFS sensors, MaxCor gap-filling for ETa, and high-accuracy Solar Diffuser BRDF measurement using the reciprocity theorem. Validation efforts cover Permafrost Active Layer estimates and HRL IMD accuracy via within-pixel sampling.

Mohammadmehdi Saberioon and Daniel Spengler
Guest Editors

remote sensing MDPI

Article

Validation of Permafrost Active Layer Estimates from Airborne SAR Observations

Andrew D. Parsekian [1,2,*], Richard H. Chen [3], Roger J. Michaelides [4], Taylor D. Sullivan [1], Leah K. Clayton [5], Lingcao Huang [6], Yuhuan Zhao [7], Elizabeth Wig [8], Mahta Moghaddam [7], Howard Zebker [8] and Kevin Schaefer [6]

1 Department of Geology & Geophysics, University of Wyoming, Laramie, WY 82071, USA; tsulli12@uwyo.edu
2 Department of Civil & Architectural Engineering, University of Wyoming, Laramie, WY 82071, USA
3 Jet Propulsion Laboratory, California Institute of Technology, Pasadena, CA 91109, USA; richard.h.chen@jpl.nasa.gov
4 Department of Geophysics, Colorado School of Mines, Golden, CO 80401, USA; michaelides@mines.edu
5 Department of Earth and Planetary Sciences, Yale University, New Haven, CT 06511, USA; leah.clayton@yale.edu
6 National Snow and Ice Data Center, Cooperative Institute for Research in Environmental Sciences, University of Colorado at Boulder, Boulder, CO 80309, USA; lingcao.huang@colorado.edu (L.H.); kevin.schaefer@colorado.edu (K.S.)
7 Viterbi School of Engineering, University of Southern California, Los Angeles, CA 90089, USA; yuhuanzh@usc.edu (Y.Z.); mahta@usc.edu (M.M.)
8 Department of Geophysics, Stanford University, Stanford, CA 94305, USA; ewig@stanford.edu (E.W.); zebker@stanford.edu (H.Z.)
* Correspondence: aparseki@uwyo.edu

Abstract: In permafrost regions, active layer thickness (ALT) observations measure the effects of climate change and predict hydrologic and elemental cycling. Often, ALT is measured through direct ground-based measurements. Recently, synthetic aperture radar (SAR) measurements from airborne platforms have emerged as a method for observing seasonal thaw subsidence, soil moisture, and ALT in permafrost regions. This study validates airborne SAR-derived ALT estimates in three regions of Alaska, USA using calibrated ground penetrating radar (GPR) geophysical data. The remotely sensed ALT estimates matched the field observations within uncertainty for 79% of locations. The average uncertainty for the GPR-derived ALT validation dataset was 0.14 m while the average uncertainty for the SAR-derived ALT in pixels coincident with GPR data was 0.19 m. In the region near Utqiaġvik, the remotely sensed ALT appeared slightly larger than field observations while in the Yukon-Kuskokwim Delta region, the remotely sensed ALT appeared slightly smaller than field observations. In the northern foothills of the Brooks Range, near Toolik Lake, there was minimal bias between the field data and remotely sensed estimates. These findings suggest that airborne SAR-derived ALT estimates compare well with in situ probing and GPR, making SAR an effective tool to monitor permafrost measurements.

Keywords: active layer; permafrost; ground penetrating radar; synthetic aperture radar

1. Introduction

Warming of the Arctic is leading to intensification of hydrologic cycles, changes to vegetation, increased river discharge, and elevated permafrost temperatures [1]. The permafrost active layer—the near-surface portion of the soil column that thaws annually in the summer—is at the nexus of change in the terrestrial Arctic system because it is a key zone for lateral groundwater flow, hosts ecological communities, and serves as the upper boundary of the permafrost [2]. Multi-scale quantification of the maximum depth that suprapermafrost soil thaws annually—the active layer thickness (ALT), primarily controlled by the maximum temperature achieved at a given location during the summer—serves as a robust indicator of climate change impacts on the Arctic [3], at many scales across Arctic landscapes. Furthermore, the high spatial heterogeneity of ALT [4] is a motivation to map ALT at high spatial resolution. This has inspired the development of a variety of tools, based on remote sensing datasets or statistical relationships, to map ALT.

The Circumpolar Active Layer Monitoring (CALM) network has acquired ALT data since the 1990s at over 200 sites in both hemispheres on 1 km^2 or 1 ha grids using direct manual push probe measurements [5]. However, the locations of CALM sites are biased towards accessible places, and there is a need to estimate ALT across the vast remote expanses of the Arctic. Other efforts have focused on making more spatially extensive maps of ALT. Statistically and physically driven models have been used to produce catchment and regional ALT maps [6,7]. Others have used statistical correlations in conjunction with regional scale measurements to map ALT [8]. Recent approaches based entirely on remote sensing measurements have shown that satellite Interferometric Synthetic Aperture Radar (InSAR) data can be used to generate maps of ALT based on the seasonal subsidence of the land surface due to freeze-thaw cycles in the active layer [9]. Alternative SAR-based methods have used backscatter signals to estimate ALT [10,11].

Recently, the possibility has arisen of using aircraft-mounted SAR systems to acquire datasets for ALT estimation similar to what had previously been observed using satellites [12]. Although airborne SAR deployments have the disadvantage that they are not placed into Earth orbit, whereas satellites in orbit for long durations can repeat temporal measurements up to every 12 days in the Arctic [13]. On the other hand, aircraft deployments have the positive characteristic that they can be tasked to meet specific time-over-target objectives, and aircraft-mounted sensors can often achieve finer ground resolution than spaceborne SAR platforms. Furthermore, aircraft missions are far less costly than satellite missions and indeed may be more cost efficient on a per-square-meter basis than field surveys for measuring ALT on the catchment- or regional-scales. In 2017, as a part of NASA's Arctic-Boreal Vulnerability Experiment (ABoVE) project, an airborne SAR dataset was acquired using the L-band Uninhabited Aerial Vehicle Synthetic Aperture Radar (UAVSAR) over Alaska and Western Canada with measurements collected in spring (April to June) and fall (September to November). The objective of ABoVE was to reveal environmental changes over large scales in the Arctic and boreal regions of North America. The airborne campaign included collection of UAVSAR data and P-band Airborne Microwave Observatory of Subcanopy and Subsurface (AirMOSS) polarimetric synthetic aperture radar (PolSAR) on 66 total flight lines covering >4 million km^2 [14].

Our purpose is to validate airborne SAR-derived ALT measurements using calibrated field geophysical data. We also present a limited comparison between remotely sensed soil volumetric water content (VWC) estimates and VWC measured in the field. Specifically, our objective is to show quantitative statistical validation of these remotely sensed products in three characteristically different permafrost regions of Alaska. This validation demonstrates the ability of the remotely sensed ALT measurements to match field-based observations within uncertainty, enabling end users to have confidence in the utility of the airborne SAR ALT product. We use probe-calibrated ground penetrating radar (GPR) data as a well-established ground truth measurement of thaw depth that can efficiently acquire data on km-scales needed to validate the remote sensing results. While the maximum seasonal thaw, i.e., ALT, is not achieved until September or October each year, there is only a slight additional downward advancement of the thaw front between late August and October [11,15], and therefore, we treat GPR- and probe-measured thaw depth as a proxy for ALT.

2. Materials and Methods

The Permafrost Dynamics Observatory (PDO) data product [16] estimates seasonal subsidence, ALT, soil moisture, and uncertainties at 30-m resolution for 66 airborne flight lines across the ABoVE domain [17] in Alaska and northwest Canada. Throughout the rest of the text, we refer to the SAR-derived ALT product as "the PDO product". The PDO retrieval uses L-band Synthetic Aperture Radar (SAR) data acquired by UAVSAR and

P-band SAR backscatter acquired by AirMOSS. As part of the ABoVE airborne campaign, NASA flew all 66 lines in June and again in August 2017 [14]. The PDO product estimates seasonal subsidence due to thawing of the active layer using InSAR of the two L-band images acquired in June and August 2017. The PDO product estimates the vertical profile of soil volumetric water content (VWC) from the seasonal subsidence and the P-band backscatter from the August flights.

We focus on three SAR swaths in Alaska across a latitudinal gradient for this validation: Utqiaġvik/Barrow (BRW), Toolik (TOO), and the Yukon–Kuskokwim Delta (YKD) (Figure 1). The abbreviations indicated in parentheses are consistent with naming schemes used in the SAR dataset (i.e., barrow, toolik, and ykdelt) [16]. The BRW swath includes a lowland coastal plain underlain by continuous permafrost and has a mean annual air temperature (MAAT) of −11 °C and <200 mm of precipitation. The TOO swath includes the rolling topography of the foothills of the Brooks Range underlain by continuous permafrost and a MAAT of −7 °C and <400 mm precipitation. The YKD swath includes delta plain lowlands underlain by discontinuous permafrost and a MAAT of −1 °C and <480 mm of precipitation. The surface environmental conditions at all swaths are similar, being dominated by moss, lichen, and forbs, and sporadic shrubs adjacent to surface water. The mechanical probing and GPR validation datasets described below for BRW and TOO are described by [18,19].

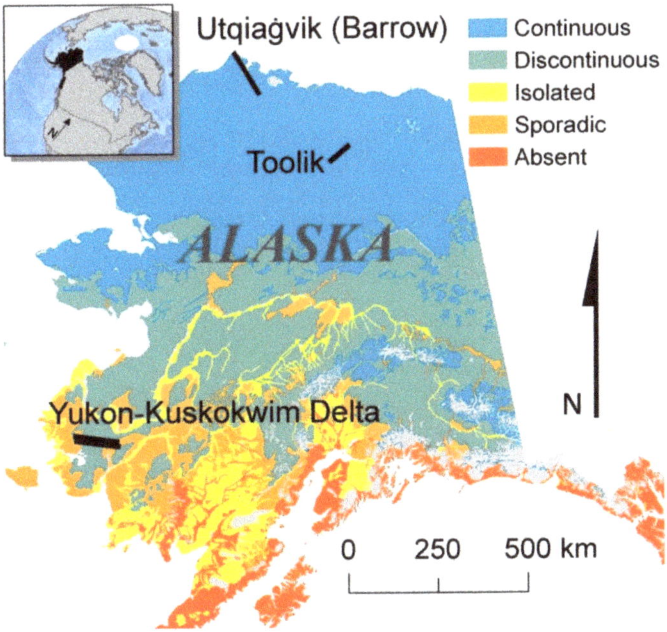

Figure 1. Airborne SAR swaths (black patches) and the permafrost classification across Alaska, USA [20].

ALT was measured using GPR calibrated to mechanical thaw probe observations. Thaw probing was done using a 1.5 m long, graduated steel rod that was inserted vertically into the active layer until refusal at the ice-bonded permafrost table, following the CALM protocol. The operator judged if the contact was permafrost or rock based on the feeling and sound of the impact. Depending on the site, thaw probe measurements were made sporadically (e.g., >100 m between probe locations) along the GPR transect to capture the large-scale spatial variability in ALT and soil characteristics; however, in some cases, high-density transects of probing at 1 m intervals were also measured to capture small-scale variability. Repeatability—the measurement of mechanical probing uncertainty—was measured occasionally by measuring in triplicate within a 0.3 m diameter circle.

All GPR data used for validation were measured using the instrumentation, settings, and protocol described by [19] with a Malå ProEx (GuidelineGeo, Stockholm, Sweden). Data were processed using ReflexW (Sandmeier Geophysical Research, Karlsruhe, Germany). The GPR transmits a radio-frequency electromagnetic (EM) pulse at a 500 MHz center frequency that propagates downward into the ground. At the permafrost table, where there is a contrast in dielectric permittivity between frozen and unfrozen materials, the EM wave is reflected back towards the instrument, and the total travel time of the wave is recorded as a waveform or 'trace'. The resulting radargram images composed of many adjacent traces are processed to remove low frequency noise and enhance late-time arrivals before being manually interpreted to extract—or 'pick'—the GPS-tagged reflection arrivals. The GPR data was acquired in 'tracks' of semi-continuous measurements automatically triggered every 0.3 s at walking speed resulting in a total of 1.9×10^5 data points across all swaths (Figure 2). The approximate spatial measurement footprint is 0.3 m^2, based on the material properties and distance to the permafrost table. If the velocity is known, travel time can be converted to ALT. Probe measurements of ALT co-located with GPR observations of travel time allow for calculation of velocity. We calculated a velocity for each co-located probe and GPR measurement within a swath and then used the average velocity for that swath to perform the time-depth-conversion for all other points in that swath. Uncertainty on ALT observations derived from GPR was estimated using the standard deviation of GPR velocity (σ_v) for each swath, i.e., supposing σ_v = 0.006 m ns^{-1}, that would correspond to ALT uncertainty of 0.065 m for a 0.5 m ALT. GPR data were measured 10–15 August 2013 for the BRW swath, 11–14 August 2014 for the TOO swath, and 13–16 August 2016 for the YKD swath. At locations where collocated GPR and thaw probe measurements were available, the calculated velocity was converted to depth-integrated VWC using an empirical equation calibrated for permafrost soils in AK [21]. We employed the depth-integrated VWC because the GPR is sensitive to the total water content throughout the active layer depth profile, in comparison to conventional soil moisture probes, e.g., time-domain reflectometry (TDR), that is limited to measurement over the length of the probe's waveguides (10 to 20 cm).

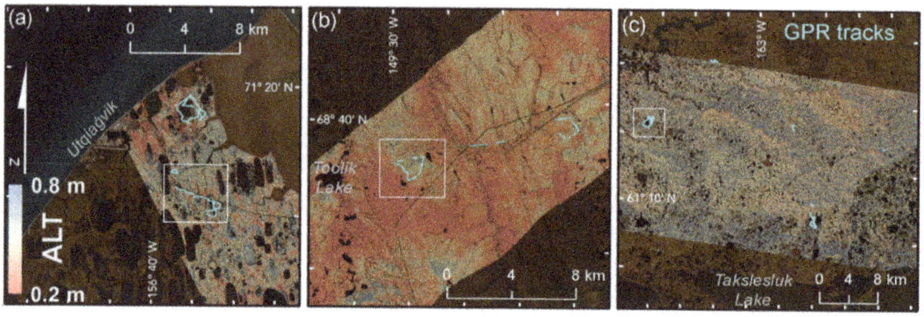

Figure 2. Ground penetrating radar (GPR) tracks displayed over the PDO product for the (**a**) BRW, (**b**) TOO, and (**c**) YKD swaths. The locations of these swaths are indicated in Figure 1. White boxes indicate locations shown in Figure 9. Latitude and longitude tick marks are in 5′ intervals. Dark patches within the terrestrial landscape are water bodies.

Airborne SAR data were collected using UAVSAR and AirMOSS radars, with two flights acquired along each swath during 2017: one near the onset of thaw, and another towards the end of the thaw season. The subsidence estimated across each swath between the L-band UAVSAR acquisitions is used to estimate ALT based on the principle that the ground surface subsides when the water in the soil melts, resulting in a decrease in the pore water volume [9]. The P-band AirMOSS backscatter is sensitive to soil dielectric properties down to ~60 cm depth. The PDO product uses both L-band InSAR and P-band backscatter to estimate seasonal subsidence, ALT, and soil volumetric water content product at 30 m pixel resolution. By estimating the water content, and assuming a typical subsurface porosity profile, the required ALT to produce the measured subsidence is calculated in a joint inversion framework [12]. Given that the SAR acquisitions are not precisely at the beginning and end of the thaw period, an accumulated degree days of thaw (ADDT) correction is applied to extrapolate the total seasonal subsidence (and therefore ALT) from the measured subsidence, based on the ADDT experienced at each swath. Full details of the postprocessing and joint retrieval can be found in [12]. The ALT product is masked to eliminate pixels with heavy forest cover and InSAR coherence < 0.35 [12].

Given the approximate two orders of magnitude scale difference between the footprint of the GPR and SAR pixel size, we averaged the GPR data in each PDO pixel. We used the same 30 m grid that the SAR results are presented on [17] and calculated the mean, standard deviation, and count of the GPR data within each pixel. Uncertainty on the averaged pixels is estimated using Gaussian error propagation where the GPR measurement error described above is added in quadrature to the scaling error (standard deviation of GPR measurements in each pixel), and representation error accounting for the difference in time between the SAR acquisition and the fieldwork, estimated as 0.045 m based on the average observed interannual variability at all three swaths (https://www2.gwu.edu/~calm/, accessed on 25 June 2021). Given the tortuous path of the GPR track across the landscape, some pixels have hundreds of ALT measurements, while others may have fewer than ten (the median count was 74 GPR points per 30 m × 30 m pixel across all sites). To ensure a representative ALT value for each validation pixel, we rejected any pixels that had fewer than 30 ALT measurements, following the Central Limit Theorem. Once the GPR dataset is calibrated for local wave velocity, upscaled to a 30 m grid, and reduced to eliminate pixels with low data count, we refer to the final product as the 'ALT validation dataset'. As detailed in [18], we use the χ^2 statistic that accounts for observational uncertainty to compare the remotely sensed product to the validation dataset. χ^2 at each pixel is calculated as:

$$\chi_n^2 = \left(\frac{r_n}{\epsilon_{0,n}}\right)^2 \tag{1}$$

where r_n is the residual between the PDO product and ALT validation dataset at pixel n, and $\varepsilon_{0,n}$ is the uncertainty in the validation dataset at pixel n. Ideal matches occur if both estimated values are within the uncertainty of each other. This means the difference between the remotely sensed and GPR ALT is smaller than uncertainty, implying the two are statistically identical. Good matches occur if the estimated ground measured value is only within the uncertainty of the SAR measurement, a marginal match is when only the uncertainties overlap, and all others are classified as 'no match'. The overall χ^2 is calculated as:

$$\chi^2 = \frac{1}{N}\sum_{1}^{N}\chi_n^2 \tag{2}$$

where N is the total number of pixels where ALT is observed.

3. Results

3.1. Calibrated GPR Dataset

Here, we focus on the dataset measured within the YKD swath as a representative example. This dataset was referred to by [22]; however, the details are first reported here. The calibrated GPR datasets measured within the BRW and TOO swaths are detailed in [19,23], respectively, and therefore we direct the reader to those publications for comprehensive explanations of those datasets. The example processed radargram (Figure 3a) shows the undulating, semi-discontinuous reflection from the permafrost table. The GPR pulse reflects at any boundary with a dielectric contrast, such as the thawed-frozen boundary at the bottom of the active layer. This is a typical image where the reflection is clearly visible along most of the line, though there are intermittent segments where no obvious reflection is present (e.g., near trace 1450) perhaps due to poor coupling or variations in the dielectric permittivity contrast [19]. Above the permafrost table reflection, there are moderately continuous to discontinuous subparallel reflections. The earliest time reflections < 10 ns may be associated with an interface between peat and mineral soil. The more horizontal arrivals above the permafrost table may also be associated with instrument noise or antenna ringing. Figure 3b shows a comparison between the probe data and GPR data after calibration to the local average wave velocity, where probe uncertainties are calculated as the standard deviation of three replicate probe measurements at the same location. A perfect match between the probe and GPR data would fall exactly along a one-to-one line, and deviation of points from the one-to-one line is primarily a result of spatial variability in soil moisture that is not accounted for when using a site-wide velocity. The average site-wide velocities for all swaths are shown in Table 1.

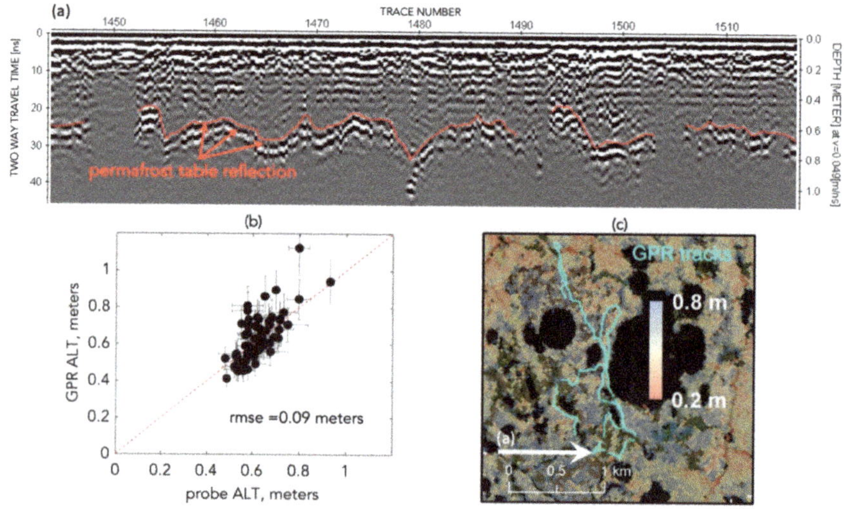

Figure 3. GPR results from within the YKD swath. (**a**) Processed radargram with time-depth-conversion applied using site-calibrated wave velocity. The interpreted reflection from the permafrost table is indicated in red. (**b**) Comparison between GPR-measured ALT and thaw probe data with one-to-one line. (**c**) Location where the radargram in (**a**) was measured (white arrow), also showing the SAR ALT product overlain on aerial imagery. The GPR track is shown in the southeast corner of Figure 2c.

Table 1. Average ($\bar{x}$) and standard deviation (σ) for the SAR and validation (VAL) datasets and for the GPR wave velocity, v.

	$\bar{x}_{ALT(SAR)}$	$\bar{x}_{ALT(VAL)}$	$\sigma_{ALT(SAR)}$	$\sigma_{ALT(VAL)}$	$\bar{x}_v$	σ_v
	Meters	**Meters**	**Meters**	**Meters**	**m ns^{-1}**	**m ns^{-1}**
all	0.50	0.50	0.14	0.15	0.046	0.006
BRW	0.48	0.38	0.14	0.07	0.042	0.006
TOO	0.43	0.47	0.09	0.09	0.048	0.005
YKD	0.56	0.67	0.13	0.08	0.049	0.006

3.2. Comparison of Validation Dataset to Airborne SAR

A direct comparison of the validation dataset with the PDO product (Figure 4a) shows BRW to have the thinnest active layer, TOO intermediate, and YKD as thickest. The TOO result falls approximately around the one-to-one line, while the BRW results indicate that the SAR product slightly overestimates ALT on average, while at YKD the SAR result slightly underestimates ALT on average. Despite these small biases, the error bars on the average SAR results do overlap the one-to-one line. Visualizing the same data as a 2D histogram (Figure 4b) illustrates that the relationship is somewhat linear when considering the more frequent occurrence of points close to the one-to-one line. The overall root mean squared error (RMSE) between the PDO product and validation dataset is 0.176 m, which translates to 20–70% ALT uncertainty.

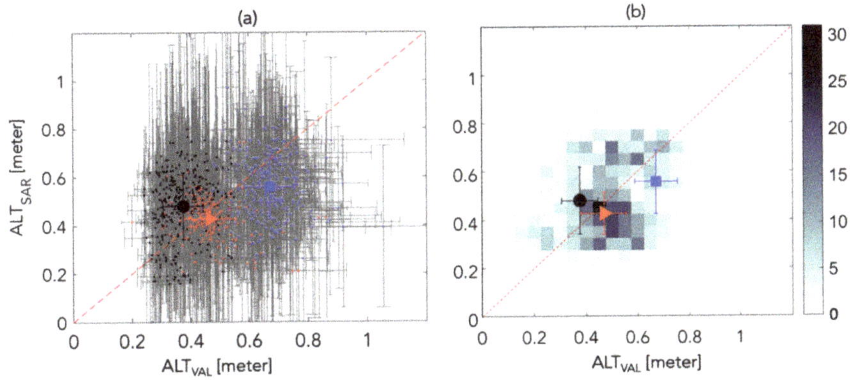

Figure 4. (**a**) ALT validation data compared against the collocated SAR result. Grey bars are individual estimated uncertainties (vertical error bars indicate posterior uncertainty for the PDO product; horizontal error bars indicate validation uncertainty from error propagation of the GPR errors, scaling errors, and representation errors); large colored markers are the average and standard deviations for each swath. (**b**) A 2D histogram, where grayscale indicates count, displaying the same data as in panel (**a**). Black, BRW; red, TOO; blue, YKD.

3.3. Validation Results Based on Airborne SAR Coherence

Including only the highest quality pixels in the validation reduces the RMSE (Figure 5). In Figure 4, we used all SAR pixels above the coherence threshold of 0.35. The magnitude of the correlation is referred to as the "coherence" [24]. As coherence increases, the uncertainty in the estimated ALT decreases, making it more difficult to reach the ideal match criteria in the χ^2 statistic. Nevertheless, the higher quality pixels with high coherence show reduced RMSE. Setting a higher coherence threshold reduces the number of pixels in the validation statistics. For example, if we set the coherence threshold to 0.65, the RMSE decreases to 0.09 m, 0.16 m, and 0.17 m for TOO, BRW, and YKD respectively, while reducing the number of usable pixels in the validation statistics to about 300. Depending on the application, a user of the PDO dataset could define their own coherence threshold to focus on the highest quality pixels with the lowest uncertainties, at the cost of fewer usable pixels per swath and less spatially continuous data coverage.

These results indicate that noise in the data influences the accuracy of the results, as opposed to a problem with the retrieval. Coherence loss is driven by noise in the interferogram, which in turn results from small differences in surface scattering between the two scenes. The pixels with the least amount of noise tend to converge towards the validation data, resulting in lower RMSE.

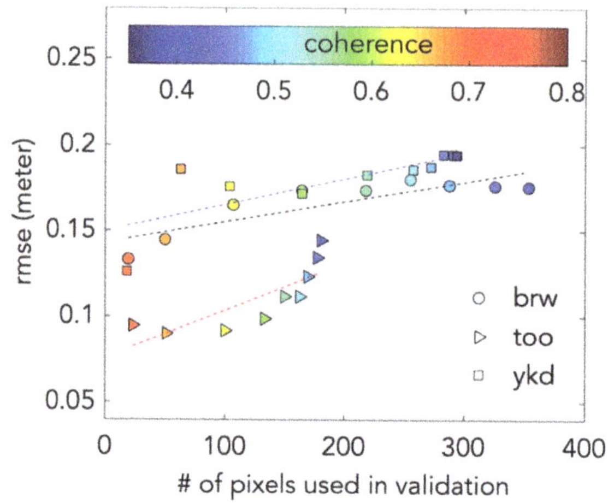

Figure 5. The relationship between the number of pixels used for validation and the resulting RMSE, showing that when SAR pixels are culled by coherence, a smaller RMSE can be achieved at the expense of a smaller statistical population.

3.4. Uncertainty Assessment

Here we address the uncertainty of each measurement method. For probing measurements, the only site where we have estimates of measurement repeatability is YKD. Given that these distributions appear to be skewed (Figure 6), particularly the SAR, we report the median rather than the mean. The probe measurements have the lowest uncertainty, likely because the primary sources of error are simply the operators' judgment of the ground surface and the ability to read the 0.01 m graduations marked on the probe. The scaled GPR data has around double the uncertainty of the probe data, due to both velocity uncertainty and scaling. SAR uncertainty is two to three times larger than GPR due to errors both in the measurement (as a result of having infrequent acquisitions from airborne platforms), as well as assumptions in the conversion from subsidence to ALT. Please refer to [12] for more details on SAR uncertainties.

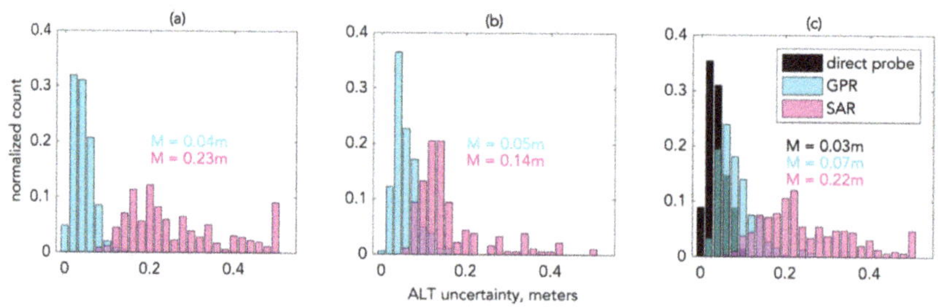

Figure 6. Frequency distributions of ALT uncertainty for the (**a**) BRW, (**b**) TOO, and (**c**) YKD swaths. Median values (M) are reported for each frequency distribution.

3.5. Evaluation of ALT Observations across Measurement Methods

Next, we illustrate a comparison of the ALT frequency distributions of each measurement for each swath. Histograms indicate the SAR and GPR data are visually similar, though statistically different (Figure 7). Although we do not expect the populations to have equivalent distributions or even to exhibit Gaussian shape, we nonetheless tested this possibility statistically using the nonparametric Kruskal–Wallis test [25]. This nonparametric one-way analysis of variance (ANOVA) test on ranks compares whether all four populations originate from the same distribution. Resulting p-values > 0.05 would indicate that they are not significantly different, however we found that the p-values were <0.05 in all cases: $p_{BRW} = <10^{-6}$, $p_{TOO} = 0.0025$, $p_{YKD} = <10^{-6}$. This mismatch is attributed to the contrasting sampling area of each measurement: ~0.01 m^2 for probing, ~0.3 m^2 for GPR and 900 m^2 for SAR that each have the capacity to detect variability in ALT on different spatial scales. At BRW (Figure 7a) there are notable shifts in the peaks of the histograms between the total GPR dataset and the 30 m scaled validation GPR dataset. Although this may seem counterintuitive given that both were derived from the same population, this difference arises because many of the GPR points were not included in the scaled validation product due to either failing to meet the 30 point-per-pixel threshold or because the GPR measurement was in a location that was masked out of the SAR swath due to low SAR coherence. In contrast, the frequency distributions for each measurement population with the TOO and YKD swaths are approximately coincident.

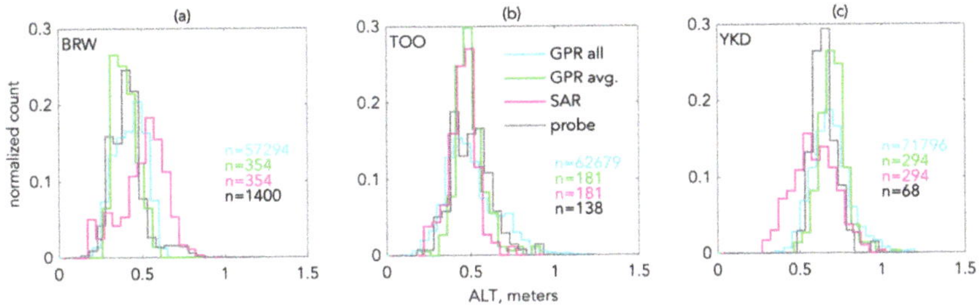

Figure 7. Histograms of ALT for the (**a**) BRW, (**b**) TOO, and (**c**) YKD swaths.

3.6. Volumetric Water Content Comparison

While our primary focus is the validation of ALT estimates, here we provide a basic comparison of VWC estimates from the joint retrieval against field measurements. We are restricted to a limited comparison for VWC because our field measurements only observe depth-integrated VWC when there are collocated probe and GPR measurements, and this only occurs sporadically every few hundred meters along each GPR track. Therefore, a single depth-integrated VWC observation that may not be representative of the local average conditions would be compared to a 30 m SAR pixel. Furthermore, at BRW where there are more probe data available, they are distributed across only 11 pixels, which is too few to make a statistical argument. The histograms at BRW (Figure 8a) are different from the SAR-estimated values substantially underestimating field measurements. Comparison of the VWC histograms at TOO and YKD (Figure 8b,c) reveals that the populations are similar with median differences of 0.03 $m^3\ m^{-3}$ and 0.08 $m^3\ m^{-3}$ respectively, suggesting a close match between SAR-derived VWC and field conditions.

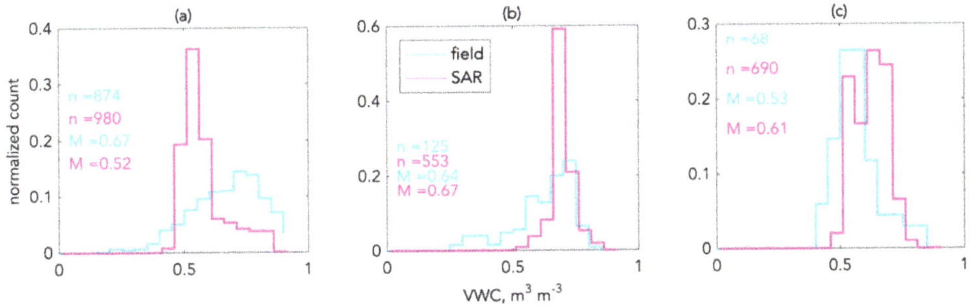

Figure 8. Frequency distributions of VWC for the (**a**) BRW, (**b**) TOO, and (**c**) YKD swaths. Population counts, n, indicate the total number of individual depth-integrated VWC observations and the number of pixels for the field and SAR datasets respectively. M indicates median.

3.7. χ^2 Classification Results

The χ^2 classification quantifies how well SAR pixels match the field data within uncertainty. These results are summarized in Table 2 for all observations and grouped by swath. The positive bias at BRW indicates SAR overestimated ALT, while the negative biases at the other swaths indicate SAR underestimated ALT. In total, 79% of all pixels were either an ideal match or a good match with the validation dataset—ideal matches being statistically identical. TOO had the highest percentage of good match or better (82%), while YKD had the smallest percentage in those categories (77%).

Table 2. Results of χ^2 classification. RMSE indicated root mean squared error, n_{SAR} is the number of SAR pixels available for validation.

	X^2	RMSE	bias	n_{SAR}	Ideal Match	Good Match	Marginal Match	No Match
	-	Meters	Meters	-	-	-	-	-
all	9.93	0.176	0.00	829	25%	54%	10%	11%
BRW	11.86	0.176	0.10	354	16%	64%	11%	9%
TOO	6.89	0.144	−0.04	181	43%	39%	9%	9%
YKD	9.49	0.194	−0.11	294	26%	52%	9%	14%

To illustrate how the χ^2 classified pixels are distributed across the landscape, we present pixels along representative GPR tracks from each swath in Figure 9a–c. The tracks do not appear spatially continuous for two reasons: (1) we rejected SAR pixels with less than 30 validation ALT measurements, and (2) SAR pixels with coherence less than 0.35 are not used in the validation. Residuals are plotted spatially in Figure 9d–f. There are no obvious spatial patterns in either the classifications or the residuals (i.e., spatial patterns χ^2 are approximately random), except that the residuals appear to be more positive in BRW and more negative in YKD, corresponding to the overall observed bias (Table 2). The approximate randomness of χ^2 spatial patterns suggests that cases of marginal match or no march likely result from random noise than a systematic bias or problems with the PDO retrieval.

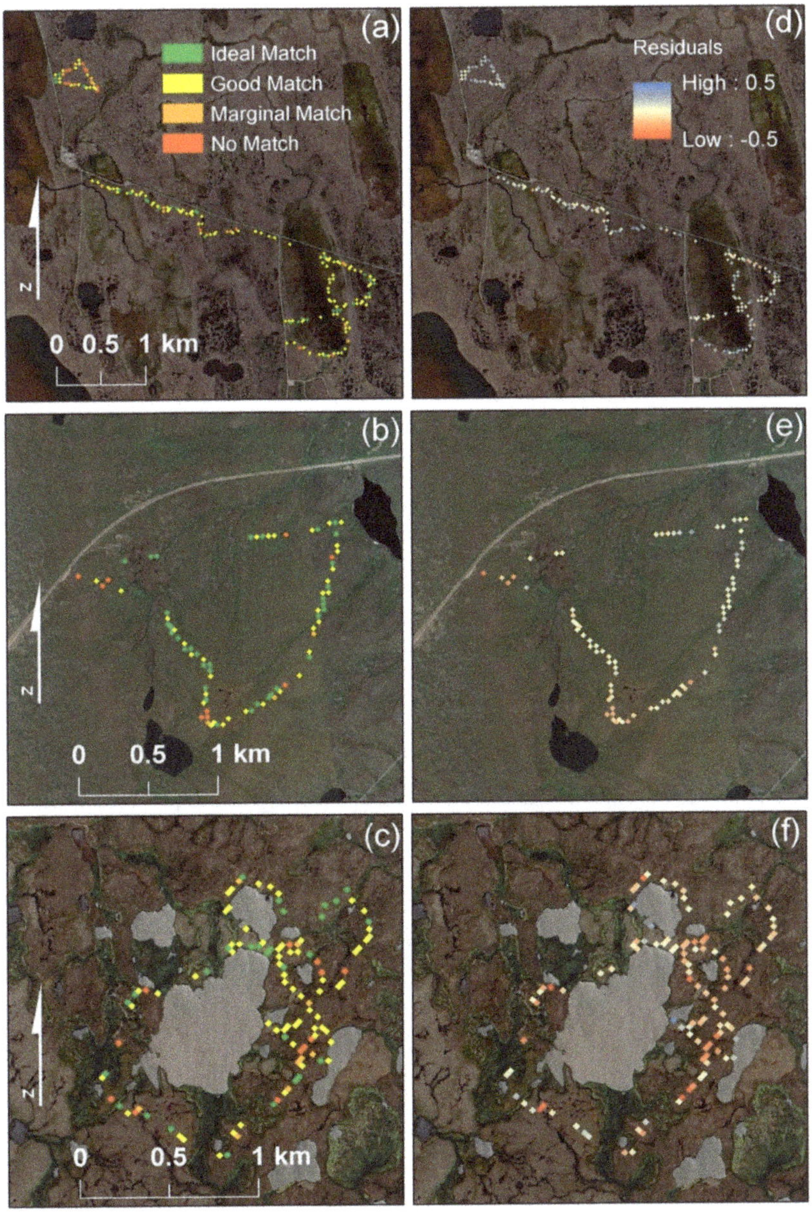

Figure 9. χ^2 classifications for (**a**) BRW, (**b**) TOO, and (**c**) YKD with corresponding residuals (**d–f**).

4. Discussion

4.1. Evaluation of Validation Data

ALT may vary substantially even over short distances on the scale of meters (e.g., Figure 3a). Using the minimum threshold of 30 GPR-measurements-per-pixel produces a statistically representative ALT observation for each pixel, corresponding to a data density of 3 measurements per 100 m^2. The median point density across all of our sites is 8 measurements per 100 m^2, and the maximum is 54 measurements per 100 m^2. For comparison, the data density at a 1 km^2 CALM grid is around 0.01 measurement per 100 m^2.

We found the median uncertainty on the direct manual thaw probe measurements in the YKD swath to be 0.03 m (Figure 7c) based on triplicate measurements within a 0.3 m radius. This is equivalent to the 0.03 m uncertainty reported within the BRW swath [18], and similar to the 0.04 m uncertainty reported for sandy arctic soils [10]. Given this consistency in mechanical probe uncertainty, we judge this level of uncertainty to be appropriate to apply to probing data for all of our study areas.

The direct comparison of probe measurements to coincident calibrated GPR observations (Figure 3b) highlights the effectiveness of GPR for noninvasively measuring an interface between thawed and frozen soil. Similar comparisons have been made [18,19] illustrating the overall linearity of the relationship, but also some expected spread of the data away from the one-to-one line due to spatial variation in VWC. While the variation in VWC (i.e., proportional to permittivity or wave velocity) is clear from Figure 3b, across the datasets from all three sites, there is relatively low variation in the mean wave velocity (Table 1). This may be useful information for other studies that cannot perform detailed calibrations.

Obtaining coincident measurements in time remains a substantial challenge, particularly when coordinating aircraft and ground teams. The aircraft may acquire data across the whole state of Alaska in the time it takes the field team to acquire validation measurements at a single site. There are related restrictions when attempting to coordinate the measurement timing with natural processes such as the maximum depth of thaw corresponding with ALT. As described, it was impossible to acquire contemporaneous field validation datasets, and indeed our field measurements were in some cases years different from the aircraft flights. Based on available timeseries data, we accounted for this representation error to the best of our ability, however, we nonetheless acknowledge that ALT experiences interannual variability that cannot entirely be accounted for in our approach, which may have resulted in some false positives and false negatives in our χ^2 classification.

While our field ALT measurements were made in August rather than the end of the thaw season in September or October, we assume these measurements approximate maximum thaw due to the deceleration of the thaw front late in the season. While we do not have detailed models or timeseries data exactly at our swaths, we consider previously presented modeling results [11,15] that modeled active layer thaw dynamics at 70°N latitude. The results of these studies indicated that between August 14 (approximately the time of measurement of our field data) and the onset of freezing, the permafrost table only advanced 0.01–0.03 m, which is within the uncertainty of both our probing and GPR data. Therefore, we judge the seasonal timing of our data collection to have a negligible effect on the validation results.

Although our VWC validation is limited due to the low data density of field measurements, we nonetheless highlight the importance of using depth-integrated VWC obtained from GPR observations. For typical soil moisture probes—e.g., TDR and similar dielectric approaches—the VWC is measured as an average along the length of the waveguides. While there is nothing physically incorrect about this, if a user is attempting to make minimally invasive observations (i.e., not digging a pit) by inserting the waveguides vertically into the ground surface this will only sample the top, partially saturated portion of the soil column. Such measurements are not representative of the soil water content throughout the active layer as demonstrated by the lack of correlation between depth-integrated VWC measurements and TDR probes [26]. In contrast, time-consuming soil pits disturb the tundra sufficiently to prevent future re-measurement, and soil pits likely influence the in situ soil moisture distribution. Furthermore, it is impractical to dig a large enough number of soil pits to achieve the spatial representativeness we seek herein.

4.2. Comparison of Large-Scale ALT Estimates

Two general approaches for making large scale maps of ALT are (1) remotely sensed observations of active layer properties and processes, or (2) discrete or indirect observations extended to large scales using spatial statistics. The first group is further divided into observations from either airborne or spaceborne sensor platforms. Given that satellites in orbit are continuously acquiring images at the same location on an approximately 12 day interval, ALT estimates from spaceborne SAR sensors have an advantage of long duration continuous deployment, meaning that ALT estimates may be derived from data measured multiple times per thaw season for several consecutive years [9]. In contrast, airborne platforms have the potential flexibility to target specific observation times and provide measurements at finer spatial resolution and possibly at multiple radar frequency bands.

Previously calculated satellite-based ALT estimates within the BRW swath were found to underestimate ALT by 0.02 m compared to field data [18], which is smaller than the 0.10 m overestimate in the PDO product we found using airborne measurements (Table 2) at similar pixel size. The difference in native spatial resolution (i.e., the intrinsic instrument resolution before spatial averaging) between airborne and spaceborne SAR could be responsible for this observed bias [12]. Furthermore, the spaceborne InSAR dataset was composed of many more SAR scenes than the airborne dataset [18]. The bias (0.10 m) observed in the airborne dataset is similar to the airborne-ALT uncertainty (+/−0.14 m), the validation dataset uncertainty (+/−0.07 m) (Table 1), and the spaceborne-ALT uncertainty (+/−0.19 m) [18]. The spaceborne-ALT results within the BRW swath were found to achieve a χ^2 category of either ideal match or good match at 74% of validation pixels, similar to the 80% of pixels in the same category for the airborne-ALT measurement (Table 2). Within the YKD swath, previously calculated satellite-ALT observations revealed that 66% of pixels were either ideal or good matches [22], compared to 78% in the same categories for the airborne-ALT observations (Table 2).

An example of a large-scale ALT estimate made using spatial statistics is available in the Yukon Flats region [8]. This area is closest to the TOO swath, although the Yukon Flats region is on the border between continuous and discontinuous permafrost and has different geologic substrate and landscape history, so we do not intend to draw a direct comparison between these two sites. Nonetheless, we observe that statistically predicted ALT around Ft. Yukon had a bias of approximately −0.09 m for a 30 m × 30 m pixel size [8], similar to the bias of −0.04 m we observed at the TOO swath (Table 2). A different statistical ALT estimate approach of the 26,000 km^2 Kuparuk River basin which partially includes the TOO swath found a bias of 0.02 m for estimates on a 300 m × 300 m pixel size, though the validation set was limited to 12 points scaled up from the 121-point CALM grids [6]. Even larger scale estimates of ALT have been attempted on the Russian Arctic drainage basin using climate inputs as drivers and assumptions about soil variables in a Stefan equation framework [7]; however, the extremely sparse direct observations make this scale of ALT product challenging to compare with our validation, though the modeled ALT is reported to be underestimated. SAR Backscatter-derived ALT estimates on the Yamal peninsula, Russia, achieved an RMSE of 0.2 m for ALT ranging from 0.8 to 1.4, or uncertainty of 14–25% [10].

4.3. Value and Limitations of Airborne SAR Estimates of ALT and VWC

There are several potential areas where airborne-SAR estimates of ALT may provide particular advantages compared to other large-scale ALT mapping methods. Perhaps most notable is the potential to retrieve subsurface VWC estimates concurrently with ALT due to the implications for developing a more complete understanding of hydrology and energy balance if both parameters are available [26]. Our present validation of VWC is limited, though it suggests promise for the accuracy of the VWC parameter (Figure 8). Currently, it is not possible to directly estimate VWC from spaceborne-ALT measurements for the whole active layer depth from spaceborne microwave instruments due to limited penetration depth, and existing spatial statistical models have not attempted to include this property directly. Another potential value of airborne-derived ALT estimates is the possibility of recovering finer-scale ALT variability. Although in this study we used 30 × 30 m pixels, this was a choice driven by the objective of using a standardized grid to enable different science datasets to be integrated and analyzed easily [17]. Using a different flight plan and a more frequent intra-seasonal measurement interval could allow for ALT and VWC retrieval at 10 m resolution.

One limitation of airborne-estimated ALT is that measurements can only be acquired when flights are tasked to do so. Therefore, SAR analysis may need to be conducted on fewer datasets than might be available from spaceborne platforms, resulting in the need to use spatial averaging and upscaling from the native resolution to achieve an acceptable signal-to-noise ratio in the SAR data. While 30 m pixel resolution may be acceptable when considering a swath that covers nearly 2500 km^2, it is also important to recognize that ALT varies substantially on the meter-scale (Figure 3a).

Although our comparison is limited to three swaths, the data (Figure 4) may suggest that airborne SAR may overestimate ALT when the true value is thinner (northern latitude) and underestimate ALT when the true value is thicker (southern latitude). While the three swaths detailed herein do represent a wide range of latitude, a detailed examination of more swaths within the latitude gradient would help to reveal if this bias is limited to BRW, TOO, and YKD or if it is systematic and linked to some aspect of the data acquisition or processing [12]. Additionally, based on the field VWC measurements (Figure 8), soil moisture is greatest in BRW and least at YKD, raising the possibility that limitations in the ability to retrieve ALT VWC may be partially responsible for the bias in Figure 4. The surface characteristics of all sites are similar, with low typical tundra vegetation and no trees, and therefore we anticipate this is not a key factor in the bias shown in Figure 4.

4.4. Future Research on Validating Remotely Sensed Active Layer Products

GPR-derived ALT datasets [19,23] have been successfully demonstrated as a field survey technique for validating SAR-estimated ALT products [18,22] due to the capability of GPR to acquire tens of thousands of ALT data points in large scale transects at acceptable uncertainty levels. Here we have further bolstered confidence in this approach and added important details related to scaling such as a minimum validation point density threshold and propagation of scaling uncertainty to the validation product. Illuminating the linkages and correspondence between SAR, GPR, and probe-measured ALT and study sites features is a key future research task. There are additional refinements that could be made to improve future validations.

First, it would be valuable to have more spatially extensive VWC field data to validate the VWC component of the joint retrieval. This is particularly challenging because each point where VWC is estimated requires 1–2 min of measurement and recording time at a minimum to complete the direct probing. So-called 'high-density' transects of 100 m total length and 1 m spacing between VWC measurements have been explored [23] with the objective of capturing some of the fine-scale spatial variability in ALT and VWC. However, in the best case, one high-density transect only could be used to validate up to three 30 m SAR pixels. Although this approach would approximately meet our criteria of validating SAR pixels only if there are >30 field data points within the pixels, the high-density surveys are a large time investment for a limited amount of validation. It would be useful to explore the spatial correlation length scales for VWC that may help to justify tolerance of <30 field data points per pixel. Furthermore, soil pits would be useful for characterizing the porosity profile [11] to allow for saturation calculations, and improved site-specific dielectric-VWC transforms may enable higher precision VWC estimates. To that end, we also emphasize the importance of acquiring depth-integrated VWC measurements, either in addition to or instead of conventional soil moisture probe (waveguide) measurements because depth-integrated VWC has been demonstrated to have the strongest correlation with active layer physical processes [27].

Another area of work would be the extension of the continuous multi-offset GPR approach for measuring ALT [27]. This simultaneously retrieves velocity and travel time, thereby eliminating the need for probe measurements beyond quality assurance/quality control. This multi-offset approach would have the additional advantage of resolving the challenge of spatially sporadic VWC measurements described above. Past efforts at this approach have been limited to simple, flat surface microtopography due to complications that arise with GPR antenna positioning on rough surfaces, and distinct subsurface layering (e.g., peat over mineral soil). It is possible that coupling a multi-layer GPR inversion [28] with the multi-offset acquisition scheme and using a ridged antenna sled that slides across the tundra vegetation may help to overcome these limitations.

4.5. Implications for Monitoring Thaw, Mapping ALT and Model Parameterizations

Monitoring ALT is valuable for understanding how permafrost landscapes are changing in response to climate warming—currently, this is primarily achieved through the network of CALM sites [4]. While this network of sites is extremely valuable for permafrost monitoring, the 1 km^2 manually-probed site scale cannot capture the dynamics of all landscape features. Therefore, the PDO product validated herein provides a useful baseline against which future observations may be compared. While it may not be possible to re-fly all swaths annually, a decadal resurvey of the measured swaths may reveal landscape-scale patchiness or other changes to ALT outside the resolution of the CALM grids.

Permafrost hydrology modeling relies on ALT because this is the primary zone for water dynamics—e.g., lateral flow—in continuous permafrost landscapes. Distributed hydrologic models use ALT as an input parameter that defines the depth to the impervious layer [29]. ALT is similarly important to permafrost carbon (C) modeling because this depth defines the boundary between bioavailable C and permafrost-sequestered C that may be released to the atmosphere under future warmer climate scenarios [30]. The validated PDO product is at an ideal scale for watershed hydrology or C cycling models—either as input parameters or to validate the results if ALT is calculated physically within the model [15,31].

5. Conclusions

We have demonstrated that 79% of the airborne-derived ALT values PDO product pixels are either an ideal match or good match χ^2 classes in comparison with the field validation dataset. Overall, the RMSE between the PDO product and validation dataset is 0.176 m, which equates to a deviation in ALT between the two datasets of 20–70%. Considering the χ^2 and RMSE results together, the airborne SAR-derived ALT products exhibit accuracies similar to previously-reported large-scale ALT estimation methods, and therefore we conclude that the airborne SAR-derived ALT products are successfully validated within uncertainty.

Author Contributions: Conceptualization, A.D.P. and K.S.; methodology, A.D.P., T.D.S., K.S., R.H.C., R.J.M.; formal analysis, A.D.P.; writing—original draft preparation, A.D.P.; writing—review and editing, K.S., R.H.C., R.J.M., T.D.S., L.K.C., L.H., Y.Z., E.W., M.M., H.Z.; project administration, K.S.; funding acquisition, K.S., M.M., H.Z., A.D.P. All authors have read and agreed to the published version of the manuscript.

Funding: This research was funded by NASA, grant numbers NNX13AM25G, NNX14A154G, NNX16AH36A, and NNX17AC59A.

Data Availability Statement: The data used in the manuscript are available at: https://doi.org/10.3334/ornldaac/1355, https://doi.org/10.3334/ORNLDAAC/1796, http://dx.doi.org/10.3334/ORNLDAAC/1265, and https://doi.org/10.3334/ORNLDAAC/1903 (all accessed on 25 June 2021).

Acknowledgments: We thank S. Schaefer, S. Ludwig, and S. Natali for assistance in the field acquiring data in the Yukon-Kuskokwim Delta. We thank the US Fish and Wildlife Service for permitting fieldwork in the Yukon-Kuskokwim Delta.

Conflicts of Interest: The authors declare no conflict of interest.

References

1. Box, J.; Colgan, W.T.; Christensen, T.R.; Schmidt, N.M.; Lund, M.; Parmentier, F.-J.W.; Brown, R.; Bhatt, U.S.; Euskirchen, E.S.; Romanovsky, V.E.; et al. Key indicators of Arctic climate change: 1971–2017. *Environ. Res. Lett.* **2019**, *14*, 045010. [CrossRef]
2. Hinzman, L.D.; Kane, D.L.; Gieck, R.E.; Everett, K.R. Hydrologic and thermal properties of the active layer in the Alas-kan Arctic. *Cold Reg. Sci. Technol.* **1991**, *19*, 95–110. [CrossRef]
3. Van Everdinger, R. *Multi-Language Glossary of Permafrost and Related Ground-Ice Terms (Revised 2005)*; National Snow and Ice Data Center/World Data Center for Glaciology: Boulder, CO, USA, 1998.
4. Hinkel, K.M. Spatial and temporal patterns of active layer thickness at Circumpolar Active Layer Monitoring (CALM) sites in northern Alaska, 1995–2000. *J. Geophys. Res. Space Phys.* **2003**, *108*. [CrossRef]
5. Abramov, A.; Davydov, S.; Ivashchenko, A.; Karelin, D.; Kholodov, A.; Kraev, G.; Lupachev, A.; Maslakov, A.; Ostroumov, V.; Rivkina, E.; et al. Two decades of active layer thickness monitoring in northeastern Asia. *Polar Geogr.* **2019**, 1–17. [CrossRef]
6. Nelson, F.E.; Shiklomanov, N.I.; Mueller, G.R.; Hinkel, K.M.; Walker, D.A.; Bockheim, J.G. Estimating Active-Layer Thickness over a Large Region: Kuparuk River Basin, Alaska, U.S.A. *Arct. Alp. Res.* **1997**, *29*, 367. [CrossRef]
7. Zhang, T.; Frauenfeld, O.W.; Serreze, M.C.; Etringer, A.; Oelke, C.; McCreight, J.L.; Barry, R.; A Gilichinsky, D.; Yang, D.; Ye, H.; et al. Spatial and temporal variability in active layer thickness over the Russian Arctic drainage basin. *J. Geophys. Res. Space Phys.* **2005**, *110*, 1–14. [CrossRef]
8. Pastick, N.J.; Jorgenson, M.T.; Wylie, B.K.; Minsley, B.J.; Ji, L.; Walvoord, M.A.; Smith, B.D.; Abraham, J.D.; Rose, J.R. Extending Airborne Electromagnetic Surveys for Regional Active Layer and Permafrost Mapping with Remote Sensing and Ancillary Data, Yukon Flats Ecoregion, Central Alaska. *Permafr. Periglac. Process.* **2013**, *24*, 184–199. [CrossRef]
9. Liu, L.; Schaefer, K.; Zhang, T.; Wahr, J. Estimating 1992–2000 average active layer thickness on the Alaskan North Slope from remotely sensed surface subsidence. *J. Geophys. Res. Space Phys.* **2012**, *117*. [CrossRef]
10. Widhalm, B.; Bartsch, A.; Leibman, M.; Khomutov, A. Active-layer thickness estimation from X-band SAR backscatter intensity. *Cryosphere* **2017**, *11*, 483–496. [CrossRef]
11. Chen, R.H.; Tabatabaeenejad, A.; Moghaddam, M. Retrieval of Permafrost Active Layer Properties Using Time-Series P-Band Radar Observations. *IEEE Trans. Geosci. Remote. Sens.* **2019**, *57*, 6037–6054. [CrossRef]
12. Michaelides, R.J.; Chen, R.H.; Zhao, Y.; Schaefer, K.; Parsekian, A.D.; Sullivan, T.; Moghaddam, M.; Zebker, H.A.; Liu, L.; Xu, X.; et al. Permafrost Dynamics Observatory (PDO)—Part I: Postprocessing and Calibration Methods of UAVSAR L-band InSAR Data for Seasonal Subsidence Estimation. *Earth Space Sci.* **2021**. [CrossRef]
13. Chen, J.; Günther, F.; Grosse, G.; Liu, L.; Lin, H. Sentinel-1 InSAR Measurements of Elevation Changes over Yedoma Uplands on Sobo-Sise Island, Lena Delta. *Remote Sens.* **2018**, *10*, 1152. [CrossRef]
14. Miller, C.E.; Griffith, P.C.; Goetz, S.J.; Hoy, E.E.; Pinto, N.; McCubbin, I.B.; Thorpe, A.K.; Hofton, M.M.; Hodkinson, D.J.; Hansen, C.; et al. An overview of ABoVE airborne campaign data acquisitions and science opportunities. *Environ. Res. Lett.* **2019**, *14*, 080201. [CrossRef]
15. Swenson, S.C.; Lawrence, D.; Lee, H. Improved simulation of the terrestrial hydrological cycle in permafrost regions by the Community Land Model. *J. Adv. Model. Earth Syst.* **2012**, *4*. [CrossRef]
16. Schaefer, K.; Michaelides, R.J.; Chen, R.H.; Sullivan, T.D.; Parsekian, A.D.; Zhao, Y.; Bakian-Dogaheh, K.; Tabatabaeenejad, A.; Moghaddam, M.; Chen, J.; et al. *ABoVE: Active Layer Thickness Derived from Airborne L-and P-band SAR, Alaska, 2017*; ORNL DAAC: Oak Ridge, TN, USA, 2021. [CrossRef]
17. Loboda, T.V.; Hoy, E.E.; Carroll, M.L. *ABoVE: Study Domain and Standard Reference Grids, Version 2*; ORNL DAAC: Oak Ridge, TN, USA, 2019. [CrossRef]
18. Schaefer, K.; Liu, L.; Parsekian, A.; Jafarov, E.; Chen, A.; Zhang, T.; Gusmeroli, A.; Panda, S.; Zebker, H.A.; Schaefer, T. Remotely Sensed Active Layer Thickness (ReSALT) at Barrow, Alaska Using Interferometric Synthetic Aperture Radar. *Remote Sens.* **2015**, *7*, 3735–3759. [CrossRef]
19. Chen, A.; Parsekian, A.D.; Schaefer, K.; Jafarov, E.; Panda, S.; Liu, L.; Zhang, T.; Zebker, H. Ground-penetrating ra-dar-derived measurements of active-layer thickness on the landscape scale with sparse calibration at Toolik and Happy Valley, Alaska. *Geophysics* **2016**, *81*, H9–H19. [CrossRef]
20. Jorgenson, M.T.; Yoshikawa, K.; Kanevskiy, M.; Shur, Y.; Romanovsky, V.; Marchenko, S.; Grosse, G.; Brown, J.; Jones, B. Permafrost characteristics of Alaska. In Proceedings of the Ninth International Conference on Permafrost, University of Alaska, Fairbanks, AK, USA, 29 June–3 July 2008; Volume 3, pp. 121–122.
21. Engstrom, R.; Hope, A.; Kwon, H.; Stow, D.; Zamolodchikov, D. Spatial distribution of near surface soil moisture and its relationship to microtopography in the Alaskan Arctic coastal plain. *Hydrol. Res.* **2005**, *36*, 219–234. [CrossRef]
22. Michaelides, R.J.; Schaefer, K.; Zebker, H.A.; Parsekian, A.; Liu, L.; Chen, J.; Natali, S.M.; Ludwig, S.; Schaefer, S.R. Inference of the impact of wildfire on permafrost and active layer thickness in a discontinuous permafrost region using the remotely sensed active layer thickness (ReSALT) algorithm. *Environ. Res. Lett.* **2018**, *14*, 035007. [CrossRef]

23. Jafarov, E.E.; Parsekian, A.D.; Schaefer, K.; Liu, L.; Chen, A.C.; Panda, S.K.; Zhang, T. Estimating active layer thickness and volumetric water content from ground penetrating radar measurements in Barrow, Alaska. *Geosci. Data J.* **2017**, *4*, 72–79. [CrossRef]
24. Rosen, P.A.; Hensley, S.; Joughin, I.R.; Li, F.K.; Madsen, S.N.; Rodriguez, E.; Goldstein, R.M. Synthetic aperture radar interferometry. *Proc. IEEE* **2000**, *88*, 333–380. [CrossRef]
25. Breslow, N. A generalized Kruskal-Wallis test for comparing K samples subject to unequal patterns of censorship. *Biometrika* **1970**, *57*, 579–594. [CrossRef]
26. Clayton, L.K.; Schaefer, K.; Battaglia, M.J.; Bourgeau-Chavez, L.; Chen, J.; Chen, R.H.; Chen, A.; Bakian-Dogaheh, K.; Grelik, S.; Jafarov, E.; et al. Active layer thickness as a function of soil water content. *Environ. Res. Lett.* **2021**, *16*, 055028. [CrossRef]
27. Gerhards, H.; Wollschläger, U.; Yu, Q.; Schiwek, P.; Pan, X.; Roth, K. Continuous and simultaneous measurement of reflector depth and average soil-water content with multichannel ground-penetrating radar. *Geophysics* **2008**, *73*, J15–J23. [CrossRef]
28. Parsekian, A.D. Inverse Methods to Improve Accuracy of Water Content Estimates from Multi-offset GPR. *J. Environ. Eng. Geophys.* **2018**, *23*, 349–361. [CrossRef]
29. Fabre, C.; Sauvage, S.; Tananaev, N.; Srinivasan, R.; Teisserenc, R.; Pérez, J.M.S. Using Modeling Tools to Better Understand Permafrost Hydrology. *Water* **2017**, *9*, 418. [CrossRef]
30. Chaefer, K.; Zhang, T.; Bruhwiler, L.; Barrett, A.P. Amount and timing of permafrost carbon release in response to climate warming. *Tellus B Chem. Phys. Meteorol.* **2011**, *63*, 168–180. [CrossRef]
31. Painter, S.L.; Coon, E.T.; Atchley, A.L.; Berndt, M.; Garimella, R.; Moulton, J.D.; Svyatskiy, D.; Wilson, C.J. Integrated surface/subsurface permafrost thermal hydrology: Model formulation and proof-of-concept simulations. *Water Resour. Res.* **2016**, *52*, 6062–6077. [CrossRef]

remote sensing

MDPI

Article

A Quick Band-to-Band Mis-Registration Detection Method for Sentinel-2 MSI Images

Tianxin Chen [1] and Yongxue Liu [1,2,*]

1 School of Geography and Ocean Science, Nanjing University, Nanjing 210046, China; chentianxin001@outlook.com
2 Jiangsu Center for Collaborative Innovation in Geographical Information Resource Development and Application, Nanjing 210046, China
* Correspondence: yongxue@nju.edu.cn

Abstract: A band-to-band mis-registration (BBMR) error often occurs in remote sensing (RS) images acquired by multi-spectral push broom spectrometers such as the Sentinel-2 Multi-spectral Instrument (MSI), leading to adverse impacts on the reliability of further RS applications. Although the systematic band-to-band registration conducted during the image production process corrects most BBMR errors, there are still quite a few images being observed with discernible BBMR. Thus, a quick BBMR detection method is needed to assess the quality of online RS products. We here propose a hybrid framework for detecting BBMR between the visible bands in MSI images. This framework comprises three main steps: first, candidate chips are captured based on Google Earth Engine (GEE) spatial analysis functions to shrink the valid areas inside image scenes as potential target chips. The redundant data pertaining to the local operation process are thus narrowed down. Second, spectral abnormal areas are precisely extracted from inside every single chip, excluding the influences of clouds and water surfaces. Finally, the abnormal areas are matched pixel by pixel between bands, and the best-fit coordinates are then determined to compare with tolerance. Here, the proposed method was applied to 71,493 scenes of MSI Level-1C images covering China and its surrounding areas on the GEE platform. From these images, 4356 chips from 442 scenes were detected with inter-band offsets among the visible bands. Further manual visual inspection revealed that the proposed method had an accuracy of 98.07% at the chip scale and 88.46% at the scene scale.

Keywords: band-to-band mis-registration (BBMR); spectral anomaly detection; multi-spectral instrument (MSI); Google Earth Engine (GEE); hybrid computation

1. Introduction

The band-to-band mis-registration (BBMR) of multi-spectral remote sensing (RS) images refers to the misalignment between bands caused by differences in the imaging time or angle between detection elements [1]. BBMR has an adverse impact on the reliability of RS applications [2,3], such as target recognition, land-cover classification, and change detection [4]. In particular, appreciably more false information is generated if BBMR occurs over areas with higher spectral complexity. For multi-spectral push broom spectrometers without a beam splitter, e.g., the Sentinel-2 Multi-spectral Instrument (MSI), different spectral band stripe filters are mounted on every single detector. This imposes a slight time offset between each spectral channel sensor during imaging [5], where long-track displacement is approximately 14 km for MSI images [6]. After systematic band-to-band registration (BBR), the inter-band geometric displacement of MSI images should be under 0.3 pixels at 99.7% confidence [7]. However, the pre-processing routine does not necessarily correct all of the displacements, as evidenced by the fact that images with visible inter-band displacement have been observed [8] and have been reported in the Sentinel-2 data quality reports [9]. MSI images have been widely used in a broad range of earth observation applications due to their high revisit frequency (five days at the equator with Sentinel-2A/B), spatial resolution (10 m for visible bands), and radiometric performance [10,11]. Therefore, a robust method is needed to identify MSI images with obvious BBMR errors.

The Sentinel-2 system is designed for high geometrical and spectral measuring performance with ongoing multi-spectral images: approximately 10,000 MSI images can be acquired daily with its twin-satellite capability, and as of July 2020, ~20 million products have been made available. With the support of Google Earth Engine (GEE)—a cloud platform that hosts petabyte-scale geospatial datasets—MSI images and their derivatives have played a crucial role in the global environment and security monitoring [12–15]. Some BBMR detection methods have been promoted targeting different types of platforms. Therefore, determining whether or not a single MSI image has obvious BBMR may not be especially difficult. However, determining the presence of BBMR in large numbers of MSI images from the Sentinel-2 archives still represents a great challenge because time efficiency and the elimination of redundant information both need to be considered due to the rapid expansion of the MSI dataset. With more than 800 embedded functions, the GEE platform offers an effective way to handle "big data." Its extensive spatial analysis functions provide an opportunity that may fulfill the lack of effective methods designed specifically for quickly detecting BBMR in a series of MSI images. Thus, the main aim of this study was to use the convenience of GEE to improve BBMR detection efficiency.

This study proposes a GEE-based algorithm to efficiently detect MSI images with mis-registration between the visible bands. The paper is organized as follows: First, we describe the spectral features of BBMR in MSI images and detail the proposed procedure in Section 2; we then present the experimental results in Section 3; and finally, we discuss the limitations and potential applications of the proposed method in Section 4.

2. Materials and Methods

2.1. Characteristics of BBMR in MSI Images

The MSI onboard Sentinel-2A/B satellites are designed for high geometrical and spectral measuring performance in a large swath with 13 spectral bands in the visible/near-infrared (VNIR) and short-wave infrared (SWIR) spectral ranges [16]. To cover the 290 km swath width, 12 MSI detectors are arranged in a staggered position on the focal planes, each with 13 spectral channel sensors, introducing along-track parallax both between detectors and between bands. In addition to the internal distortion of sensor detection elements, the relative offsets between bands can also be caused by platform jitters or terrain [17]. While it is difficult to determine the exact source of error, it is practical to detect and correct inter-band offsets based on each image's abnormal spectral features.

In MSI true-color images (TCI, R: Band 4; G: Band 3; B: Band 2, 10 m resolution), an inter-band displacement over one pixel can be discernable, creating a "rainbow" effect [18]. This usually comprises a bright-green stripe on one side of the geo-feature's edge and a magenta stripe on the other side (Figure 1c). The parallax can be useful in some cases; for example, the "rainbow" effect of objects moving at high speed caused by inter-band parallax can be used to capture small targets such as vehicles, ships, and planes in high-spatial-resolution images. It can also track the motion of moving clouds and water [19,20]. Meanwhile, the blurring of object boundaries may lead to geometric deviations in target recognition or land-cover misclassifications, which affect the reliability of RS applications. As shown in Figure 1c1–c3, the green stripes show relatively high values in the differences between Bands 3 and 4, whereas the magenta stripes appear to show relatively low values. This indicates the existence of an offset between Bands 3 and 4. The offset between Bands 2 and 3 can also be observed in the same way. However, the difference between Bands 2 and 4 only displays the features of geo-objects, rather than the "rainbow" stripe, which implies there are no apparent offsets between them. According to the analysis, the displacement of Band 3 is, therefore, the leading cause of the "rainbow" effect.

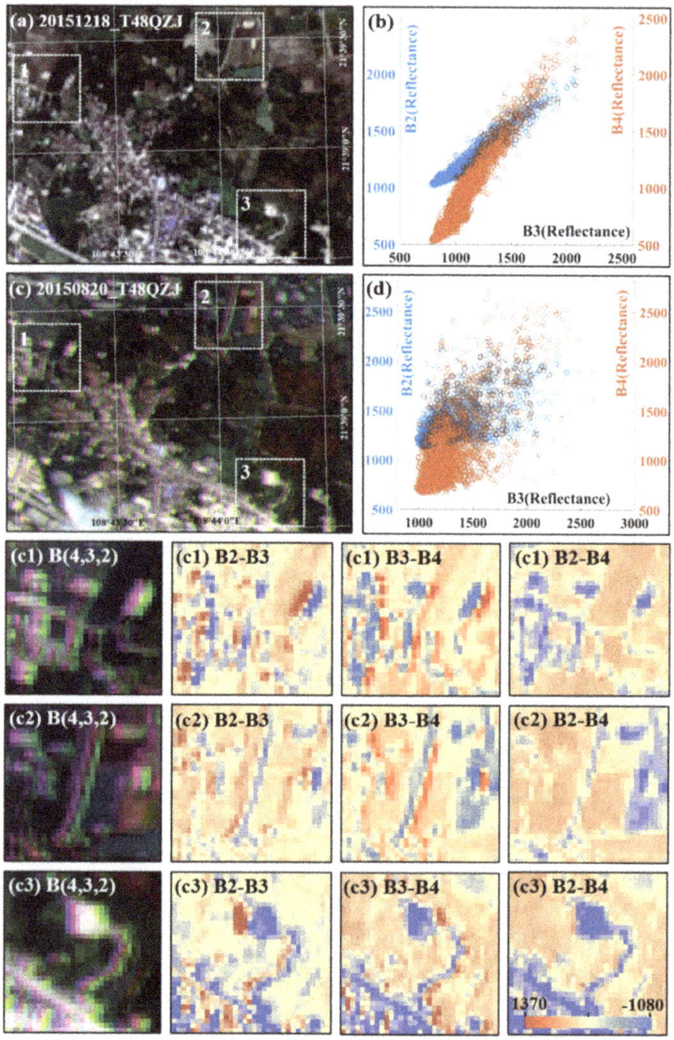

Figure 1. MSI images with and without discernible BBMR in the visible bands (Bands 2–4). (**a**) A subset of an MSI true-color image (48QZJ) acquired on 18 December 2015, without BBMR; (**c**) a subset of an MSI true-color image acquired on 20 August 2015, with BBMR in the same area as (**a**); and (**b**–**d**) the correlation between the reflectance of Bands 2–3 and Bands 4–3 inside three square image chips in (**a**–**c**). (Note that the reflectance values in (**b**,**d**) were scaled by 10,000.) (**c1**–**c3**) The true-color image and difference value between the visible bands of the three chips.

Comparing MSI images with and without BBMR over the same area, we found that the relationships among the visible bands differed significantly (Figure 1a–d). Figure 1b clearly shows a higher correlation among the visible bands' reflectance compared to that in the image without BBMR. This suggested that the inter-band offset could be captured by the bright-color feature (i.e., bright green: $\rho_{B3} - \rho_{B2} > 0.05$ and $\rho_{B3} - \rho_{B4} > 0.05$ in most images, ρ stands for the reflectance value) and could be determined by the high correlation between the visible bands. The following method was therefore proposed based on this core idea. Note that this study only focuses on the offset between the green band and the other two visible bands.

2.2. Inter-Band Mis-Registration Detection Method

There are two types of most commonly used methods for BBMR detection: sensor based and feature based. The former focuses on building a mathematical model between the interference factors during imaging and the inter-band offset, such as reproducing the imaging process in a laboratory to calibrate the interior of the camera and relative orientation parameters [17]. The latter analyzes the spectral features of images, mostly based on the cross-correlations between bands [21,22] and processed with phase correlation or the quantile matching method [23]. The detailed procedure varies depending on the method used to locate feature points. For example, image processing methods such as quad-tree decomposition [24] and singular value decomposition [25] have been used to calculate the BBMR inside an image chip. Meanwhile, other methods have been used to select the region of interest based on entropy [26] or to construct triangulation methods [27]. For the optimal results of band-to-band registration, height difference displacement analysis [28], cloud masks, and low-rank analysis [29] have been used to provide different levels of strategies for registration. The methods mentioned above mostly targeted charge-coupled devices (CCDs) and some sensors carried on an unmanned aerial system (UAS). Note that there are few BBMR detection methods for MSI images, and a known one is processed with phase correlation calculation on a sliding window basis for a whole image [18]. However, to be applied to a series of spatiotemporal multi-spectral images, the BBMR detection method requires a more efficient determination strategy and a larger storage space. Moreover, the spectral characteristics of surface objects in RS images cannot be effectively expressed over a large spatial scale (~110×110 km^2 for each scene). Therefore, filtering redundant data in the image and narrowing the size of the image participating in a single operation process are both priorities. We propose the following hybrid framework to detect the BBMR in MSI images, and it comprises three main steps (Figure 2): candidate chip detection based on GEE, spectral abnormal areas extraction, and feature matching. The first step shrinks the valid areas inside image scenes as potential targets to download, the second step precisely extracts pixels with spectral abnormal features inside every single chip, and the third step determines the best-fit coordinates in the accumulation cross-correlation matrix of all abnormal areas.

2.2.1. Candidate Chip Detection Based on GEE

The computation priority on GEE is distributed by the geometrical range and complexity of the operation to guarantee efficiency in general, which means that it is not realistic to conduct the complete data calculation process online. Our strategy, which maximizes the advantages of the analysis functions of GEE, involves simplifying the online spectral abnormal area detection process and downloading all image chips containing valid spectral information from a whole scene as candidates for further determination in a local computation environment. This approach could narrow the size of an image to approximately one-thousandth that of the original scene and thus could significantly reduce local data storage requirements.

First, as described in Section 2.1, the abnormal spectral feature can be used to capture inter-band mis-registration candidates. In this experiment, our aim was to detect anomalies in Band 3 (i.e., $\rho_{B3} - \rho_{B2} > 0.05$ and $\rho_{B3} - \rho_{B4} > 0.05$) that presented bright-green features in true-color images. Then, a rough filtering of false reflectance anomalies in Band 3 needed to be conducted before downloading the image chips to minimize excessive network transferring. False anomalies along the seam-lines of two adjacent granules (caused by moving clouds) and those inside traffic networks (caused by moving vehicles) can be effectively excluded using a similar method to that used in a previous study [8]. The M-estimator sample and consensus method is used to find the linear-distributed candidates in every scene. Moving vehicles can be frequently observed along roads (especially along highways). Therefore, a frequency image is used to better exclude them in a time series of MSI images. Moreover, the sun-glint noise that appears on water surfaces with high reflectance or moving ships may also lead to false detections.

Here, we use a global water map provided by the Environmental Systems Research Institute (ESRI) to filter chips that are fully covered by water surfaces as invalid data. After filtering the false positives, the detected abnormal spectral pixels are divided into different areas according to connectivity. The centroids of every area are calculated and saved as candidates in coordinate files. Finally, visible band chips (301 × 301 pixels) of MSI images centering on each candidate (pixel) with a 150-pixel extension in 4 neighborhoods could then be downloaded from the Sentinel-2 Collection on GEE.

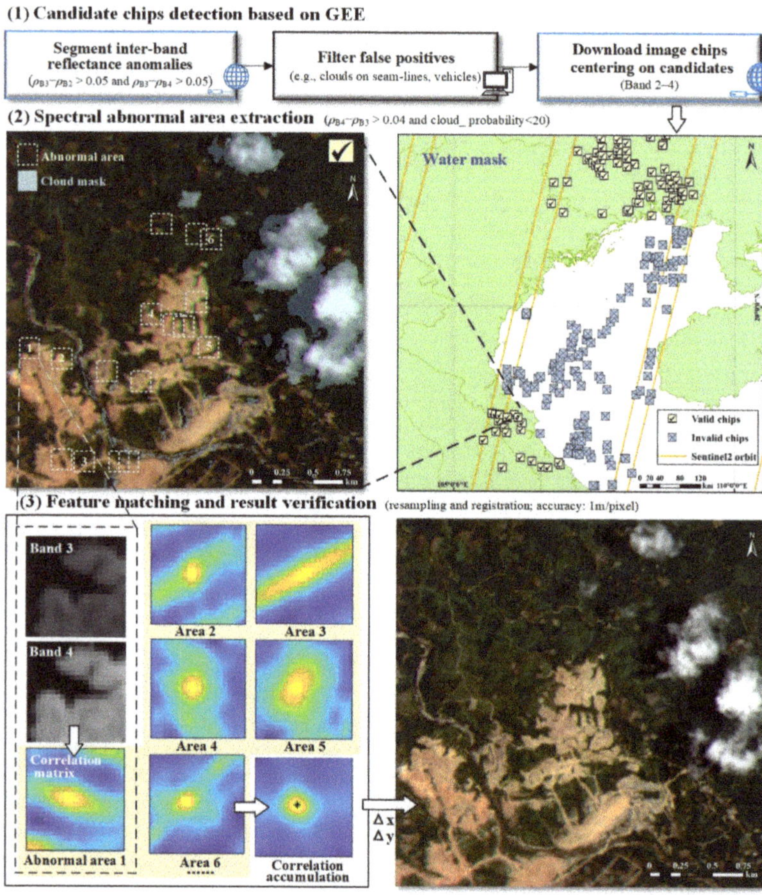

Figure 2. The hybrid framework of BBMR detection in an MSI image chip. The valid candidate chips (301 × 301 pixels) were selected after spectral anomaly detection and water masks. With the filtration of clouds, the abnormal areas inside the chips were extracted to calculate the best-fit location in the correlation matrix between visible bands. (The sample chip used in this figure was subsetted from scene 20150810_T48QXE.).

2.2.2. Spectral Abnormal Area Extraction

The abnormal area detection methods used in image registration can be divided into two main types: area based and feature based. To obtain a more solid result while also reducing algorithm complexity, we chose the latter to locate the inter-band mis-registration points inside the chip, depending on the spectral difference between bands. The specific steps are as follows:

(i) Spectral feature extraction. Inside each image chip, the threshold of the abnormal spectral point needs to be expanded to extract more features in areas with relatively low heterogeneity. The spectral reflectance peak around the green light band that vegetation has may lead to false positives if we simply narrow the threshold using the same criterion as that applied during the detection of candidate chips. Instead, we focus on bright-magenta features ($\rho_{B4} - \rho_{B3} > 0.04$) that appear in pairs with green ones in order to avoid vegetation interference. Only focusing on the BBR of the surface objects in the experiment, the "rainbow" effect appearing at the edge of moving clouds also needs to be excluded to obtain more convincing candidate points. Our primary choice of cloud mask is the cloud probability product (10 m resolution) processed by the s2cloudless open-source algorithm on GEE, because it ensures data coverage, avoids image resampling, and has significant high confidence compared to Band QA60. In this experiment, points with cloud probability of >20% were removed from the candidates. Note that the thresholds were the optimal choices determined by trial-and-error analysis and succeeded in extracting reflectance anomalies in Band 3 of MSI images covering different backgrounds. Although the performance was supported by the experiment result, the transfer to other datasets may still need more tests for adjustment.

(ii) A random sampling of abnormal spectral points. If the number of candidate points inside a chip is less than three, it will be filtered as an accidental error or a flying airplane; if the number is more than 24, the chip will be divided into a 3×4 grid for random sampling inside each grid in order to obtain a roughly geometrical uniform distribution of 24 points inside the chip. As the chips are downloaded centering on every abnormal area, we considered that the potential mis-registration inside a chip is spatial continuous in most cases. Therefore, this shrinks the computation complexity, while maintaining the main spectral feature.

2.2.3. Feature Matching and Result Verification

The high relativity among visible bands makes it possible to assess inter-band mis-registration by cross-correlation. In this case, we only evaluate the correlation between Bands 3 and 4 because as the bright-green and red stripes appear symmetrically, the existing offset patterns between Bands 3–4 and Bands 3–2 in the experimental images are essentially the same. With each abnormal spectral point, a small window (21×21 pixels) is built to calculate the statistical best-fit position between Bands 3 and 4. The window is centered on the abnormal spectral pixel with a 10-pixel extension in 4 neighborhoods. The window size is designed to meet the demand of surface feature representation and reduce the number of pixels participating in every single operation at the same time. Taking the Band 3 window as the reference image and moving the Band 4 window as the sensed image in the range of 10×10 pixels, the image correlation is calculated at every position to establish a 21×21 cross-correlation matrix. This matrix is resampled into 105×105, and the maximum correlation value position is calculated as the best-fit position at the improved accuracy of 0.2 pixels. To enhance the abnormal spectral feature detected in different areas and reduce the error caused by object texture, we accumulate all of the cross-correlation matrices and obtain a general best-fit position of all of the candidate windows. Meanwhile, we use a k-means clustering result of all the best-fit positions to ensure reliability, because based on the assumption that the inter-band offsets are similar inside the chip, the offset at each point should also have a similar direction. If the accumulation position and the clustering position are both outside of the 0.37×0.37 window, with a distance within 0.75 pixels between each position, then this image chip will be considered with inter-band mis-registration. For the mis-registration result, all three bands are resampled to 0.1 pixels and Band 3 is moved back to the best-fit position for manual visual inspection to check the result. The improvement in the true-color image composition is used to verify the effect.

3. Results

3.1. Detection Results

The proposed method was applied to 71,493 scenes of MSI Level-1C top of atmosphere (TOA) reflectance images covering China and its surrounding areas (58°43′37″N–0°23′15″S, 142°29′40″E–67°8′44″E) acquired from 21 July 2015 to 28 February 2016, the first few months after the launch of Sentinel-2A. We use the candidate chip products provided by a former study into airplane detection [8], as the detection method for this experiment shares similarities in spectral features with the flying airplane detection method based on the parallax between MSI visible bands. After the reflectance anomaly detection and false-positive filter, a total of 130,784 MSI image chips from 19,842 scenes were downloaded from GEE. In the local environment, 4356 chips from 442 scenes were detected with BBMR. From this, 4272 chips from 391 scenes were confirmed with existing mis-registration by manual visual inspection with an accuracy up to 98.07% at the chip scale and 88.46% at the scene scale. The average number of downloaded chips per image was 6.59, and it only accounted for 0.49% of the total pixels in an image. The reduction of downloaded images significantly improved the computation performance. According to the test statistic of 15,388 chips from 25 July 2015 to 15 August 2015, the average operation time for a chip was 0.76 s. Therefore, the method could be considered efficient for application on a sizeable set of image data. However, the limitation of the dataset brings possibilities of more chips with BBMR that may have remained undetected, because some areas with clustered abnormal points were filtered before being downloaded. Thus, this result only presents the temporal and spatial distribution of images with BBMR to a certain extent.

As shown in Figure 3, the chips with confirmed BBMR mainly appeared across 12 days from 29 July to 3 September 2015, during the early period following the launch of Sentinel-2A. Later in the study interval, BBMR only showed up occasionally on four days (24 November, 25 November, 11 December 2015, and 3 February 2016), and other detection results were mostly determined as false positives (53 chips from 41 scenes in 27 days). The confirmed mis-registrations not only showed up in multiple chips inside a single day but often in multiple scenes as well. The peak appeared on 10 August 2015, when there were 1431 chips with BBMR captured from 123 scenes, followed by 20 August and 3 September 2015 with 56 scenes and 15 August 2015 with 52 scenes. Although there was no apparent temporal pattern regarding the repetitive spatial appearances of BBMR, they did occur on some of the scenes over several days: scenes T50TML, T50TLL, and T50TLK were each detected to have BBMR four times, with 9 scenes detected three times and 40 scenes detected twice. Most offset distances were in the range of 0.37–3, with 77.40% of chips under 1.5 pixels, 90.67% of chips under 2.0 pixels, and 95.25% of chips under 2.5 pixels. The proportion of these images is not quite high due to the fine quality of the Sentinel-2 product; hence mis-registration could be an easily neglected problem. Therefore, we believe this method is essential and necessary for further remote sensing applications.

The BBMR distances and directions not only varied between different chips acquired on the same day (Figure 3) but also varied between different chips inside the same scene. For example, 10 chips detected to have confirmed BBMR (Figure 4b) in scene 20150810_T48RZP showed different offset distances and directions (Figure 4e) without representing a distinct spatial clustering pattern. This indicates that the offset was not homogeneous inside the range of the scene. Therefore, images need to be defined into smaller units for more precise BBMR detection and correction. In this research, we only focused on quick BBMR detection. The discontinuity between different chips will not affect the reliability of the result, because the detection will not accurately extract the boundary of the mis-registration area. However, based on the detection result, the requirement of fine registration for further application may need a continuous area border. The other method mentioned in Section 4.3 may provide a potential support to solve this problem with a flexible window.

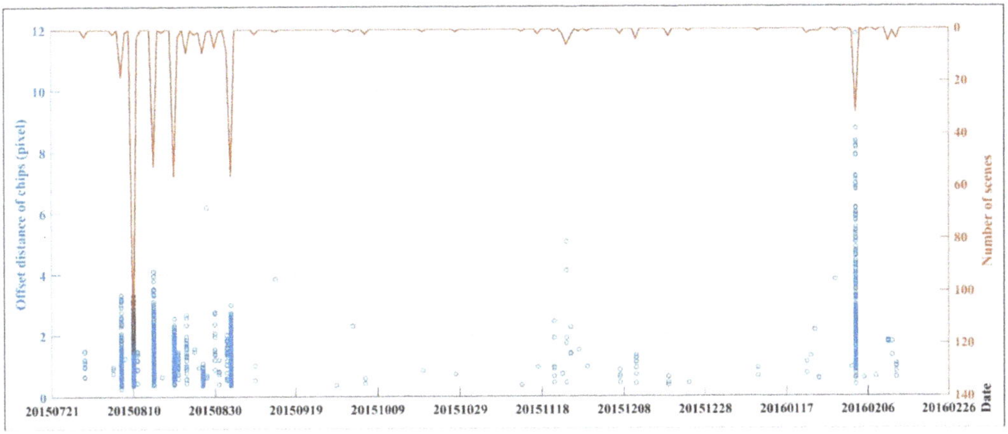

Figure 3. Temporal distribution of chips and scenes with an inter-band offset. The chips are represented with blue circles by the offset distance (pixel); the orange line represents the trend of scenes.

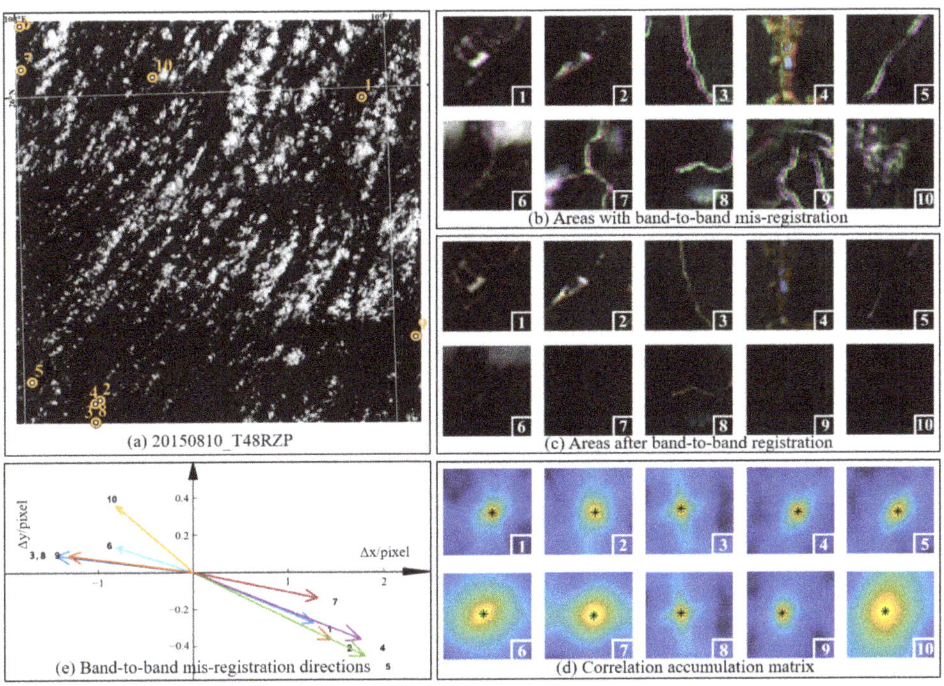

Figure 4. Inter-band offset directions and distances of different chips inside one scene. (**a**) Positions of 10 chips in scene 20150810_T48RZP, (**b**) original true-color images of these chips, (**c**) BBR results of these chips, and (**d**,**e**) different offset directions and distances of these chips. Note that image chips in (**b**,**c**) were zoomed in (40 × 40 pixels) to better present the spectral feature.

To analyze the performance of this method, a test area from scene T48PYC (10,000 × 10,000 m^2, centered on 15.5157°N, 107.8083°S) was tracked from 21 July 2015 to 28 February 2016 with manual visual inspection to check the detection result. From 45 scenes acquired from 18 days during this period (including all versions of images with different processing dates), the image acquired on 20 August 2015, processed on 26 April 2016, was detected and confirmed with BBMR. In the rest of the negative images, 24 were confirmed without BBMR and 20 were fully covered by clouds. This could support the reliability of the proposed method with a low probability of mis-detection.

3.2. Spatial Distribution of Detected MSI Chips with BBMR

To better analyze the spatial distribution of the detection results, all of the geographical positions were plotted on a map in Figure 5a. The images acquired on 10 August, 20 August, 23 August, 27 August, and 3 September 2015 appeared in two relative orbits across mid-eastern China, making up the main clustering area. The detections from other days showed up sporadically in other areas. Furthermore, the inter-band offset distance did not show a clear spatial pattern related to the trajectory of the satellite or to the geographical location of the image. To further analyze the inter-band offset direction pattern, a rectangular coordinate was established, as shown in Figure 5b. Based on the random sample consensus algorithm, we constructed a fitting straight line through the origin ($y = -0.1784x$) to represent the general trend of the clustering offsets in the second and fourth quadrants; 96.06% of offsets were inside the boundary of an ellipse centering on the origin with the same orientation angle as this fitting line (major axis: 8, minor axis: 2). The spatial distribution of the inter-band offset directions along the fitting line shows a possible connection between the offset direction and the orbital motion direction of the satellite, because the angle between the minor axis of the ellipse and x-axis (100.11°) is close to the orbit inclination of Sentinel-2A (98.62°).

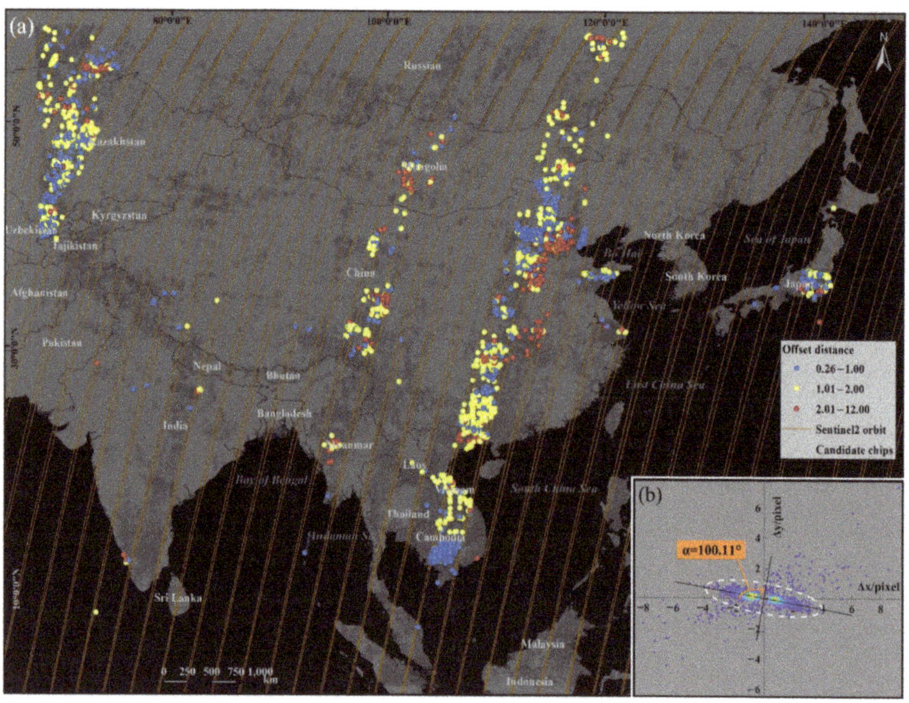

Figure 5. Spatial distribution of image chips with detectable BBMR. (**a**) Distribution of chips around China and its neighboring areas. The chips detected with BBMR were graded by the offset distance with different colors, and other candidate chips were represented by gray circles. (**b**) Sets the offset position of 4356 detected chips to a rectangular coordinate with a fitting ellipse (covered 96.06% chips) centered on the origin.

4. Discussion

4.1. Limitations

4.1.1. Limitations of Introducing Additional Data

Noticeable spectral differences between visible bands can be widely observed near objects with high reflectivity features, such as water surfaces or clouds, leading to a large amount of false detection. Due to the limitation of spectral features provided by visible bands, it is challenging to filter this kind of error solely by analyzing the texture features of each chip. Therefore, we introduced additional data into the experiment to create both water masks and cloud masks to reduce these particular false detections. However, the practical data and method both still have some flaws.

(i) Cloud masks. The machine-learning-based cloud detection product achieves much higher accuracy in cloud filtering than Band QA60 (Figure 6b,c). In some cases, however (Figure 6(b3)), it might show a false high probability in areas without clouds. This could lead to the omission of some significant abnormal spectral points, which in turn may affect the BBR accuracy. To guarantee the filtration of clouds in most chips, the threshold of the cloud probability cannot be set too low. Thus, we adjusted the threshold while eroding the edge of the logical operation matrix to lessen the mis-filtering of ground objects. For comparison, we used chips acquired on 15 August, 20 August (with multiple confirmed BBMRs), and 1 December 2015 (without BBMRs) as experimental data to test the effect of introducing the s2cloudless product (Table 1). While a few mis-detections appeared on 15 August and 20 August 2015, the number of false detections increased significantly, leading to a significant drop in accuracy from ~90% to ~40%. Meanwhile, all of the newly added detections were confirmed as false positives on 1 December 2015. This indicates that introducing a cloud mask leads to some mis-detections on days when images with known BBMR appear. However, in most cases (especially on days without confirmed BBMR images), the increment in detection accuracy was notable. Considering that BBMR is known to occur regularly (it tends to appear in multiple chips and scenes inside one day), missed detections will not greatly affect the general temporal tendency compared to the significant number of false positives that would be detected without the cloud mask. The crucial role that the mask plays in the filtration of moving clouds cannot currently be substituted. Thus, we regard this error as acceptable in the context of our experiment.

(ii) Water masks. Although the movements of seam-lines and the edges of lakes and rivers did not represent notable differences in kilometer-scale TOA images, the global water map still could not filter all of the chips that were capturing water surfaces, as there were some inevitable errors during the process of rasterization and the resampling of the water shapefile data; these errors may affect the accuracy of water masking. If using the Sentinel water surface classification product, additional errors may also be caused by interpretations such as the cloud mask, which may bring about new problems. According to the results, the proportion of false-positive detections caused by water was extremely low (5 of 4363 chips, 0.11%), and we consider that it is of no great significance to introduce the water surface classification product to optimize the result.

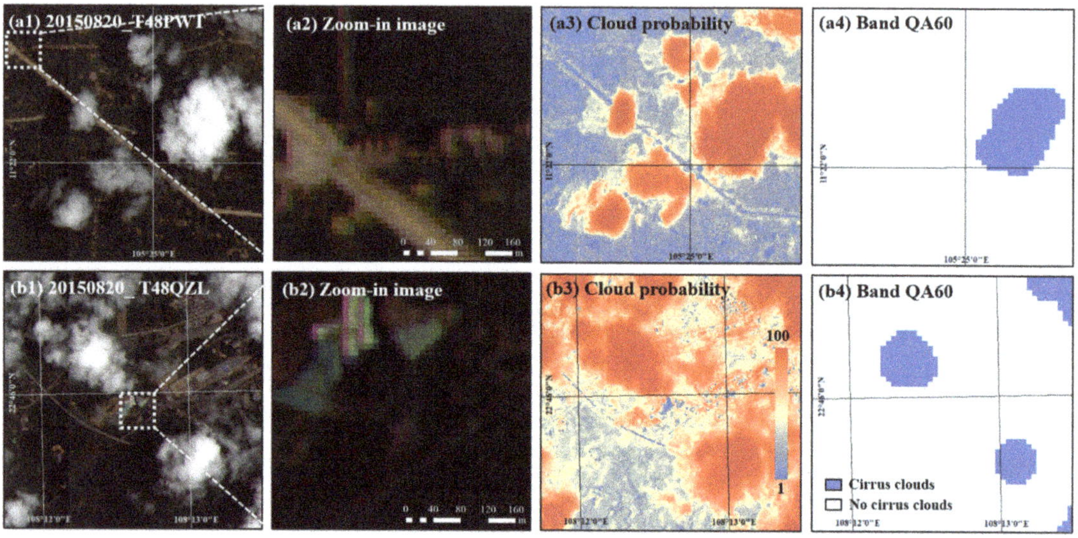

Figure 6. Influence of detection results with the introduction of cloud masks. (**a1**–**b4**) Compares the accuracy of the s2cloudless product and Band QA60 in two scenes with BBMR. The s2cloudless product identifies the range of the cloud more precisely than Band QA60 in most cases like (**a3**,**a4**), but sometimes also presents a false high value in areas like (**b3**), leading to the mis-detection of BBMR.

Table 1. Influence of introducing a cloud mask on detection results.

Image Acquired Date		**2015/08/15**		**2015/08/20**		**2015/12/01**	
Image scale		Chip	Scene	Chip	Scene	Chip	Scene
Image quantity		3335	165	2925	160	2940	158
With cloud masks	Detected	285	52	372	56	0	0
	Confirmed	281	49	370	54	0	0
	Accuracy (%)	98.60%	94.23%	99.46%	96.43%	100%	100%
Without cloud masks	Detected	1598	127	1220	111	368	72
	Confirmed	392	59	490	67	0	0
	Accuracy (%)	24.53%	46.46%	40.16%	60.36%	0%	0%

4.1.2. The Limitation of Application in Exceptional Cases

(i) Homogeneous areas. The main limitation of the proposed method is that it is based on the assumption that there is sufficient variance inside each chip and that there will be a unique and clear maximum in every correlation matrix, corresponding to the offsets between bands. However, areas such as deserts and grasslands, which have repetitive or almost homogeneous textures may have a rather low variance in a small image chip. This homogeneity could lead to an inaccurate offset direction. In our experiment, there were few image chips in which we could not distinguish the difference before and after BBR visually, due to a lack of discernable features outstanding from the image background. This indicates that the accumulation correlation matrix can mitigate the influence of homogeneous areas in most cases, and the false-positive results that arise occasionally will only have a slight impact on the detection accuracy.

(ii) Areas with complex inter-band offsets. Our method assumes that the feature of mis-registration between bands shows spatial consistency within the scale of one chip

(301 × 301 pixels). However, the situation is much more complicated in practical terms. We found out that in some chips, both the range and the spatial features of the mis-registration could not fit the previously expected ideal state. These kinds of cases often show up in areas with obvious topographic relief in the experiment images. As shown in Figure 7a, although mis-registration was successfully detected and corrected in part of the chip (Figure 7(a2)), some areas (Figure 7(a1)) without visual errors showed new mis-registrations after registration. Segmenting chips into smaller regions or re-performing a spectral anomaly detection on the image after BBR may filter this kind of error.

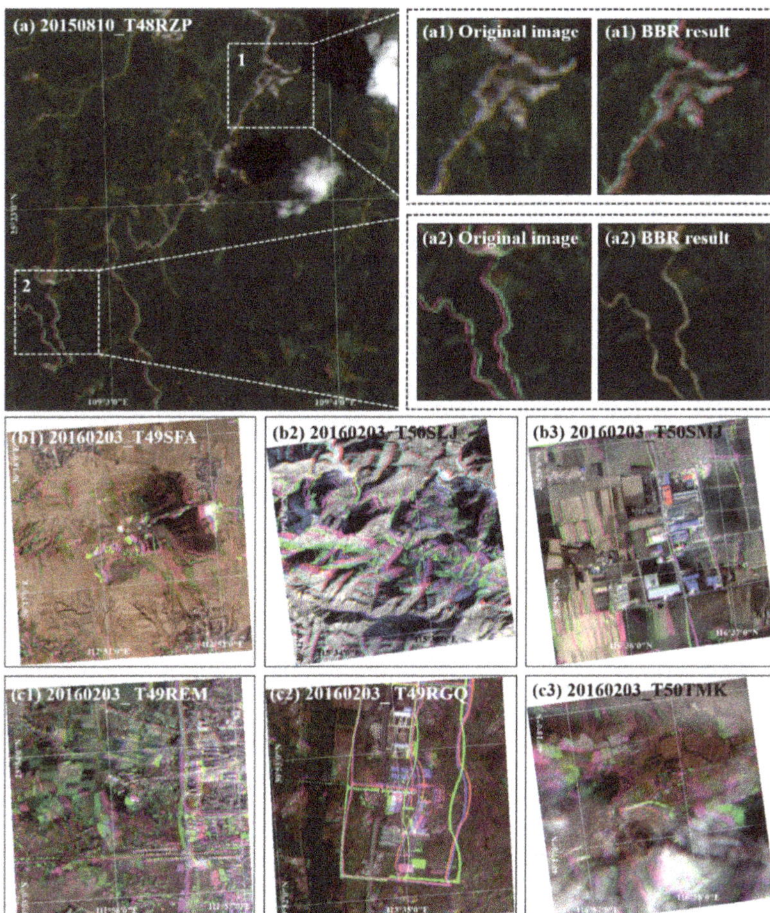

Figure 7. Examples of areas with complex inter-band offset. (**a**) The inter-band offset only appeared in part of the chip of 20150810_T48RZP, and the BBR of (**a2**) leads to a new error in (**a1**). (**b1–c3**) Some complex offset acquired on 3 February 2016; the offset in (**b1–b3**) can be detected, but the registration effect is limited; the offset in (**c1–c3**) cannot be detected.

In few cases, mis-registration can become dramatically complicated (Figure 7b,c): chips represent more than two types of inter-band offset errors, including translation, rotation, and mirror symmetry, and offsets can sometimes occur between three bands at the same time (Figure 7(b2),(c2)). Due to the spatial and spectral limitation of this method, although some chips can be captured (Figure 7(b1)–(b3)), a small number of chips still cannot be effectively detected (Figure 7(c1)–(c3)). These cases indicate that rigid registration methods, such as the simple translation provided, cannot fix complicated instances of BBMR within the scale of the chip. This demonstrates the need for better BBR strategies to apply to broader contexts. Nonetheless, this kind of rare abnormal situation only occurred on one day during our experiment. Therefore, considering the primary purpose of identifying mis-registrations, we consider it to be an accidental situation.

4.1.3. Limitations of the Image Dataset

The method for capturing candidate chips shares similarities in spectral features with a previously published airplane detection method based on the parallax between MSI visible bands [8]; the latter differs in that it focuses more on the separate single target. Actually, many BBMRs observed in the image chips during the parallax research inspired us to propose a method to trace the inter-band mis-registration over time. GEE keeps updating image data with more accurate registration results, meaning that much fewer inter-band offsets could be observed in the images that were re-downloaded in 2019 for the purposes of this study. Thus, considering the unity and integrity of the original features, we decided to use the existing data products downloaded from 21 May 2019 to 27 June 2019 for this experiment. The limitations of the experimental image dataset may therefore have caused some mis-detections due to filtering of the clustered spectral abnormal area. However, the detection result could still be valuable as a reference to reveal the general tendency of MSI BBR accuracy development during the early period following the launch of Sentinel-2A.

4.2. Transferability to Other MSI Bands

The MSI visible bands were used in this experiment, but the core of this method could be applied to other MSI bands or satellite data based on GEE. Taking Bands 8A (NIR), 11 (near-SWIR), and 12 (far-SWIR) of the MSI images as an example, in scene 20180816_T10SEJ, the true-color image did not represent an apparent inter-band offset among the visible bands (Figure 8a). However, in a false-color image (R: 12, G: 11, B: 8A) with 20 m spatial resolution, there was a nearly 11-pixel dislocation around the edge of the lake (Figure 8b). These bands could be used to build a spectral thermal anomaly index to detect high-temperature thermal anomalies. Therefore, the inter-band mis-registration may have led to many false positives during thermal anomaly detection and could have introduced unreliability in subsequent work. This dislocation was determined to appear between Band 8A and the other bands, following difference analysis among the three bands. By performing a procedure similar to that carried out during BBMR detection for visible bands, the anomaly pixels in this scene could also be captured by a particular feature ($\rho_{B8A} > 0.21$, $\rho_{B12} < 0.01$, $\rho_{B11} < 0.01$) and ameliorated to a certain extent, as shown in Figure 8(c1),(c2).

Due to the complexities of the spectral features in these bands, the establishment of a generally applicable extraction method for anomalies requires further research. Similarly, the abnormal spectrum area can also be captured by the spectrum differences between other target bands, while discriminant conditions might be more complicated depending on their inter-band spectrum covariance. Thus, the method proposed here might exhibit a certain generality and superiority regarding the detection of long-term inter-band mis-registration.

4.3. Comparison with Existed BBMR Detection Method

In comparison to another existed BBMR detection method promoted by Skakun et al. (2017) [18] for MSI images, the main improvement in this paper is the change in calculation objects. Differentiating from a calculation traversing the whole image, the method promoted in this paper is target originated, which considerably reduces the number of pixels that participate in the operation.

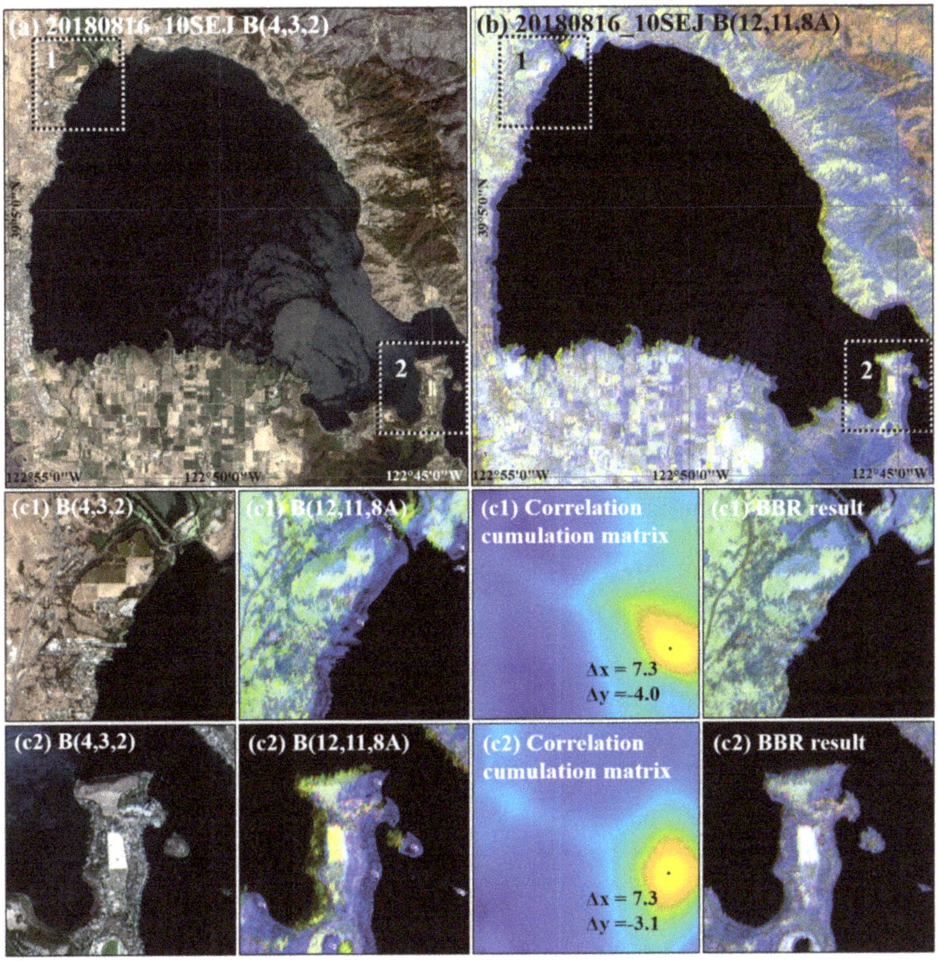

Figure 8. Apparent BBMR exists among MSI Band 8A and SWIR bands (Bands 11 and 12), while there is hardly discernible BBMR among visible bands (Band 2–4) of the same acquisition. (**a**) A subset of true-color MSI image (10SEJ) acquired on 16 August 2018, (**b**) false-color image (Bands 8A, 11, and 12) of the same extent as (**a**), and (**c1**,**c2**) examples of a rough BBR process in two image chips (Bands 12, 11, 8A; size: 168 × 168 pixels) cut from scene 10SEJ.

The method promoted by Skakun et al. was based on the phase correlation calculation on a sliding window basis. With the adjustment of window size and step length, this method could be applied to detect moving targets such as planes or clouds. However, this paper is focused on the BBMR detection of static objects, which fail to meet the required threshold of multispectral band registration. Therefore, moving targets such as airplanes, ships, and clouds need to be eliminated from detecting objects. The potential application of Skakun et al.'s method does not quite match our needs. In this case, the sliding window is more suitable for further spatial analyzation of BBMR in a detected area to refine the result. For example, in some cases, the BBMR did not show spatial consistency within the scale of one chip (301 × 301 pixels), as we discussed in Section 4.1, and therefore, a fine boundary of BBMR area is needed to avoid the occurrence of new mis-registration in other areas after the correction. It could also be used in areas with complex BBMR to determine different directions of offsets between bands in the adjacent parts inside an image chip. Moreover, this method may also have potential application in the analysis for spatially continuously changing events such as cloud motion or hydrological events.

As shown in Figure 9, scene 20150820_48PZC (processed on 26 April 2016) was detected with serious BBMR. The mis-registration areas were widely distributed inside this image. We used a sliding window basis (window size: 16 × 16 pixels, step length: 9 pixels) for BBMR detection, and the result is shown as a contour (Figure 9b). Targeting the detection of the BBMR area, the result is effective on land (Figure 9c1) but still has trouble filtering the water surface with sun-glint (Figure 9c2) or other noise (Figure 9c3). Moreover, in many other images, BBMR only showed up in the local area with a small proportion of pixels. Therefore, if we use a sliding window to scan the whole image, the invalid calculation times will tremendously increase in each determination process. Although the efficiency can be improved by the adjustment of the size and step length of a sliding window, it still requires considerable operations to acquire a subpixel level offset result. With the introduction of a pre-processing procedure to cut the original image into chips, the method promoted in this paper diminishes the image data involved in the calculation. The conditional random sampling step to choose feature points inside each image chip makes a further data reduction.

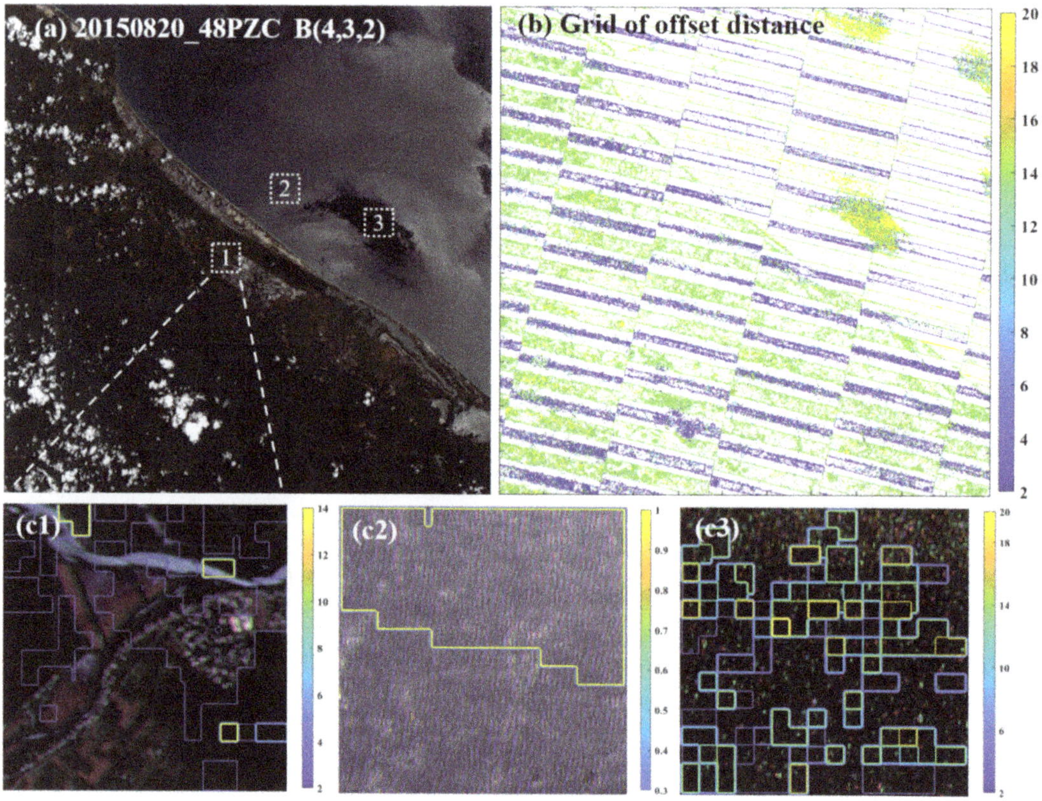

Figure 9. A true-color image with BBMR and the detection result using the phase correlation calculation on a sliding window basis (window size: 16 × 16 pixels, step length: 9 pixels). The result is presented as a grid with a colored border, divided by the inter-band offset distance (pixels) in (**b**). (**c1–c3**) The zoom-in image clips (150 × 150 pixels) from the marked areas in the image (**a**) with a contour result overlay.

To summarize, the method promoted in this paper is more suitable for quick BBMR detection in a large number of images due to the improved computing model based on chips. The existed BBMR method could acquire a more accurate BBMR boundary, and it could be applied to optimize our results in some cases.

5. Conclusions

The recent development of cloud sharing and computing technology has made it easier for researchers to access massive amounts of geographic image data. Though this provides convenience, it also brings higher demands on retrieval methods and data quality to ensure the reliability and efficiency of further applications. As for MSI images, determining unqualified image with band-to-band mis-registration is crucial for subsequent RS application. It is more challenging because time efficiency and the elimination of redundant information both need to be carefully considered to handle the ever-increasing MSI images. Here, we proposed a hybrid framework for detecting BBMR between the visible bands in MSI images, including rough GEE online pre-processing and a precise local operation. To enhance the efficiency of handling massive MSI images, we adopted online feature filtration to extract spectral abnormal areas in the MSI images, potentially caused by the band-to-band mis-registration; and to guarantee the robustness, we conducted band correlation to determine the presence of BBMR, only based on image chips containing spectral abnormal areas derived from the online operation.

The proposed BBMR detection framework was applied to 71,493 scenes, and the results were used to reproduce the spatial and temporal distributions of images with BBMR during a seven-month period at both chip and scene scales. Although limitations remained, leading to some mis-detections, the results were still able to reveal trends in BBR accuracy during the early period after the launch of Sentinel-2A. The hybrid computation framework proposed here may also be transferable to BBMR detection in other bands or with other satellite data. This process would demand the adjustment of spatial features and the optimization of the geometry registration to address more complicated situations. Moreover, by combining both online and local computation advantages, this method provides an efficient perspective for users to conveniently evaluate image quality when dealing with a large amount of data. This idea may have further potential applications in other image-processing problems.

Author Contributions: Conceptualization, T.C. and Y.L.; methodology, resources, investigation, writing—review and editing, supervision, Y.L.; software, writing—original draft preparation, visualization, T.C. All authors have read and agreed to the published version of the manuscript.

Funding: This research was supported by the Key Research and Development Program of China (grant no. 2019YFA0606601) and the Jiangsu Provincial Natural Science Foundation (grant no. BK20160023).

Data Availability Statement: The authors are grateful to the European Space Agency (ESA) and Google Earth Engine (GEE) for providing Sentinel-2 MSI images and online computation.

Conflicts of Interest: The authors declare no conflict of interest.

References

1. Fang, Z.; Cao, C.; Jiang, W.; Ji, W.; Xu, M.; Lu, S. Multi-Spectral Image Inter-Band Registration Technology Research. In Proceedings of the 2012 IEEE International Geoscience and Remote Sensing Symposium, Munich, Germany, 22–27 July 2012; pp. 4287–4290.
2. Xie, Y.; Xiong, X.; Qu, J.J.; Che, N.; Wang, L. Sensitivity analysis of MODIS band-to-band registration characterization and its impact on the science data products. *SPIE Opt. Eng. Appl.* **2007**, *6679*, 667908.
3. Xu, H.; Sun, R.; Zhang, L.; Tang, Y.; Liu, S.; Wang, Z. Influence on Image Interpretation of Band to Band Registration Error in High Resolution Satellite Remote Sensing Imagery. In Proceedings of the 2012 2nd International Conference on Remote Sensing, Environment and Transportation Engineering, Nanjing, China, 1–3 June 2012; pp. 1–4.
4. Belward, A.S.; Skoien, J.O. Who launched what, when and why; trends in global land-cover observation capacity from civilian earth observation satellites. *ISPRS J. Photogram. Remote Sens.* **2015**, *103*, 115–128. [CrossRef]
5. ESA. Sentinel-2 User Handbook. Available online: https://sentinel.esa.int/documents/247904/685211/Sentinel-2_User_Handbook (accessed on 24 April 2020).

6. ESA. Sentinel-2 Technical Guides. Available online: https://earth.esa.int/web/sentinel/technicalguides/sentinel-2-msi/performance (accessed on 24 April 2020).
7. Gascon, F.; Bouzinac, C.; Thépaut, O.; Jung, M.; Francesconi, B.; Louis, J.; Lonjou, V.; Lafrance, B.; Massera, S.; Gaudel-Vacaresse, A.; et al. Copernicus Sentinel-2A Calibration and Products Validation Status. *Remote Sens.* **2017**, *9*, 584. [CrossRef]
8. Liu, Y.; Xu, B.; Zhi, W.; Hu, C.; Dong, Y.; Jin, S.; Lu, Y.; Chen, T.; Xu, W.; Liu, Y.; et al. Space eye on flying aircraft: From Sentinel-2 MSI parallax to hybrid computing. *Remote Sens. Environ.* **2020**, *246*, 111867. [CrossRef]
9. ESA. Sentinel-2 L1C Data Quality Report. Available online: https://sentinel.esa.int/documents/247904/685211/Sentinel-2_L1C_Data_Quality_Report/6ad66f15-48ca-4e65-b304-59ef00b7f0e0?version=1.56 (accessed on 6 February 2020).
10. Berger, M.; Moreno, J.; Johannessen, J.A.; Levelt, P.F.; Hanssen, R.F. ESA's sentinel missions in support of Earth system science. *Remote Sens. Environ.* **2012**, *120*, 84–90. [CrossRef]
11. Malenovský, Z.; Rott, H.; Cihlar, J.; Schaepman, M.E.; García-Santos, G.; Fernandes, R.; Berger, M. Sentinels for science: Potential of Sentinel-1, -2, and -3 missions for scientific observations of ocean, cryosphere, and land. *Remote Sens. Environ.* **2012**, *120*, 91–101. [CrossRef]
12. Hass, E.; Hill, J.; Stoffels, J.; Frantz, D.; Uhl, A. Improvement of the Fmask algorithm for Sentinel-2 images: Separating clouds from bright surfaces based on parallax effects. *Remote Sen. Environ. Interdiscip. J.* **2018**, *215*, 471–481.
13. Munyati, C. The potential for integrating Sentinel 2 MSI with SPOT 5 HRG and Landsat 8 OLI imagery for monitoring semi-arid savannah woody cover. *Int. J. Remote Sens.* **2017**, *38*, 4888–4913. [CrossRef]
14. Pahlevan, N.; Balasubramanian, S.V.; He, J.; Sarkar, S.; Franz, B.A. Sentinel-2 MultiSpectral Instrument (MSI) data processing for aquatic science applications: Demonstrations and validations. *Remote Sen. Environ. Interdiscip. J.* **2017**, *201*, 47–56. [CrossRef]
15. ED Chaves, M.; CA Picoli, M.; D Sanches, I. Recent Applications of Landsat 8/OLI and Sentinel-2/MSI for Land Use and Land Cover Mapping: A Systematic Review. *Remote Sens.* **2020**, *12*, 3062. [CrossRef]
16. ESA. Sentinel-2 Products Specification Document. Available online: https://sentinel.esa.int/documents/247904/349490/S2_MSI_Product_Specification (accessed on 24 April 2020).
17. Ping, J.J.; Yeou, R.J.; Ying, H.C. Band-to-band registration and ortho-rectification of multilens/multispectral imagery: A case study of MiniMCA-12 acquired by a fixed-wing UAS. *ISPRS J. Photogramm. Remote Sens.* **2016**, *114*, 66–77.
18. Sergii, S.; Eric, V.; Jean-Claude, R.; Christopher, J. Multi-spectral misregistration of Sentinel-2A images: Analysis and implications for potential applications. *IEEE Geosci. Remote Sens. Lett.* **2017**, *14*, 2408–2412.
19. Amos, C.; Petropoulos, G.P.; Ferentinos, K.P. Determining the use of Sentinel-2A MSI for wildfire burning & severity detection. *Int. J. Remote Sens.* **2019**, *40*, 905–930. [CrossRef]
20. Kääb, A.; Leprince, S. Motion detection using near-simultaneous satellite acquisitions. *Remote Sens. Environ.* **2014**, *154*, 164–179. [CrossRef]
21. Berenstein, C.A.; Kanal, L.N.; Lavine, D.; Olson, E.C. A Geometric Approach to Subpixel Registration Accuracy. *Comput. Graphics Image Proc.* **1987**, *40*, 334–360. [CrossRef]
22. Yang, K.; Fleig, A.J.; Electric, I.o.; Engineer, E. MODIS band-to-band registration. In Proceedings of the IGARSS 2000, IEEE 2000 International Geoscience and Remote Sensing Symposium. Taking the Pulse of the Planet: The Role of Remote Sensing in Managing the Environment. Proceedings (Cat. No.00CH37120), Honolulu, HI, USA, 24–28 July 2000; pp. 887–889.
23. Kim, W.; Jung, S.; Moon, Y.; Mangum, S.C. Morphological Band Registration of Multispectral Cameras for Water Quality Analysis with Unmanned Aerial Vehicle. *Remote Sens.* **2020**, *12*, 2024. [CrossRef]
24. Simper, A.; Electric, I.O.; Engineer, E. Correcting general band-to-band misregistrations. In Proceedings of the 3rd IEEE International Conference on Image Processing, Lausanne, Switzerland, 19 September 1996; pp. 597–600.
25. Hao, L.; Tuanjie, L.; Haiqing, Z. High-accuracy band to band registration method for multi-spectral images of HJ-1A/B. *J. Electron.* **2012**, *29*, 361–367.
26. Du, Q.; Bruce, L.M.; Orduyilmaz, A.; Raksuntorn, N. Automatic Registration and Mosaicking for Airborne Multispectral Image Sequences. *Photogramm. Eng. Remote Sens. J. Am. Soc. Photogramm.* **2008**, *74*, 169–181. [CrossRef]
27. Pan, J.; Zhu, Y.; Wang, M.; Zhang, B.X. Parallel Band-to-Band Registration for HJ-1A1B CCD Images Using OpenMP. In Proceedings of the 2011 International Symposium on Image and Data Fusion, Yunnan, China, 9–11 August 2011; pp. 1–4.
28. Hassanpour, M.; Javan, F.D.; Azizi, A. Band to Band Registration of Multi-Spectral Aerial Imagery—Relief Displacement and Miss-Registration Error. *ISPRS Int. Arch. Photogramm. Remote Sens. Spat. Inform. Sci.* **2019**, *XLII-4/W18*, 467–474. [CrossRef]
29. Zhao, X.; Gao, Z.; Sun, W.; Wen, F. A Coarse-To-Fine Band Registration Framework for Multi/Hyperspectral Remote Sensing Images Considering Cloud Influence. *ISPRS Ann. Photogramm. Remote Sens Spat. Inform. Sci.* **2020**, *V-3-2020*, 201–208. [CrossRef]

remote sensing

Article

On-Orbit Radiometric Performance of GF-7 Satellite Multispectral Imagery

Hongzhao Tang [1,2], Junfeng Xie [2,*], Xinming Tang [2], Wei Chen [3] and Qi Li [1]

1 School of Earth and Space Sciences, Peking University, Beijing 100871, China; tanghz@pku.edu.cn (H.T.); liqi@pku.edu.cn (Q.L.)
2 Land Satellite Remote Sensing Application Center, Ministry of Natural Resources, Beijing 100094, China; tangxm@lasac.cn
3 College of Geoscience and Surveying Engineering, China University of Mining and Technology, Beijing 100083, China; chenw@cumtb.edu.cn
* Correspondence: xiejf@lasac.cn; Tel.:+86-10-68412293

Abstract: China's first civilian, sub-meter, high-resolution stereo mapping satellite, GF-7, launched on 3 November 2019. Radiometric characterization of GF-7 multispectral imagery has been performed in this study. A relative radiometric accuracy evaluation of the GF-7 multispectral imagery was performed using several large uniform scenes, and the results showed that the accuracy is better than 2%. The absolute radiometric evaluation of the GF-7 satellite sensor was conducted at the Baotou and Dunhuang calibration sites, using the reflectance-based vicarious approach. The synchronous measurements of surface reflectance and atmospheric parameters were collected as the input for the radiative transfer model. The official radiometrically calibrated coefficient of the GF-7 multispectral imagery was evaluated with the predicted top-of-atmosphere (TOA) radiance from the radiative transfer model. The results indicated that the absolute radiometric accuracy of GF-7 multispectral imagery is better than 5%. In order to monitor the radiometric stability of the GF-7 satellite multispectral sensor, a relative and absolute radiometric accuracy assessment campaign should be performed several times a year.

Keywords: GF-7 satellite; radiometric performance; multispectral sensor; reflectance-based

1. Introduction

The Gao-Fen 7 (GF-7) satellite is China's first civilian, sub-meter, high-resolution stereo mapping satellite, and has the highest mapping accuracy standard of China's Gao-Fen series of satellites [1]. The GF-7 satellite was launched into orbit on 3 November 2019. The GF-7 satellite carried two optical cameras, which can obtain high-spatial-resolution optical stereo imagery pointing at forward and backward angles. The inclination angle of the forward-looking camera is +26° and the backward-looking camera is −5° from the nadir. The GF-7 satellite was designed to satisfy China's need for sub-meter, high-resolution satellite imageries, which were primarily used for 1:10,000 scale mapping. The GF-7 satellite can not only obtain accurate geometric information and structural information of ground features, but also multispectral information of features, when combined with the high-resolution multi-spectral camera, which plays a great role in China's natural resource and other industrial applications.

Most studies have focused on geometric calibration of the GF-7 satellite, and far less attention has been paid to radiometric calibration and validation [2]. Since the launch of the GF-7 satellite, GF-7 products have successfully played an important role in China's environmental, agricultural, and other quantitative applications, which require accurate and reliable radiometric information. The radiometric calibration accuracy of the GF-7 satellite is critical in these studies. The Land Satellite Remote Sensing Application Center (LASAC) of the Chinese Ministry of Natural Resources delivered the radiometrically corrected and georectified images (the standard level 1A products) of the GF-7 satellite to government and industry users. The radiometric calibration accuracy of the GF-7 satellite standard products must be evaluated.

The GF-7 satellite was designed to meet uniform relative radiometric responses across the scene. The pixel-to-pixel variation in the detector response of the GF-7 satellite sensor was eliminated by relative radiometric correction. LASAC performed a relative radiometric accuracy evaluation of the GF-7 imagery, using several large uniform scenes, such as a uniform ocean scene and the Libya invariant test site. The nonuniform variations in the GF-7 imagery could be detected in this study.

The absolute radiometric calibration of the GF-7 multispectral imagery is critical for providing highly accurate quantitative measurements of the Earth's surface. LASAC performed an absolute radiometric assessment of the GF-7 satellite sensor using the following two calibration sites: Dunhuang and Baotou. Reflectance-based vicarious radiometric calibration was commonly recognized as one of the most reliable approaches for on-orbit calibration and validation of an optical satellite sensor. Two independent absolute radiometric assessment campaigns of the GF-7 satellite, at different calibration sites, both used this reflectance-based approach. The surface reflectance and atmospheric parameters of the calibration sites were synchronously measured at the GF-7 satellite overpass time. These synchronous measurements were taken as the inputs of the radiative transfer model (such as MODTRAN 6.0), to predict the top-of-atmosphere (TOA) radiance, which was used to evaluate the GF-7 satellite sensor radiometric calibration. In order to ensure the accuracy and reliability of the GF-7 satellite multispectral imagery, the on-orbit radiometric performance must be regularly evaluated every year.

2. GF-7 Satellite Background

The GF-7 satellite is positioned on a 500 km sun-synchronous orbit, and covers the global region every 59 days. The designed life expectancy of this satellite is 5 years. It can survey the Earth at ±84° latitude, with a 5-day revisit time. Table 1 shows the technical specification of the GF-7 satellite. The GF-7 satellite carries two optical cameras (DLC) that point forward and backward, as shown in Figure 1. These two cameras are arranged at inclinations of +26° (forward-looking camera, FWD) and −5° (backward-looking camera, BWD) from the nadir. The FWD is a panchromatic camera, while the BWD is a multispectral imager with five bands, which include a panchromatic band and four multi-spectral bands (blue, green, red, and near infrared). The GF-7 DLC has about 20 km of ground swath. The GF-7 satellite is mainly used for mapping 1:10,000-scale topographic maps and geographic information products.

Table 1. The technical specification of the GF-7 satellite.

GF-7 Satellite	Technical Specification
Launch date	3 November 2019
Mission duration	5 Years
Orbit	500 km sun-synchronous orbit
Equator crossing time	10:30 AM
Repeat cycle time	59 days
Revisit cycle time	5 days
Spatial resolution of DLC	Panchromatic band: FWD: 0.80 m (GSD) BWD: 0.65 m (GSD) Multispectral band: BWD:2.60 m (GSD)
Swath width of DLC	20 km

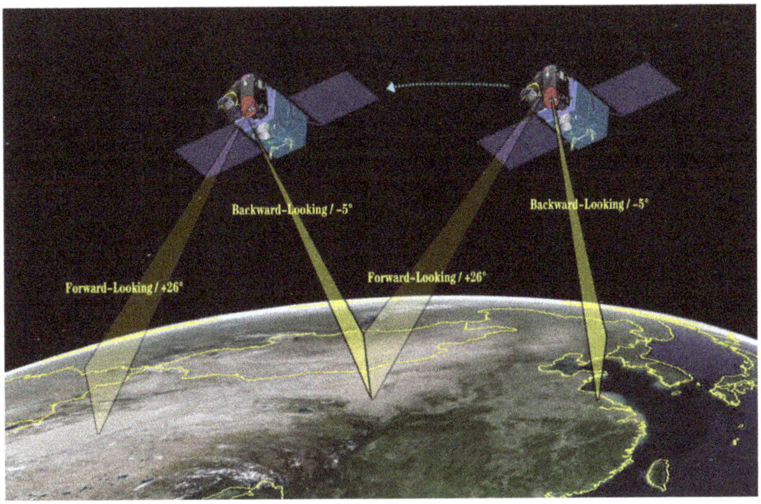

Figure 1. The DLC (forward-looking and backward-looking cameras) of the GF7 satellite.

2.1. GF-7 Multispectral Sensor

The GF-7 satellite carried two optical cameras (FWD and BWD), which can obtain high-spatial-resolution optical stereo imagery, pointing at forward and backward angles. The inclination angle of FWD is +26° and BWD is −5° from the nadir. The FWD is a panchromatic camera with a spatial resolution of 0.8 m, and the BWD is an imager that provides the repetitive acquisition of panchromatic (BWD-PAN, 450–920 nm) imagery with a spatial resolution of 0.65 m, and four-band multispectral imagery (BWD-MUX) of blue (460–530 nm), green (510–590 nm), red (620–690 nm), and near infrared (750–890 nm) with a spatial resolution of 2.6 m. Each camera has its own lens and consists of three single CCD lines located in an across-track dimension.

The spectral response function (SRF) of the GF-7 BWD is shown in Figure 2. Several of the spectral response properties of the GF-7 BWD are listed in Table 2.

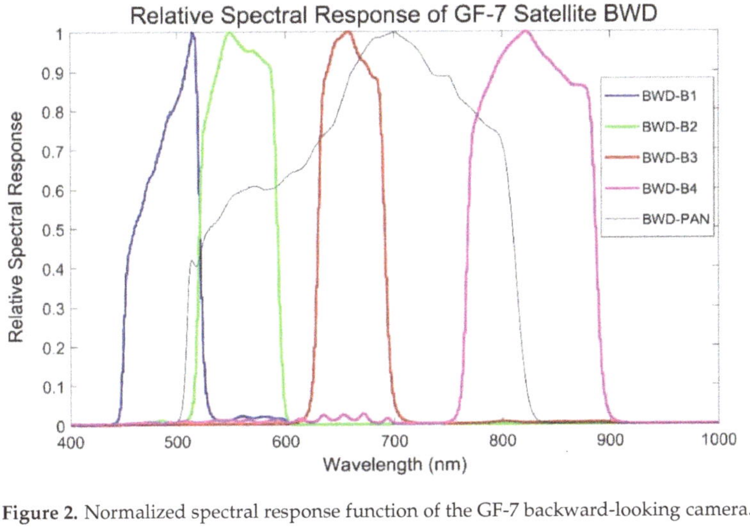

Figure 2. Normalized spectral response function of the GF-7 backward-looking camera.

Table 2. The spectral response properties of the GF-7 BWD.

Description	Band	Spectral Range (nm)	Center Wavelength (nm)	Specified Spectral Range at 50% Transmittance Points (nm)	Bandwidth (nm)
BWDPAN	PAN	450–920	746	556.2–912.5	470
BWDMUX	Blue	460–530	490	462.3–512.2	70
	Green	510–590	563	515.7–587.1	80
	Red	620–690	676	625.8–693.0	70
	NIR	750–890	807	763.9–885.4	140

2.2. LASAC GF-7 Data Products

Several types of GF-7 data products are delivered by LASAC to China's government and different industrial users. The standard level 1A products were both radiometrically and geometrically corrected (without using ground control points (GCPs)) and projected to a CGCS2000 datum. The standard level 1A products were orthorectified to improve the spatial accuracy with GCPs using Geomatica-PCI software [3]. In this study, all the radiometric analysis was performed on the standard level 1A products. The two cameras of the GF-7 satellite are formed by several time-delayed integral imaging (TDI) CCD linear arrays; the GF-7 imagery needs to be stitched in accordance with the geometric and radiometric characteristics. Evaluation and verification of the radiometric accuracy of the GF-7 imagery is essential for the application of satellite-retrieved data.

3. Relative Radiometric Accuracy Assessment

The GF-7 satellite BWD multispectral imager is a typical push-broom sensor, which consists of three single CCD arrays located in an across-track dimension. The linear detector array of the GF-7 BWD constructs an image one row at a time as the satellite moves [4,5]. Each CCD array of the BWD has 3072 detectors. There are two overlap regions between three CCD arrays, and each overlap region contains 108 detectors. Figure 3 shows the three CCD arrays of the GF-7 satellite multispectral sensor and the corresponding overlap region when imaging. Variations between the detectors of the GF-7 BWD always appear in the along-track scanning direction. Noise was apparent in the GF-7 satellite raw data before the relative radiometric correction. The radiometric nonuniformity always caused banding and striping, or dark lines, in the along-track scanning direction on the imagery [6–8]. The on-orbit detector-to-detector radiometric variation of the GF-7 multispectral imagery was corrected using relative radiometric calibration coefficients.

A relative radiometric accuracy evaluation of the GF-7 satellite multispectral imagery, using uniform scenes (such as desert and ocean), was performed in this study. Several large uniform scenes, including the Libya and Algeria pseudo-invariant calibration sites (PICs), have been used for evaluating the relative radiometric uniformity. The nonuniform variations in the GF-7 imagery could be detected using these large uniform scenes. Table 3 shows the worldwide uniform sites chosen for the relative radiometric accuracy assessment in this study.

Table 3. The uniform sites for relative radiometric accuracy assessment.

Num	Site	Latitude	Longitude	Area
1	Saharan	25.9° N	14.4° W	60 km × 60 km
2	Algeria	5.86° N	31.1° E	23 km × 20 km
3	Libya 4	28.52° N	23.39° E	23 km × 24 km

Figure 4 shows the GF-7 BWD-MUX blue band imagery of the uniform Libya 4 PICs. The columns of each detector (pixel) in the uniform imagery were summed and normalized. Figure 5 shows the mean normalized DN value curve of each detector in the uniform Libya 4 PICs on the GF-7 multispectral imagery, indicating that the relative differences in all the detectors are less than 2% for each band. We can observe that the array transitions between the different CCD arrays appear as step discontinuities in Figure 5. The largest relative difference appears in the array transition region. Similar results were obtained from the GF-7 multispectral imagery of the other larger uniform scenes in this study. The relative radiometric accuracy of the GF-7 multispectral imagery is better than 2% for all the bands.

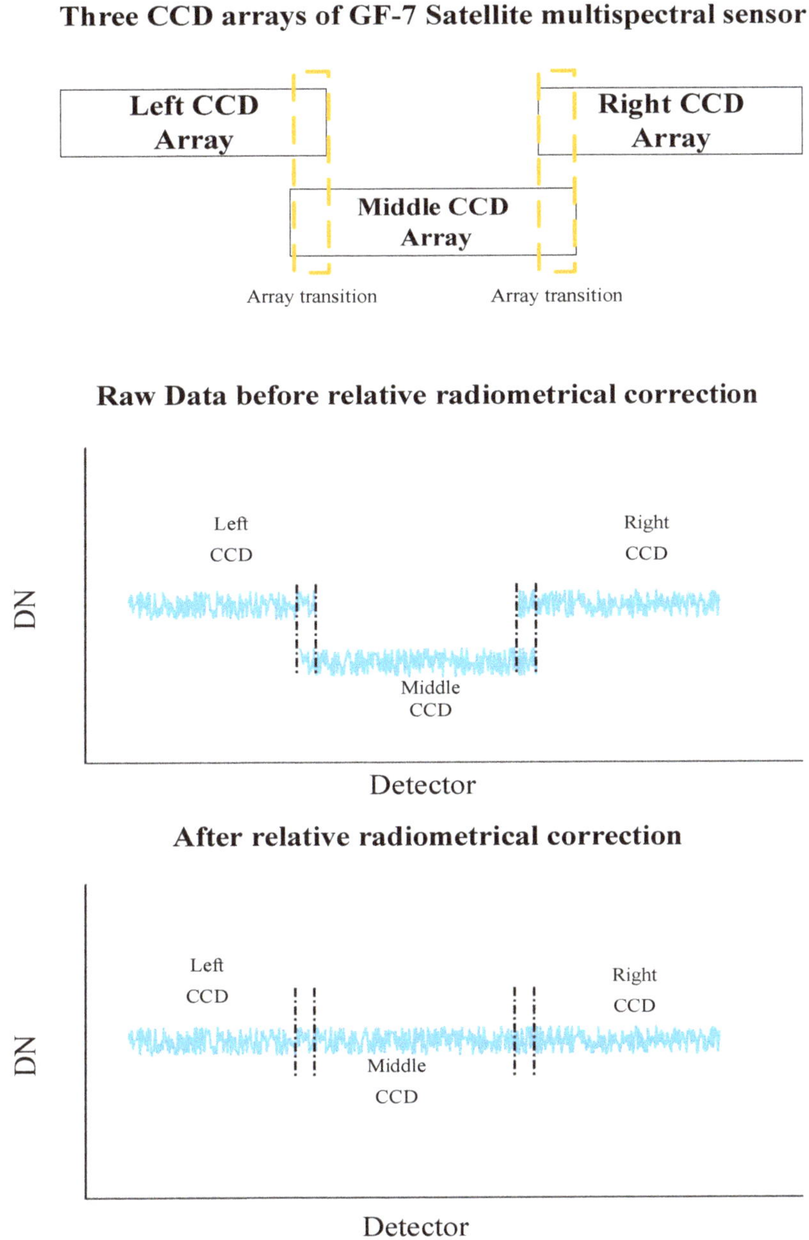

Figure 3. The three CCD arrays of the GF-7 satellite multispectral sensor.

Figure 4. GF-7 BWD-MUX blue band imagery of the uniform Libya 4 PICs (8 August 2020).

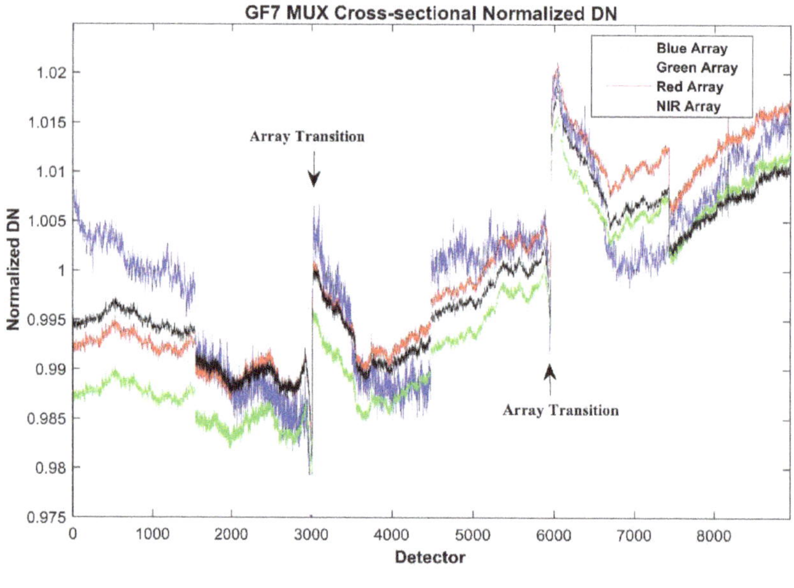

Figure 5. The normalized DN in the along-track direction of the uniform Libya 4 PICs.

4. Absolute Radiometric Accuracy Assessment

4.1. Reflectance-Based Radiometric Calibration Approach

Two independent absolute radiometric assessment campaigns of the GF-7 satellite at different calibration sites were performed by LASAC, using a reflectance-based approach. Reflectance-based radiometric calibration was commonly recognized as one of the most reliable approaches [9–12]. The surface spectral reflectance and atmospheric parameters of the calibration sites were synchronously measured during the GF-7 satellite overpass time. A radiative transfer model, such as MODTRAN, is commonly used for radiometric calibration and validation [13,14]. The predicted TOA radiance of the target was computed by MODTRAN using these synchronous measurements, and the sensor-measured TOA radiance was radiometrically calibrated from DN with absolute radiometric calibration coefficients. The relative difference between the predicted TOA radiance and sensor-measured TOA radiance was recognized as the accuracy of the radiometric calibration [15–19]. Figure 6 shows a detailed flow chart of the reflectance-based absolute radiometric evaluation approach.

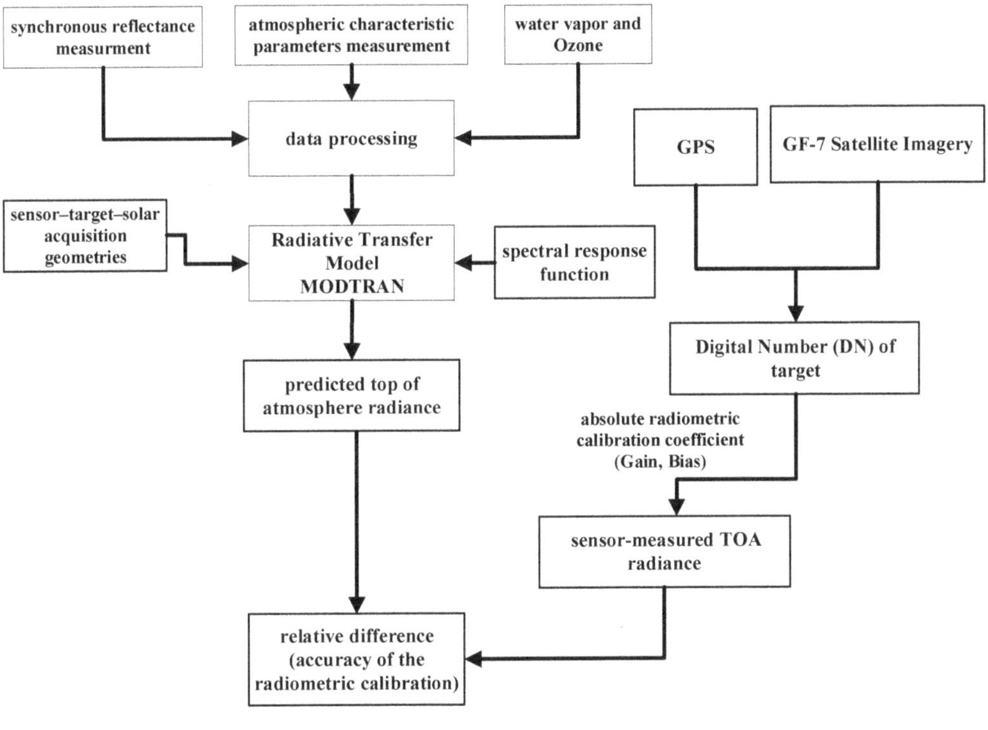

Figure 6. Detailed flow chart of the reflectance-based approach.

The TOA reflectance (known as apparent reflectance) of the GF-7 satellite multispectral imager $\rho^*(\mu_s, \phi_s; \mu_v, \phi_v)$ can be expressed as:

$$\rho^*(\mu_s, \mu_v, \phi_s - \phi_v) = \rho_a(\lambda) + \frac{\rho}{1 - S(\lambda)\rho} T_{\theta_s}(\lambda) T_{\theta_v}(\lambda) \tag{1}$$

The predicted TOA radiance $L_{predicted}(\theta_s, \phi_s; \theta_v, \phi_v)$ can be determined as follows:

$$L(\theta_s, \phi_s; \theta_v, \phi_v) = \frac{\mu_s E_s \rho^*}{\pi d^2} = \frac{\mu_s E_s}{\pi d^2} \left[\rho_a(\lambda) + \frac{\rho}{1 - S(\lambda)\rho} T_{\theta_s}(\lambda) T_{\theta_v}(\lambda) \right] \tag{2}$$

where θ_s is the solar zenith angle, θ_v is the viewing zenith angle, ϕ_s is the solar azimuth angle, and ϕ_v is the viewing azimuth angle. $\mu_s = \cos\theta_s$ and $\mu_v = \cos\theta_v$ are the cosine values. $T_{\theta_s}(\lambda)$ is the total transmittance from the solar to the Earth, and $T_{\theta_v}(\lambda)$ is the total transmittance from the Earth to the satellite sensor. $\rho_a(\lambda)$ is the atmospheric path reflectance, $S(\lambda)$ is the atmospheric spherical albedo, E_s is the exoatmospheric solar irradiance (ESUN), and d is the solar–Earth distance factor [20].

The average DN values of the targets were extracted from the GF-7 satellite multispectral imagery. The sensor-measured TOA radiance $L_{measured}$ was radiometrically calibrated from DN with absolute calibration coefficients.

$$L_{measured} = Gain \times DN + Bias \tag{3}$$

where $L_{measured}$ is the sensor-measured TOA radiance, expressed in units of $W \times m^{-2} \times sr^{-1} \times \mu m^{-1}$. $Gain$ and $Bias$ are the absolute radiometric calibration coefficients. DN is the digital number from the satellite imagery.

$\Delta L\%$ is the relative difference between the predicted TOA radiance and the sensor-measured TOA radiance, and this relative difference is recognized as the accuracy of the absolute radiometric calibration.

$$\Delta L\% = \frac{L_{measured} - L_{predicted}}{L_{predicted}} \times 100\% \tag{4}$$

4.2. Baotou Site Absolute Radiometric Calibration

4.2.1. Baotou Calibration Site

In 2013, the Committee on Earth Observation Satellites (CEOS) Working Group on Calibration and Validation (WGCV) Infrared and Visible Optical Sensors Subgroup (IVOS) established the radiometric calibration network (RadCalNet), which consists of four international calibration and validation sites located in the USA, France, Namibia, and China. The RadCalNet provides the automated in situ measurements of surface spectral reflectance, atmospheric parameters, and the corresponding nadir-viewing TOA reflectance using the radiative transfer model [21–23]. The Baotou site is located in the center of Inner Mongolia, China, and is currently operated by the Aerospace Information Research Institute, Chinese Academy of Science [24,25]. The Baotou site covers a flat area of 300 km^2, dominated by sand and bare soil. A series of targets and infrastructure have been built in the Baotou site to provide effective supports for the radiometric calibration and validation of high-resolution satellite optical sensors. Figure 7 shows the gray-scale permanent artificial targets in the Baotou site. The gray-scale permanent artificial targets are composed of four uniform gravel square targets (one black, one gray, and two white). The white, gray, and black targets, with known spectral reflectance [56%, 18% and 7%], have fairly flat spectral reflectance.

4.2.2. Surface Reflectance

The reflectance measurements of the targets were collected within about 30 min before and after the overpass time of the GF-7 satellite. The reflectance was measured by an SVC HR-1024i spectroradiometer, which covered the wavelength range from 350 nm to 2500 nm. SVC HR-1024i is a high-resolution, field-portable spectroradiometer, and the spectral resolution in the 400–1000 nm wavelength range is 1.5 nm. The reflectance measurement was interpolated at 1 nm intervals in the 350–2500 nm wavelength range by the SVC software. The size of each gray-scale permanent artificial target is 48 m × 48 m; it covers about 15 cross-track pixels and 15 along-track pixels of the GF-7 multispectral imagery. We took 3 × 3 pixels as one sample, and 25 samples were collected in each single target. The parameters of the SVC spectroradiometer were configured to measure five spectra per sample, and then a total of 100 samples and more than 500 spectra were collected. Figure 8 shows the averaged reflectance measurement from the gray-scale permanent artificial targets in the range of 400–1000 nm.

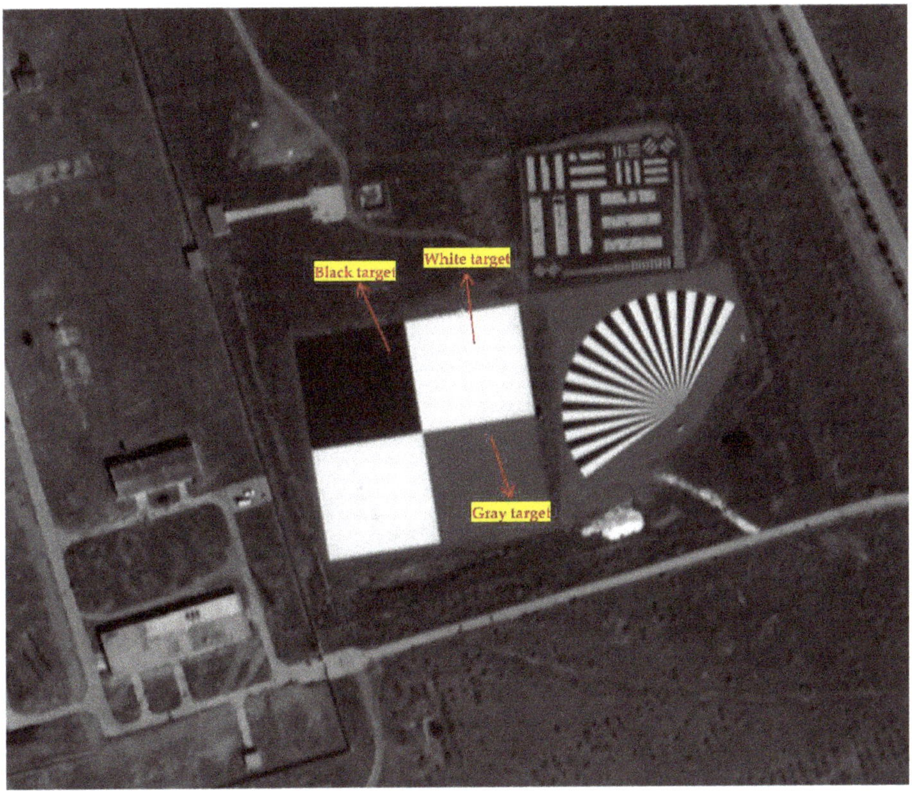

Figure 7. The gray-scale permanent artificial targets at the Baotou site of GF-7 BWD imagery (15 September 2020).

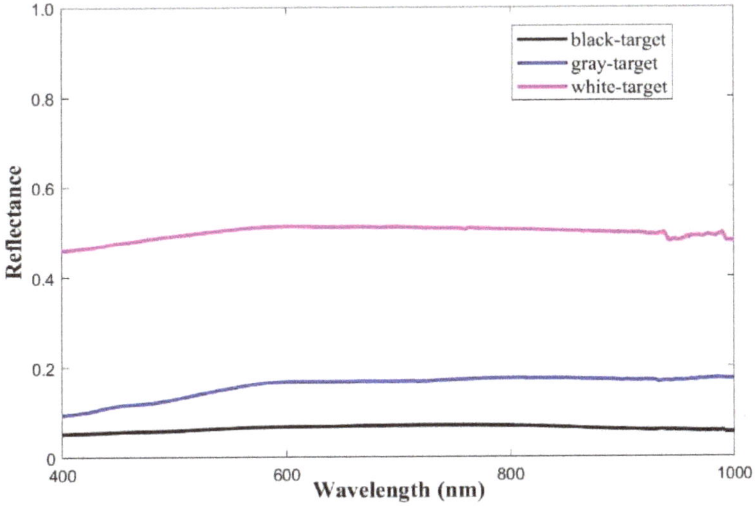

Figure 8. Spectral reflectance measurement data of gray-scale permanent targets.

In this study, a multi-angle bidirectional reflectance distribution function (BRDF) measurement instrument was used to quantify the bidirectional reflectance effect of the gray-scale permanent targets, as shown in Figure 9. The BRDF characteristics of the gray-scale permanent targets were measured according to the angle of incidence and reflection, using the SVC spectroradiometer with the multi-angle instrument [26]. The measured BRDF values of the targets were interpolated at 1 nm intervals in the 400–1000 nm wavelength range.

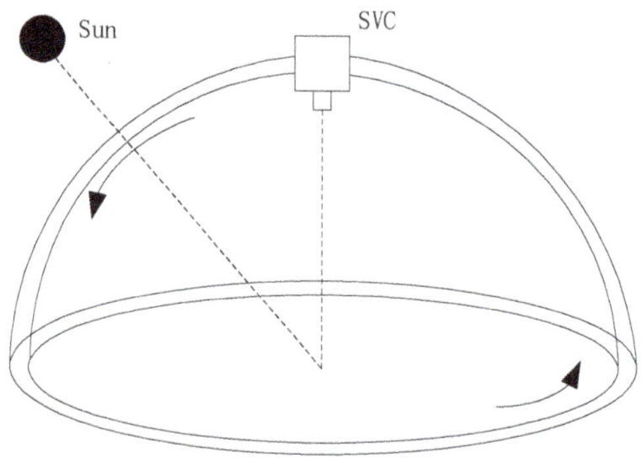

Figure 9. The BRDF measurement instrument.

The anisotropy factor (ANIF) was used to interpret the BRDF effects of elevation and azimuth direction in this study [27,28].

$$\mathrm{ANIF}(\theta_{\mathrm{i}}, \varphi_{\mathrm{i}}, \theta_{\mathrm{v}}, \varphi_{\mathrm{v}}, \lambda) = \frac{\mathrm{BRDF}(\theta_{\mathrm{i}}, \varphi_{\mathrm{i}}, \theta_{\mathrm{v}}, \varphi_{\mathrm{v}}, \lambda)}{\mathrm{BRDF}_N(\theta_{\mathrm{i}}, \varphi_{\mathrm{i}}, \theta_{\mathrm{N}}, \varphi_{\mathrm{v}}, \lambda)} \tag{5}$$

where $\mathrm{BRDF}(\theta_{\mathrm{i}}, \varphi_{\mathrm{i}}, \theta_{\mathrm{v}}, \varphi_{\mathrm{v}}, \lambda)$ is the BRDF for the reflected radiance in the azimuth (φ_{v}) and zenith (θ_{v}) directions at wavelength (λ), which comes from the incident radiance in the azimuth (φ_{i}) and zenith (θ_{i}) directions. $\mathrm{BRDF}_N(\theta_{\mathrm{i}}, \varphi_{\mathrm{i}}, \theta_{\mathrm{N}}, \varphi_{\mathrm{v}}, \lambda)$ is measured in the nadir direction ($\theta_{\mathrm{N}} = 0^{\circ}$). Figure 10 shows the ANIF values of the gray target from the BRDF measurement.

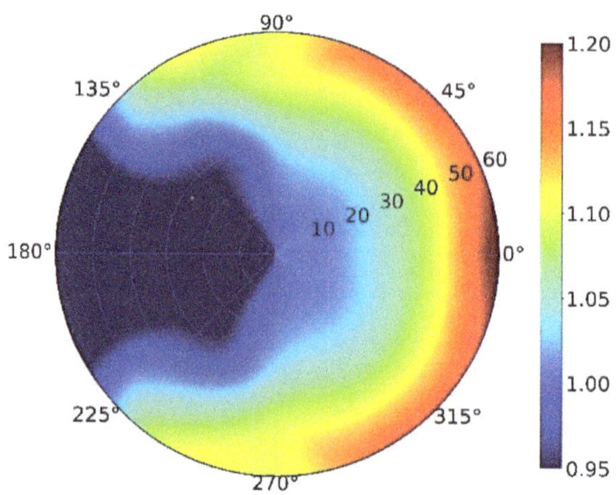

Figure 10. The ANIF values of gray target from BRDF measurement.

These ANIF values are used to better understand the BRDF effects of the gray-scale permanent targets for the GF-7 radiometric calibration. The SVC reflectance measurement in the nadir direction $\rho(\theta_i, \varphi_i, \theta_N, \varphi_v, \lambda)$ of the gray-scale permanent targets was modified by Formula (6), where $\rho^*(\theta_i, \varphi_i, \theta_v, \varphi_v, \lambda)$ is the reflectance of the GF-7 for the relative sensor–target–solar geometry condition [29].

$$\rho^*(\theta_i, \varphi_i, \theta_v, \varphi_v, \lambda) = \frac{\rho(\theta_i, \varphi_i, \theta_N, \varphi_v, \lambda)}{\mathrm{ANIF}(\theta_i, \varphi_i, \theta_v, \varphi_v, \lambda)} \tag{6}$$

4.2.3. Atmospheric Measurements

The atmospheric parameters at the Baotou site were measured by a Cimel CE318 sun photometer. The aerosol optical depth (AOD), aerosol microphysical information, and column amounts of water vapor can be collected automatically every day at this site. Figure 11 shows the synchronous measurement of aerosol optical depth at the Baotou site during the overpass of the GF-7 satellite on 23 July 2020. These atmospheric parameter measurements serve as inputs in MODTRAN, to predict the TOA radiance.

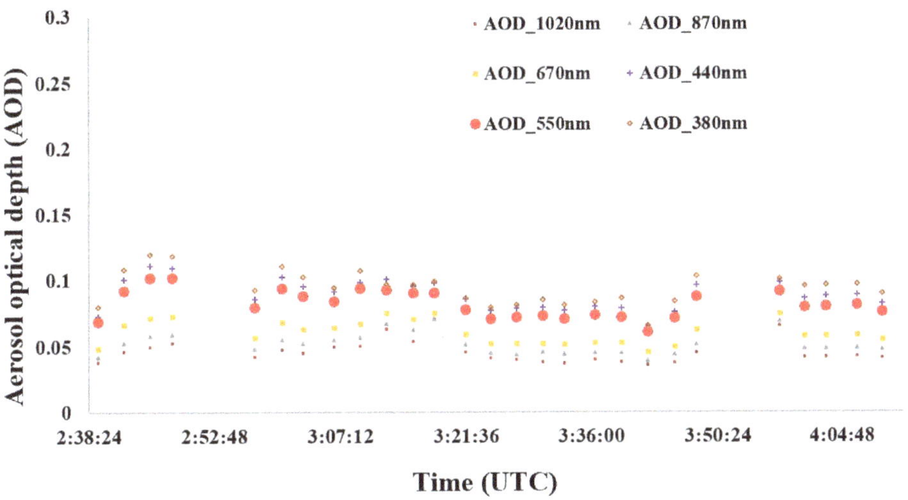

Figure 11. Synchronous measurement of the atmospheric parameters of the Baotou site on 23 July 2020.

4.2.4. Reflectance-Based Radiometric Calibration Campaign

Vicarious reflectance-based radiometric calibration and evaluation campaigns were performed at the Baotou site on the following four separate days: 23 July, 28 July, 15 September, and 20 September 2020. There was no cloud cover on any of the days. Table 4 shows the solar–target–satellite acquisition geometric conditions of the GF7 satellite at the Baotou site. The GF-7 satellite acquired imagery on 23 June at 03:46 (UTC time) at a viewing elevation angle of 84.484° and a viewing azimuth angle of 177.117°, as shown in Figure 12. The viewing elevation angles of the other three imageries acquired on 28 July, 15 September, and 20 September are all more than 82°. Tables 5 and 6 summarize the absolute radiometric calibration result of the GF-7 satellite at the Baotou site on 23 and 28 July, and 15 and 20 September 2020. The comparison between the sensor-measured TOA radiance and the predicted TOA radiance in Tables 5 and 6 showed that the relative difference is less than 5% for all the bands of the GF-7 multispectral imagery, indicating that the accuracy of the absolute radiometric calibration is better than 5%.

Table 4. Solar–target–satellite acquisition geometric conditions of the GF7 satellite at the Baotou site.

Date	Overpass Time (UTC)	Solar Elevation (°)	Viewing Elevation (°)	Solar Azimuth (°)	Viewing Azimuth (°)
23 July 2020	03:46:26	65.412	84.484	143.097	177.117
28 July 2020	03:44:57	64.285	82.554	143.641	146.680
15 September 2020	03:48:11	50.564	84.614	160.732	191.945
20 September 2020	03:47:13	48.724	83.789	161.678	160.266

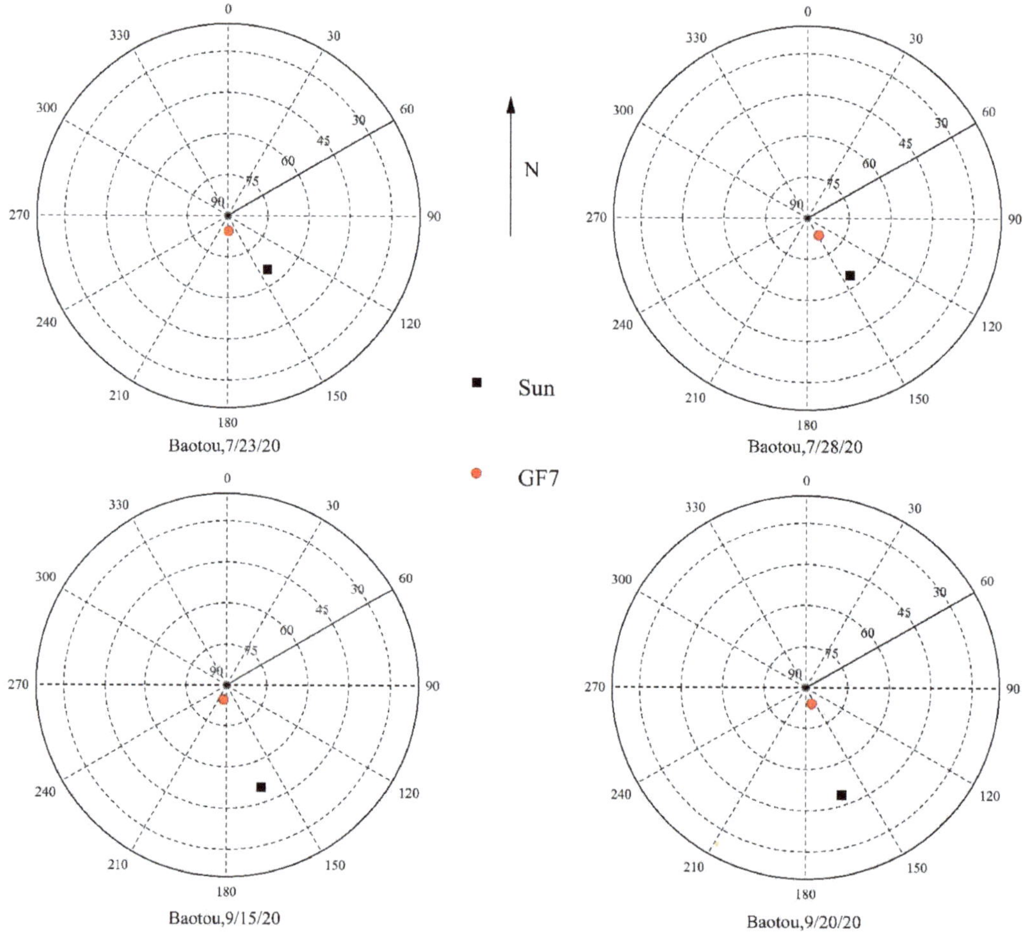

Figure 12. The GF-7 satellite Baotou site acquisition geometry.

Table 5. The absolute radiometric calibration result of the GF-7 satellite at the Baotou site on 23 and 28 July 2020.

Date	Band	Target	$L_{measured}$	$L_{predicted}$	$\Delta L\%$
			$W \times m^{-2} \times sr^{-1} \times \mu m^{-1}$		
23 July 2020	Blue	Black	58.588	56.472	3.61%
		Gray	90.581	87.675	3.21%
		White	233.324	221.986	4.86%
	Green	Black	45.366	43.142	4.90%
		Gray	85.648	82.794	3.33%
		White	211.256	217.673	−3.04%
	Red	Black	34.575	35.878	−3.77%
		Gray	72.852	75.672	−3.87%
		White	180.602	188.231	−4.22%
	NIR	Black	21.524	22.004	−2.23%
		Gray	51.116	53.168	−4.01%
		White	124.943	127.552	−2.09%
28 July 2020	Blue	Black	58.002	55.789	3.97%
		Gray	89.675	86.614	3.53%
		White	230.734	219.766	4.99%
	Green	Black	44.912	43.051	4.32%
		Gray	84.792	81.792	3.67%
		White	209.143	215.039	−2.74%
	Red	Black	34.229	35.444	−3.43%
		Gray	72.123	74.756	−3.52%
		White	178.796	185.953	−3.85%
	NIR	Black	21.309	21.738	−1.97%
		Gray	50.605	52.525	−3.66%
		White	123.694	126.009	−1.84%

Table 6. The absolute radiometric calibration result of the GF-7 satellite at the Baotou site on 15 and 20 September 2020.

Date	Band	Target	$L_{measured}$	$L_{predicted}$	$\Delta L\%$
			$W \times m^{-2} \times sr^{-1} \times \mu m^{-1}$		
15 September 2020	Blue	Black	50.438	51.879	−2.78%
		Gray	76.954	79.030	−2.63%
		White	179.236	185.092	−3.16%
	Green	Black	40.598	39.558	2.63%
		Gray	70.735	73.713	−4.04%
		White	165.577	160.149	3.39%
	Red	Black	30.954	29.991	3.21%
		Gray	65.673	62.680	4.78%
		White	138.312	134.854	2.56%
	NIR	Black	19.338	18.627	3.82%
		Gray	45.292	44.029	2.87%
		White	98.884	94.357	4.80%

Table 6. *Cont.*

Date	Band	Target	$L_{measured}$	$L_{predicted}$	$\Delta L\%$
			$W \times m^{-2} \times sr^{-1} \times \mu m^{-1}$		
20 September 2020	Blue	Black	49.934	51.251	−2.57%
		Gray	76.184	78.074	−2.42%
		White	177.444	182.852	−2.96%
	Green	Black	40.192	39.079	2.85%
		Gray	70.028	72.821	−3.84%
		White	163.921	158.211	3.61%
	Red	Black	30.644	29.628	3.43%
		Gray	65.016	62.040	4.80%
		White	136.929	133.222	2.78%
	NIR	Black	19.145	18.402	4.04%
		Gray	44.839	43.496	3.09%
		White	97.895	93.394	4.82%

4.3. Dunhuang Site Absolute Radiometric Calibration

The Dunhuang site was China's national radiometric calibration site, which has been used to calibrate China's satellites (such as the Feng-Yun series, Zi-Yuan series, and Gao-Fen series of satellites) since the 1990s. The Dunhuang site was selected as one of the "instrumented sites" (LandNet) by CEOS WGCV [30,31]. The Gobi Desert at the Dunhuang calibration site is large, spatially uniform, and homogeneous. The spectral reflectance of the Gobi Desert is stable, with an annual variation of less than 2%. Figure 13 shows the Gobi Desert at the Dunhuang site.

Figure 13. The Dunhuang calibration site.

The Dunhuang site is located about 25 km west of Dunhuang city, Gansu Province, China. The central region of the Gobi Desert is located at the coordinates 94.41° E, 40.09° N, with an elevation of 1280 m above sea level. The size of the central region is 1000 m × 1000 m; it covers more than 360 cross-track pixels and 360 along-track pixels of the GF-7 multispectral imagery.

The methodology took 30 × 30 pixels as one sample, and more than 150 samples were collected in the central region of the Gobi Desert and more than 600 spectra were collected at the Dunhuang calibration site. Figure 14 shows the average and standard deviations of the surface reflectance measurement of the Gobi Desert in the range of 400–1000 nm. Since the Gobi Desert is stable and homogeneous, the standard deviation of more than 600 spectra is less than 1%, as shown in Figure 14. The atmospheric parameters at the Dunhuang site were measured by a Cimel CE318 sun photometer. The daily AOD at 550 nm varied from 0.11 to 0.18 during the campaigns at the Dunhuang test site, and the synchronous AOD measurements at the GF-7 satellite overpass time were less than 0.15.

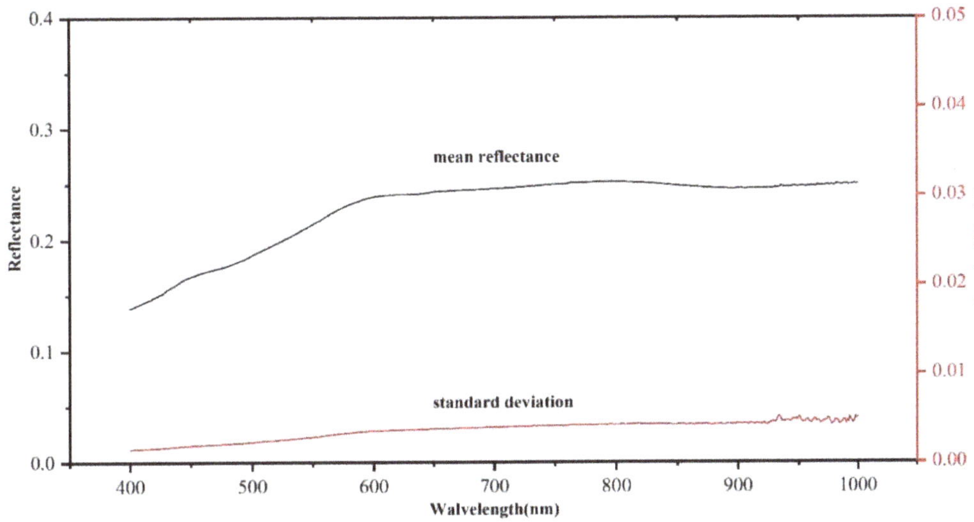

Figure 14. Surface reflectance and standard deviation of the Gobi Desert at the Dunhuang test site.

The BRDF effect of the Gobi Desert at the Dunhuang test site was also measured by the multi-angle measurement instrument, which was exactly the same as described in Section 4.2.

The vicarious radiometric calibration campaign was performed at the Dunhuang test site on the following two separate days: 10 July and 25 July 2020. There was also no cloud cover on either day. Table 7 shows the solar–target–satellite acquisition geometric conditions of the GF7 satellite at the Dunhuang test site. The GF-7 satellite acquired imagery on 10 June at 03:46 (UTC time) at a viewing elevation angle of 84.622° and a viewing azimuth angle of 187.648°, as shown in Figure 15. The other imagery was acquired on 25 July 2020 at a viewing elevation angle of 77.835° and a azimuth angle of 126.181°, as shown in Figure 15. The viewing elevation angle of the other imagery acquired on 25 July was less than 80°. Table 8 summarizes the absolute radiometric calibration result of the GF-7 satellite at the Dunhuang test site on 10 July and 25 July 2020. The comparisons between the measured TOA radiance and the predicted TOA radiance in Table 8 showed the relative differences on July 10 for the blue band (4.43%), green band (2.02%), red band (1.17%), and NIR band (2.22%), as well as the relative differences on 25 July for the blue band (4.68%), green band (3.88%), red band (3.78%), and NIR band (3.26%), which were less than 5% in all the bands. The calibration results of the Dunhuang site agreed well with the results at the Baotou site.

Table 7. Solar–target–satellite acquisition geometric conditions of the GF7 satellite at the Dunhuang site.

Date	Overpass Time (UTC)	Solar Elevation (°)	Viewing Elevation (°)	Solar Azimuth (°)	Viewing Azimuth (°)
10 July 2020	04:47:03	67.835	84.622	139.761	187.648
25 July 2020	04:42:11	65.128	77.835	140.366	126.181

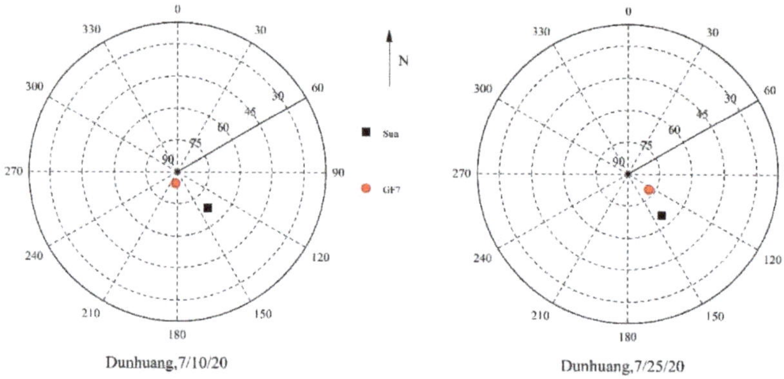

Figure 15. The GF-7 satellite Dunhuang test site acquisition geometry.

Table 8. The absolute radiometric calibration result of the GF-7 satellite at the Baotou site.

Date	Band	Target	$L_{measured}$ ($W \times m^{-2} \times sr^{-1} \times \mu m^{-1}$)	$L_{predicted}$ ($W \times m^{-2} \times sr^{-1} \times \mu m^{-1}$)	$\Delta L\%$
10 July 2020	Blue	Gobi	75.490	72.143	4.43%
	Green	Gobi	82.732	81.060	2.02%
	Red	Gobi	88.205	87.171	1.17%
	NIR	Gobi	72.180	70.580	2.22%
25 July 2020	Blue	Gobi	76.399	72.822	4.68%
	Green	Gobi	86.290	82.945	3.88%
	Red	Gobi	90.273	86.861	3.78%
	NIR	Gobi	73.906	71.499	3.26%

5. Conclusions

The relative and absolute radiometric accuracy of GF-7 satellite imagery is critical for the quantitative applications of natural resources, environment, agriculture, and other industries. In this study, a relative radiometric accuracy evaluation of GF-7 multispectral imagery, using several uniform scenes (such as the uniform Libya, Algeria, and Saharan pseudo-invariant calibration test sites), was performed by LASAC. Nonuniform variations in the original GF-7 imagery were detected before relative radiometric calibration, and the relative radiometric accuracy of the GF-7 multispectral imagery is better than 2% in these large uniform scenes. The evaluation of the absolute radiometric accuracy of the GF-7 satellite sensor at Baotou and Dunhuang, two different calibration sites, was performed by a reflectance-based approach. The BRDF of the targets was measured by a multi-angle instrument with an SVC spectroradiometer. The synchronous measurement of the surface spectral reflectance was modified by the surface BRDF model, and the atmospheric parameters were coupled with the exoatmospheric solar irradiance spectrum and relative spectral response of the sensor as inputs in MODTRAN, to predict the TOA radiance. The predicted TOA radiance was compared with the sensor-measured TOA radiance to evaluate the absolute radiometric accuracy of the GF-7 multispectral imagery. The two independent absolute radiometric assessment campaigns of the GF-7 satellite at the Baotou and Dunhuang sites were performed by LASAC. The radiometrically calibrated accuracy of the GF-7 multispectral imagery was evaluated using different targets at two different calibration sites. The results at the two calibration sites showed that the absolute radiometric accuracy is better than 5%. Considering the results shown in this study, the relative and absolute radiometric accuracy assessment campaign should be performed several times a year, to monitor the radiometric stability of the GF-7 satellite multispectral sensor. This study can also supply a reference to the on-orbit radiometric performance of Chinese high-resolution satellites.

Author Contributions: H.T. developed the research plan and supervised the work, J.X., X.T., W.C. and Q.L. participated in the literature review, method selection, data acquisition and discussions. All authors have read and agreed to the published version of the manuscript.

Funding: This work was supported by the National Key R&D Program of China, 2018YFB0504800, 2018YFB0504801.

Institutional Review Board Statement: Not applicable.

Informed Consent Statement: Not applicable.

Conflicts of Interest: The authors declare no conflict of interest.

References

1. Cao, H.Y.; Liu, F.Q.; Zhao, C.G.; Dai, J. The study of high resolution stereo mapping satellite. *Natl. Remote Sens. Bull.* **2021**, *25*, 1400–1410.
2. Xie, J.; Liu, R.; Zhao, C.G.; Tang, H.Z.; Dou, X.H.; Yang, C.; Xi, S. Design and Data Processing of China's First Spaceborne Laser Altimeter System for Earth Observation: GaoFen-7. *IEEE J. Sel. Top. Appl. Earth Obs. Remote Sens.* **2020**, *13*, 1034–1044. [CrossRef]
3. Cheng, P.; Chaapel, C. WorldView-2 Pansharpening and Geometric Correction. *GeoInformatics Mag.* **2010**, *6*, 1–6.
4. Pagnutti, M.; Ryan, R.E.; Kelly, M.; Holekamp, K.; Zanoni, V.; Thome, K. Radiometric characterization of IKONOS multispectral imagery. *Remote Sens. Environ.* **2003**, *88*, 53–68. [CrossRef]
5. Henderson, B.G.; Krause, K.S. Relative radiometric correction of QuickBird imagery using the side-slither technique on orbit. *Proc. SPIE -Int. Soc. Opt. Eng.* **2004**, *5542*, 1–12.
6. Ron, M.; Julia, B.; Raviv, L.; Brian, M.; Esad, M.; Lawrence, O.; Pat, S.; Kelly, V. Landsat-8 Operational Land Imager (OLI) Radiometric Performance On-Orbit. *Remote Sens.* **2015**, *7*, 2208–2237.
7. Markham, B.; Barsi, J.; Kvaran, G.; Kaita, E.; Ong, L.; Haque, M.O.; Mishra, N.; Czaplamyers, J.; Pahlevan, N.; Helder, D. Landsat-8 Operational Land Imager Radiometric Calibration and Stability. *Remote Sens.* **2014**, *6*, 12275–12308. [CrossRef]
8. Pesta, F.; Bhatta, S.; Helder, D. Radiometric Non-Uniformity Characterization and Correction of Landsat 8 OLI Using Earth Imagery-Based Techniques. *Remote Sens.* **2015**, *7*, 430–446. [CrossRef]
9. Thome, K.J. Absolute Radiometric Calibration of Landsat 7 ETM+ Using the Reflectance-Based Method. *Remote Sens. Environ.* **2001**, *78*, 27–38. [CrossRef]
10. Czapla-Myers, J.; McCorkel, J.; Anderson, N.; Thome, K.; Biggar, S.; Helder, D.; Aaron, D.; Leigh, L.; Mishra, N. The Ground-Based Absolute Radiometric Calibration of Landsat 8 OLI. *Remote Sens.* **2014**, *7*, 600–626. [CrossRef]
11. Biggar, S.F.; Dinguirard, M.C.; Gellman, D.I.; Henry, P.J.; Jackson, R.D.; Moran, M.S.; Slater, P.N.; Guenther, B.W. Radiometric Calibration of SPOT 2 HRV-a Comparison of Three Methods. *Proc. SPIE* **1991**, *1493*, 155–162.
12. Dinguirard, M.; Slater, P.N. Calibration of Space-Multispectral Imaging Sensors: A Review. *Remote Sens. Environ.* **1999**, *68*, 194–205. [CrossRef]
13. Gao, H.L.; Gu, X.F.; Tao, Y. The Research Overview on Visible and Near-infrared Channels Radiometric Calibration of Space-borne Optical Remote Sensors. *Remote Sens. Inf.* **2010**, *4*, 12.
14. Xiong, X.; Barnes, W.; Xie, X.; Salomonson, V. On-orbit performance of Aqua MODIS on-board calibrators. *Proc. SPIE* **2005**, *5978*, 75–80.
15. Ponzoni, F.J.; Junior, J.Z.; Lamparelli, R. In-flight absolute calibration of the CBERS-2 CCD sensor data. *An. Da Acad. Bras. De Ciências* **2008**, *80*, 799–804. [CrossRef]
16. Naughton, D. Absolute radiometric calibration of the RapidEye multispectral imager using the reflectance-based vicarious calibration method. *J. Appl. Remote Sens.* **2011**, *5*, 3544. [CrossRef]
17. Hu, X.; Liu, J.; Sun, L.; Rong, Z.G.; Li, Y.; Zhang, Y.; Zheng, Z.; Wu, R.; Zhang, L.; Gu, X.F. Characterization of CRCS Dunhuang test site and vicarious calibration utilization for Fengyun (FY) series sensors. *Can. J. Remote Sens.* **2010**, *36*, 566–582. [CrossRef]
18. Pinto, T.; Ponzoni, F.; Ruy, C.; Larry, L.; Nischal, M.; David, A.; Helder, D. First in-Flight Radiometric Calibration of MUX and WFI on-Board CBERS-4. *Remote Sens.* **2016**, *8*, 405. [CrossRef]

19. Markham, B.; Thome, K.J.; Barsi, J.A.; Kaita, E.; Helder, D.L.; Barker, J.L.; Scaramuzza, P.L. Landsat-7 ETM+ on-orbit reflective-band radiometric stability and absolute calibration. *IEEE Trans. Geosci. Remote Sens.* **2004**, *42*, 2810–2820. [CrossRef]
20. Chen, W.; Yan, L.; Li, Z.; Xiong, X.X. In-flight absolute calibration of an airborne wide-view multispectral imager using a reflectance-based method and its validation. *Int. J. Remote Sens.* **2013**, *34*, 1995–2005. [CrossRef]
21. Bouvet, M.; Thome, K.; Berthelot, B.; Bialek, A.; Czapla-Myers, J.; Fox, N.P.; Goryl, P.; Henry, P.; Ma, L.; Marcq, S.; et al. RadCalNet: A Radiometric Calibration Network for Earth Observing Imagers Operating in the Visible to Shortwave Infrared Spectral Range. *Remote Sens.* **2019**, *11*, 2401. [CrossRef]
22. Liu, Y.K.; Ma, L.L.; Qiu, S.; Qian, Y.G.; Li, C.R. On-orbit radiometric calibration of the optical sensors on-board SuperView-1 satellite using three independent methods. *Opt. Express* **2020**, *28*, 8.
23. Liu, Y.K.; Ma, L.L.; Wang, N.; Qian, Y.G.; Li, C.R. Vicarious radiometric calibration/validation of Landsat-8 operational land imager using a ground reflected radiance-based approach with Baotou site in China. *J. Appl. Remote Sens.* **2017**, *11*, 1.
24. Li, C.R.; Ma, L.L.; Tang, L.L.; Gao, C.X.; Qian, Y.G.; Wang, N.; Wang, X.H. A comprehensive calibration site for high resolution remote sensors dedicated to quantitative remote sensing and its applications. *Natl. Remote Sens. Bull.* **2021**, *25*, 198–219.
25. Li, C.R.; Ma, L.L.; Gao, C.X.; Tang, L.L.; Liu, Y.K. Permanent target for optical payload performance and data quality assessment: Spectral characterization and a case study for calibration. *J. Appl. Remote Sens.* **2014**, *8*, 083498. [CrossRef]
26. Feingersh, T.; Dorigo, W.; Richter, R.; Dor, E.B. A new model-driven correction factor for BRDF effects in HRS data. In Proceedings of the 4th EARSeL Workshop on Imaging Spectroscopy, Warsaw, Poland, 26–29 April 2005; Volume 4, pp. 565–576.
27. Kim, J.; Jin, C.; Choi, C.; Ahn, H. Radiometric characterization and validation for the KOMPSAT-3 sensor. *Remote Sens. Lett.* **2015**, *6*, 529–538. [CrossRef]
28. Jong-Min, Y.; Jisoo, H.; Jae-Heon, J.; Kwon-Ho, L.; Chang-Suk, L. Initial Radiometric Characteristics of KOMPSAT-3A Multispectral Imagery Using the 6S Radiative Transfer Model, Well-Known Radiometric Tarps, and MFRSR Measurements. *Remote Sens.* **2017**, *9*, 130.
29. Jin, C.; Ahn, H.; Seo, D.; Choi, C. Radiometric Calibration and Uncertainty Analysis of KOMPSAT-3A Using the Reflectance-Based Method. *Sensors* **2020**, *20*, 2564. [CrossRef]
30. Badawi, M.; Helder, D.; Leigh, L.M.; Jing, X. Methods for Earth-Observing Satellite Surface Reflectance Validation. *Remote Sens.* **2019**, *11*, 1543. [CrossRef]
31. Jing, X.; Leigh, L.M.; Pinto, T.; Helder, D. Evaluation of RadCalNet Output Data Using Landsat 7, Landsat 8, Sentinel 2A, and Sentinel 2B Sensors. *Remote Sens.* **2019**, *11*, 541. [CrossRef]

remote sensing

MDPI

Technical Note

Parsimonious Gap-Filling Models for Sub-Daily Actual Evapotranspiration Observations from Eddy-Covariance Systems

Danlu Guo *, Arash Parehkar, Dongryeol Ryu , Quan J. Wang and Andrew W. Western

Department of Infrastructure Engineering, The University of Melbourne, Parkville, VIC 3010, Australia; aparehkar@student.unimelb.edu.au (A.P.); dryu@unimelb.edu.au (D.R.); quan.wang@unimelb.edu.au (Q.J.W.); a.western@unimelb.edu.au (A.W.W.)

* Correspondence: danlu.guo@unimelb.edu.au

Abstract: Missing data and low data quality are common issues in field observations of actual evapotranspiration (ET_a) from eddy-covariance systems, which necessitates the need for gap-filling techniques to improve data quality and utility for further analyses. A number of models have been proposed to fill temporal gaps in ET_a or latent heat flux observations. However, existing gap-filling approaches often use multi-variate models that rely on relationships between ET_a and other meteorological and flux variables, highlighting a critical lack of parsimonious gap-filling models. This study aims to develop and evaluate parsimonious approaches to fill gaps in ET_a observations. We adapted three gap-filling models previously used for other meteorological variables but never applied to infill sub-daily ET_a or flux observations from eddy-covariance systems before. All three models are solely based on the observed diurnal patterns in the ET_a data, which infill gaps in sub-daily data with sinusoidal functions (Sinusoidal), smoothing functions (Smoothing) and pattern matching (MaxCor) approaches, respectively. We presented a systematic approach for model evaluation, considering multiple patterns of data gaps during different times of the day. The three gap-filling models were evaluated together with another benchmarking gap-filling model, mean diurnal variation (MDV) that has been commonly used and has similar data requirement. We used a case study with field measurements from an EC system over summer 2020–2021, at a maize field in southeastern Australia. We identified the MaxCor model as the best gap-filling model, which informs the diurnal pattern of the day to infill by using another day with similar temporal patterns and complete data. Following the MaxCor model, the MDV and the Sinusoidal models show comparable performances. We further discussed the infilling models in terms of their dependence on data availability and their suitability for different practical situations. The MaxCor model relies on high data availability for both days with complete data and the available records within each day to infill. The Sinusoidal model does not rely on any day with complete data, which makes it the ideal choice in situations where days with complete records are limited.

Keywords: actual evapotranspiration; latent heat; infill; data-driven; missing data; flux tower; data quality; maize field

1. Introduction

Actual evapotranspiration (ET_a) is an important component of the global water balance, accounting for about 62% of global precipitation over land [1]. Understanding and measuring ET_a can provide useful information for various water resources management applications, such as for catchment water yield, urban water supply and irrigation management [2]. In the context of irrigated fields, ET_a consists of both the evapotranspiration from crop surface and soil evaporation, which can take a total of 50–95% of surface irrigation water [3].

The eddy-covariance (EC) technique is considered to be one of the best techniques to obtain continuous, high-frequency field measurements of ET_a [4]. The technique measures sensible heat and latent heat fluxes, H and λE, where the latter corresponds to ET_a. However, missing and low-quality data are commonly seen in EC-based measurements due to instrument malfunctions, power failures and unfavorable weather conditions [4–7]. These data gaps limit the utility of these datasets when complete ET_a records are necessary, such as for water balance calculations. Therefore, effective gap-filling techniques are important to improve the completeness of ET_a data obtained from EC systems.

Several approaches have been developed to infill gaps in the λE flux observations or the ET_a observations directly, which have also been tested over a large range of field conditions including various climates zones and land cover types. Table S1.1 in the Supplementary provides a detailed summary of existing literature in these infilling approaches. These approaches largely rely on meteorological and/or other flux variables measured at the same location and over the same period as the variable to infill [7]. These gap-filling methods can be categorized into: (1) infilling λE (or ET_a) gaps with available records from neighboring time steps or with similar meteorological conditions [4,8–10], (2) building regression models between λE (or ET_a) and meteorological data, which sometimes also require further monitoring data for soil moisture and vegetation conditions such as leaf area index (LAI) [4,11–14]; and (3) predicting λE (or ET_a) with complex statistical models such as Kalman filter, multiple imputation or machine-learning algorithms developed based on meteorological conditions [11,12,14,15].

Although numerous gap-filling models were developed for λE flux and ET_a, they share a common limitation in the high model complexity and data requirement, highlighting a critical lack of parsimonious gap-filling models that operate only with the variable to infill itself. With the exception of two models, all other existing gap-filling approaches are dependent on relationships between λE flux or ET_a and other driving variables, such as additional flux and meteorological variables. The two exceptions are the mean diurnal variation (MDV) [10] and the analogue period (AP) [16] methods in which gap-filling relies solely on the variable to infill itself. Thus, the applicability and performances of most of these gap-filling methods are highly dependent on the quality and availability of those additional variables. For example, the feasibility of such methods is limited when the required meteorological/flux data are also missing [7,16]. Further, the infilling may be affected by spurious relationships produced from changing meteorological conditions and/or outliers and low-quality records within the meteorological observations [4].

This study therefore aims to develop and evaluate parsimonious models to infill gaps in ET_a observations derived from EC systems; we focus on parsimonious models that require only ET_a itself, and thus having no reliance on data for other variables. Compared to the wide range of existing methods to fill data gaps in ET_a and λE flux, the models presented here are easier to implement and are more applicable in data-limited situations. The simpler model structures also remove the dependence on the quality of measurements other than ET_a (e.g., flux variables and meteorological data), which are often used as model predictors in existing infilling models. We adapt three parsimonious models that have never been used to infill sub-daily ET_a data before. Two models are fitted to the observed diurnal patterns of ET_a data in days with gaps, and one model utilizes days with complete ET_a records to identify a matching temporal pattern to infill each day with gaps. These new models were compared with the mean diurnal variation (MDV) model, which is an existing parsimonious model that fills gaps in sub-daily data by averaging values recorded at the same time step within a short time window around the gap [10]. The MDV model has been widely used to infill gaps in flux variables [4,8,14,17].

We present a systematic approach to evaluate different infilling models considering multiple patterns of data gaps during different times of the day. We used a case study using field measurements from an EC system over summer 2020–2021 at a maize field in southeastern Australia. We discuss the relative performances of the four models along with their dependence on data availability, from which recommendations are made for different practical conditions. These gap-filling models and model recommendations presented will be highly valuable for improving the completeness and utility of ET_a measurements from EC systems in future studies. We made the R codes of all models evaluated in this study publicly available on GitHub https://github.com/DanluGuo/ETinfilling/blob/main/4_ETinfilling_Models_V2.R (accessed on 1 February 2022) along with example data.

2. Materials and Methods

The three new parsimonious gap-filling models, along with the existing model, MDV, were evaluated with field monitoring data from an EC system. The monitoring site and instrumentation is introduced in Section 2.1. The gap-filling models are introduced in Section 2.2. Section 2.3 describes the evaluation process, including (1) data resampling to represent typical missing/erroneous data at different times in a day; and (2) performance assessment and comparison for the four infilling models.

2.1. Monitoring Site for the Eddy-Covariance System and Data

To evaluate different ET_a gap-filling models, we used monitoring data from an eddy-covariance system installed at a maize field over the 2020–2021 summer cropping season. The study field was within the Goulburn-Murray Irrigation District in southeast Australia (field centered at −36.18S, 145.04E). The typical cropping season for summer maize in this region spans from November to May, while the study field was sown on 11 December 2020 and harvested on 23 April 2021. The field is located between temperate and arid steppe climate regions [18], with an annual mean rainfall of 447 mm, based on records at the closest public weather station (Kyabram, Australian Bureau of Meteorology #80091, 19 km away), from 1964 to 2021.

We continuously monitor the in-field weather condition alongside CO_2, H_2O and sensible heat fluxes using an eddy-covariance system, between 19 December 2020 and 12 April 2021. We monitored the air temperature, solar radiation, relative humidity and wind speed at 2 meters' height from a standing weather station. The CO_2 and H_2O fluxes were measured using an open path infrared gas analyser (LI-7500, Lincoln, NE. LI-COR, Inc.) and a three-dimensional (3D) sonic anemometer (CSATS3, Campbell Scientific Australia) as the core of the eddy-covariance system. All flux variables were monitored and recorded at 20 Hz frequency and subsequently processed and aggregated to 30-min interval with EddyPro® Software (Version 7.0) [19].

Figure 1 shows a photograph of the set of full monitoring equipment in field. The EddyPro® Software processes raw eddy covariance data to compute biospheric/atmosphereic fluxes of CO_2, H_2O and sensible heat including applying raw data filtering, calibration, and other algorithms for calculating and correcting fluxes [20]. The remaining energy balance components were monitored with Kipp and Zonen CNR-1 radiometer, HFT3-L REBS soil heat flux plates and TCAV soil temperature thermocouples. The solar radiation was monitored at a 30-min interval, while air temperature, wind speed and relative humidity were monitored every 5-min; therefore, the ET_a observations processed by EddyPro® was in a 30-min time step. The maize field is rectangular with cropping rows along the east–west gradient. Within the field, the weather station and the flux tower were placed next to each other at 10 m away from the northern field boundary and over 100 m away from the other three boundaries. Therefore, the vast majority of the target footprint was located at the southern side of the monitoring stations.

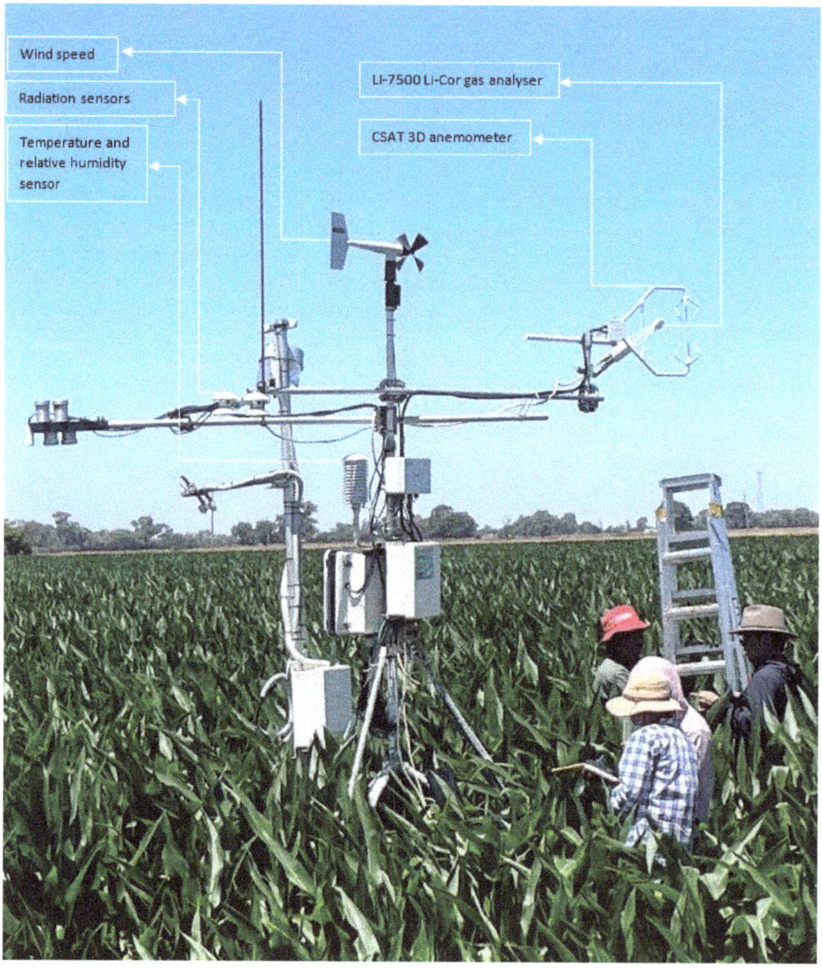

Figure 1. The eddy-covariance system and the weather station for monitoring fluxes and weather conditions at the study field. The purposes for different parts of the monitoring stations are labelled. Note that the CSAT 3D anemometer was measuring the 3D wind speed and direction, while the wind speed sensor on the top provided a second set of wind speed measures and the dominant direction within the horizontal plane for validation. The soil moisture down to 90 cm and the canopy temperature and NDVI were also monitored but observations were not used in this study.

Thirty-minute ET_a data were estimated with EddyPro using the observations from the EC systems. The reference evapotranspiration (ET_0) data at corresponding time steps was estimated using the weather observations and the FAO-56 Penman–Monteith model [21]. The key quality issue for the ET_a data occurred during periods when the wind was blowing from the north due to the limited fetch across the crop and a flux source footprint unrepresentative of the crop; therefore, the corresponding flux measurements may not be representative of the field, leading to inaccurate ET_a estimates and poor energy balance closures. When wind was from southerly directions between 112.5 and 247.5 degrees (i.e., ESE to WSW), the sum of sensible and latent heat fluxes accounted for a median of 90% of available energy, suggesting a good energy closure (Figure 2). The gaps and potential errors in the ET_a measurements prompted the need to develop effective gap-filling approaches.

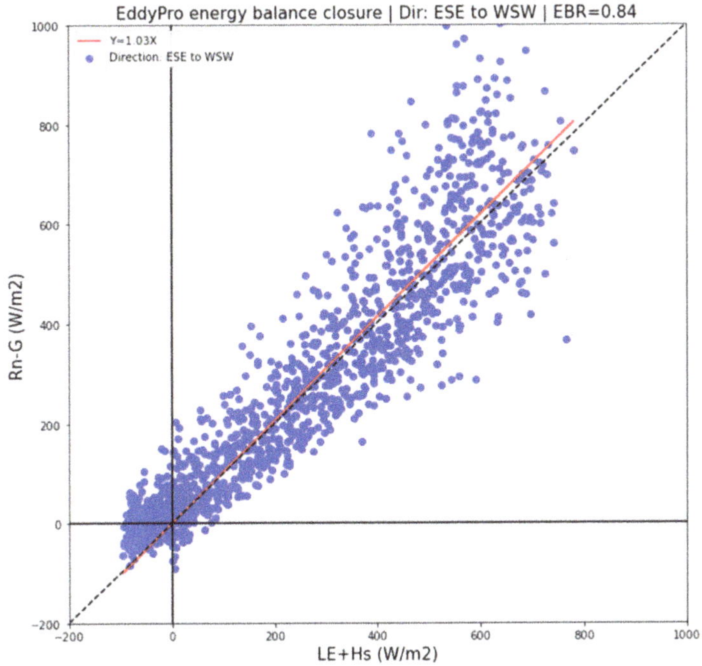

Figure 2. When wind directions are between 112.5 and 247.5 degrees (i.e., ESE to WSW), the heat fluxes (the sum of sensible and latent heat fluxes) accounted for a median of 90% of available energy, suggesting a good energy closure.

2.2. Three Gap-Filling Models for Sub-Daily ET_a

We adapted three models to infill gaps in sub-daily ET_a and evaluated their performance along with a benchmarking model using our field observations. Suppose that each day within the observation period can be categorized by data availability as either a full day (FUL), a partial day (PAR), or a sparse day (SPA) as follows:

- Full day (FUL)—where data within the day is complete or mostly (≥80%) complete;
- Partial day (PAR)—where part of the data within the day is missing but a substantial portion (30–80%) is still available;
- Sparse day (SPA)—where data within the day is mostly (>70%) missing/erroneous.

None of the three gap-filling models that we adapted and evaluated require data other than ET_a records themselves. The models use the available records differently to fill gaps in the daytime 30-min ET_a records within each PAR day (i.e., day to infill). The Sinusoidal model (*Daily sinusoidal functions of ET_a*) describes the diurnal patterns of ET_a with a sinusoidal function of the time in a day, which have been widely applied to fill gaps in time-series of meteorological data, soil heat flux data, and even gaps in daily evapotranspiration [22–24]. The Smoothing model (*Daily smoothing functions of ET_a*) describes the diurnal patterns of ET_a with polynomial functions of the time in a day, which is adapted from a common gap-filling approach for meteorological time-series [25]. The MaxCor model (*Daily temporal pattern matching for ET_a*) fills gaps in a day based on another day with complete records that is selected as having the most similar diurnal pattern with the day to infill. This model is conceptually similar to the analogue period (AP) model which fills a gap by searching the full dataset for an 'analogue period' that that has similar temporal patterns with the data surrounding the gap [16]. However, the implementation of MaxCor is much simplified as the search is based on a daily time-step, rather than the more flexible, user-defined time step as implemented in AP; specifically, MaxCor searches for an 'analogue day', while AP allows any length of analogue period and recommends case-specific investigation to determine the optimal length. None of the three models have been used to infill sub-daily ET_a data to our knowledge.

As a benchmark to the three abovementioned new models, we also included the mean diurnal variation (MDV) model in our evaluation, as this has been a widely used parsimonious gap-filling model for λE and carbon fluxes [4,8,14,17]. Similar to the three models introduced in this study, MDV requires the variable to infill as the only input. The MDV model was originally developed to fill gaps in laten heat flux observations [10]. The model fills any gap in sub-daily records by averaging values measured at the same time step on days adjacent to (both before and after) the gap, within a time window usually between 4 and 15 days. A shorter averaging period is considered insufficient to determine a reasonable mean value, while a longer average period might introduce errors due to potential non-linear impacts from other environmental variables. More details on the MDV model are included in its original paper [10]. For the model evaluation in this study, we implement MDV to infill any missing 30-min ET_a records by the average of all values recorded at the identical 30-min time slot, within the adjacent 14 days (i.e., 7 days before and 7 days after the day where gap presents).

Common to all infilling models, a specific daytime period is defined for each day to infill, based on solar time angles estimated with the latitude and longitude of the study site and the ordinal dates within the record period, following Chapter 3 of the FAO-56 guidelines [21]. Across the season, the ranges of sunrise and sunset times are between 5:30 a.m. and 6:30 a.m. and 5:30 p.m. and 7 p.m., respectively; solar noon is between 12 p.m. and 12:15 p.m. Any ET_a for times outside of daytime is treated as night-time ET_a and assumed negligible. Any day that belongs to the SPA set is not filled because the available data is considered insufficient to be filled reliably. The threshold chosen to define the SPA set (having >70% of the 30-min ET_a data missing) implies that a day can be infilled with a minimum of 15 out of 48 records available. This is a relatively low data requirement which could lead to unreliable gap filling. However, since our primary aim is to present and evaluate gap-filling models, a more important consideration in choosing the threshold was to enable a reasonable number of days remaining to be used for model evaluation (see Figure S1.1 in the Supplementary Materials for an assessment of data availability across the season). Details on how the FUL and PAR datasets were used for model evaluation is included in Section 2.3.

The three infilling models are detailed subsequently:

- *Sinusoidal—Daily sinusoidal functions of ET_a*: This model uses all available daytime 30-min ET_a records on the day to be infilled (each day in the PAR set) to fit a sinusoidal function between ET_a and time of the day, which has a period specific to that day. The fitted sinusoidal curve is then used to estimate all 30-min daytime ET_a while infilling the missing time steps. We chose the sinusoidal function because of its simplicity and ability to represent the overall diurnal patterns of ET_a, which we concluded from a visual assessment of ET_a for the FUL days within our records (days with >80% complete data, see details in Section 2.3.1). The sinusoidal function used takes the form of:

$$ET_{a,H} = Amp \times \sin\left(\frac{2\pi H}{P}\right) \tag{1}$$

Equation (1) describes the diurnal pattern of daytime 30-min ET_a with the positive half of a sine curve. H is the time since sunrise in decimal hours at the start of each 30-min slot (e.g., 1, 1.5, 2 h, and $H = 0$ at sunrise). Ideally, the full period of the sinusoidal curve, P, should be equal to twice of the daytime length (time between sunrise and sunset) of each day, which enables the representation of all daytime 30-min ET_a with half of the sine curve where the daily peak occurs halfway between the sunrise and sunset. However, we found via a preliminary analysis on all FUL days that the daytime ET_a peaks around 1:30 p.m., and the ET_a diurnal patterns seem to follow only part of the half-sine curve (Figure 3). To represent these diurnal patterns in daytime ET_a more accurately, the period P in Equation

(1) is defined for each day as four times the hour difference between the sunrise and the hour of peak daytime ET_a (1:30 p.m.), and the modelled ET_a from Equation (1) after sunset of each day is zeroed; this adjustment of the sinusoidal function ensures that the model best resembles the observed asymmetrical diurnal patterns in ET_a. *Amp* is the only model parameter to be calibrated, which represents the amplitude of the sine curve; it is fitted by minimizing the sum of squared residuals from the available data on the day to be infilled.

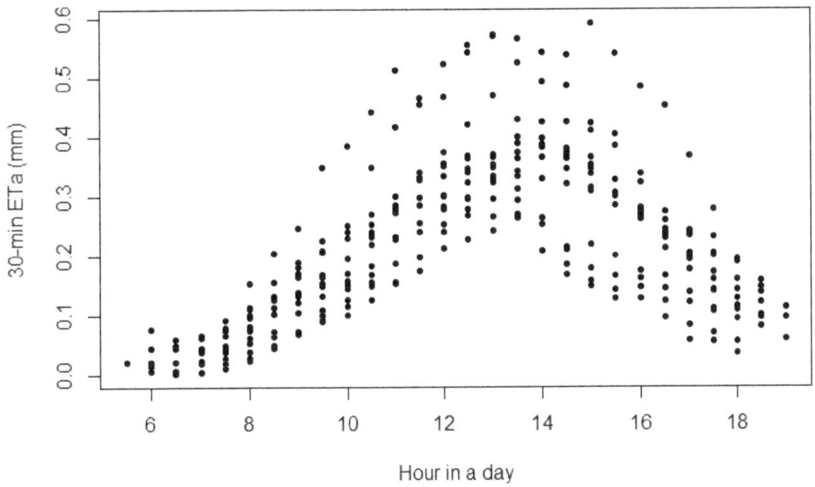

Figure 3. The diurnal patterns of 30-min daytime ET_a across all days in the season with complete data (FUL days). The values of 30-min ET_a generally peak around 1:30 p.m. and follow the shape of an incomplete half-sine curve.

- *Smoothing—Daily smoothing functions of ET_a*: This model uses all available daytime ET_a data on the day to be infilled to fit a second-order polynomial smoothing function between ET_a and time of the day. The fitted smoothing function is then used to infill ET_a for the missing time steps. The second-order polynomial smoothing function takes the form of:

$$ET_{a,H} = AH^2 + BH + C \tag{2}$$

In Equation (2), *H* is the time since sunrise in decimal hours at the start of each 30-min slot; *A*, *B* and *C* are the model parameters to be calibrated.

- *MaxCor—Daily temporal pattern matching for ET_a*: For each day in the PAR set, this model first calculates the linear correlation (i.e., Pearson correlation coefficient) between the daytime ET_a records in the current day and each day in the FUL set. The correlation calculation only considers the common timeslots where data are available in both the current day and each FUL day. Based on these correlations, the FUL day that has the maximum correlation with the day to infill is selected. Within this 'matching FUL day', all individual 30-min daytime ET_a values (ET_a_FUL) are divided by their sum ($ET_a\_tot\_FUL$) to calculate the proportions of 30-min ET_a to the daily total, ET_{a_prop}. This is described in Equation (3), where $H = 0, 0.5, 1, \ldots, 24$, denoting the time since sunrise in decimal hours:

$$ET_{a\_prop,H} = \frac{ET_a\_FUL_H}{ET_a\_tot\_FUL} = \frac{ET_a\_FUL_H}{\sum_{H=0}^{24} ET_a\_FUL_H} \tag{3}$$

To infill the data gaps in the PAR day, we first estimate the daily total ET_a of this day ($ET_a\_tot\_complete\_PAR$) by dividing the sum of all available ET_a records (*ETa_tot_avail_PAR*) by the sum of ET_{a_prop} values corresponding to these timeslots with available data. This is described in Equation (4), where *Hpar* are the time since sunrise (decimal hours) for all timeslots with available records in the PAR day:

$$ET_a\_tot\_complete\_PAR = \frac{ET_a\_tot\_avail\_PAR}{\sum_{i \in Hpar} ET_{a\_prop,i}} = \frac{\sum_{i \in Hpar} ET_a\_PAR_i}{\sum_{i \in Hpar} ET_{a\_prop,i}} \quad (4)$$

We can then estimate each 30-min ET_a value for the PAR day (ET_a_PAR) with the estimated total ET_a for the day ($ET_a\_tot\_complete\_PAR$) and all proportions of 30-min ETa, ET_{a_prop}, which enables us to fill the ET_a gaps (Equation (5)).

$$ET_a\_PAR_H = ET_a\_tot\_complete\_PAR \times ET_{a\_prop,H} \quad (5)$$

We made the R codes which implement all four abovementioned models available on GitHub https://github.com/DanluGuo/ETinfilling/blob/main/4_ETinfilling_Models_V2.R (accessed on 1 February 2022) along with the data used in this study.

2.3. Model Evaluation Process

Although the ultimate goal of the above-mentioned gap-filling models is to infill sub-daily data for days with partially missing data (i.e., the PAR set, as detailed in Section 2.2), our model evaluation was based on days within the FUL set only to understand the performance of individual infilling models. Specifically, we divided days within the FUL set into training and evaluation sets. We added artificial gaps to data (i.e., assign an NA value to some of the data) in the evaluation set to represent typical types of missing data from the field observations.

2.3.1. Classifying Daily Data Completeness

We first classified all days in our monitoring period into the FUL, PAR, or SPA sets by the completeness of data in each day, as highlighted by different colors in Figure 4.

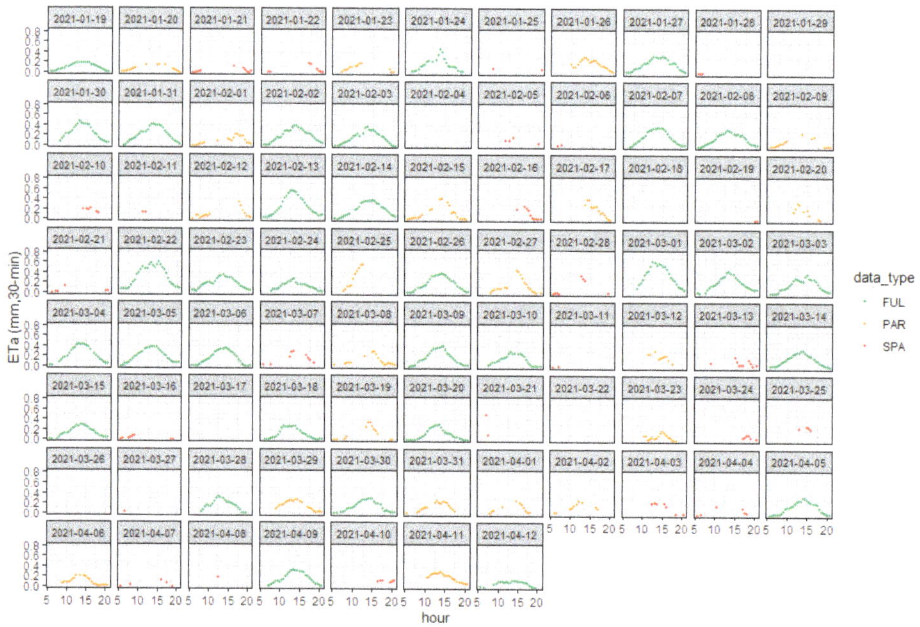

Figure 4. Classification of the completeness of 30-min ET_a data for each day within the observation period. The colors differentiate days within the FUL, PAR and SPA sets. See the below text for the explanation of individual categories and their utility.

The data were then used as follows.

1. The FUL set (green in Figure 4) contains days with complete/near complete (≥80%) records. These data will be further divided for training and evaluation of the four infilling models (Section 2.3.2).
2. The PAR set (orange in Figure 4) contains days with partially complete (30–80%) records. These data were then used to summarize the typical patterns of missing data. We identified three typical patterns of missing data as:
    - A: with most missing data in the morning (sunrise to 10 a.m.);
    - B: with most missing during mid-day (10 a.m. to 3 p.m.);
    - C: with most missing during afternoon (3 p.m. to sunset).

The days highlighted in red in Figure 4 are classified into the SPA set, where data for most (>70%) of the day were missing. As discussed in Section 2.2, these days are not recommended to be infilled because of significant lack of 'ground truth'.

2.3.2. Building the Training and Evaluation Datasets

Figure 4 identified 32 days within the FUL set, which were then randomly divided into:

1. A training set (60%, 19 days); and
2. An evaluation set (40%, 13 days).

The training dataset was used to represent days with complete data (the FUL set). Within the evaluation set, we added artificial gaps to each day to mimic each of the three typical patterns of missing data (A, B and C, Section 2.3.1). This led to three separate evaluation sets representing each type of missing data to be infilled with the four models. For each day in each evaluation set, data points corresponding to the gaps were held off (assigned as NA) and used for evaluating the performance of infilling models. For example, Figure 5 shows the training and evaluation sets that represent missing data Type A (missing morning, where red points are artificial gaps), in which all data between sunrise and 10 a.m. in the evaluation set were held off. Data splits for the other two types of missing data (B and C) are shown in Figures S1.2 and S1.3 in the Supplementary Materials.

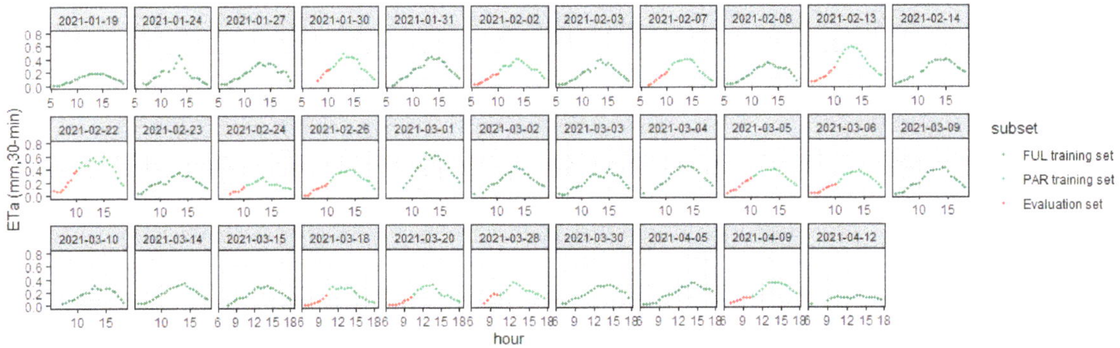

Figure 5. Split of the training and evaluation subsets to represent missing data Type A, i.e., missing morning.

2.3.3. Comparing the Model Infilling Performances

With the training and evaluation datasets created following Section 2.3.2, the four gap-filling models (three proposed models and MDV, see Section 2.2) were compared by their evaluation performance on infilling the missing 30-min ET_a data. The root-mean-squared-error (RMSE) and the r-sqaured (R^2) were used to assess the model performance against the evaluation data; the former represents the average error in infilling the 30-min ET_a relative to true observations and the latter represents the proportion of variance in infilling the 30-min ET_a observations that can be explained by the model. As a further reference to infilling performances, the RMSE values for the daily total ET_a were also calculated for days that contain missing data.

3. Results

Figure 6 summarizes the RMSE of 30-min ET_a for the four infilling models for the three types of missing data (A: missing morning; B: missing mid-day; C: missing afternoon), respectively. The MaxCor model consistently has the lowest model errors across all three situations, with RMSE values between 0.03 and 0.07. Following MaxCor, the next best models are the MDV (with RMSE ranging from 0.04 to 0.1) and the Sinusoidal (with RMSE ranging from 0.05 to 0.1) models, which have comparable magnitudes of errors. The Smoothing model has the worst performance with the highest errors for both Type A (missing morning, RMSE = 0.18) and Type C (missing afternoon, RMSE = 0.2). Considering the variance explained, the Sinusoidal and the MaxCor have higher ability to explain the observed variance, with R^2 values ranging from 0.7 to 0.87 and 0.68 to 0.82, respectively. The MDV model struggles to explain variance for Type B (missing mid-day, R^2 = 0.25) while the Smoothing model again shows limited performance for Types A and C (missing morning and afternoon, R^2 = 0.45 and 0.37, respectively).

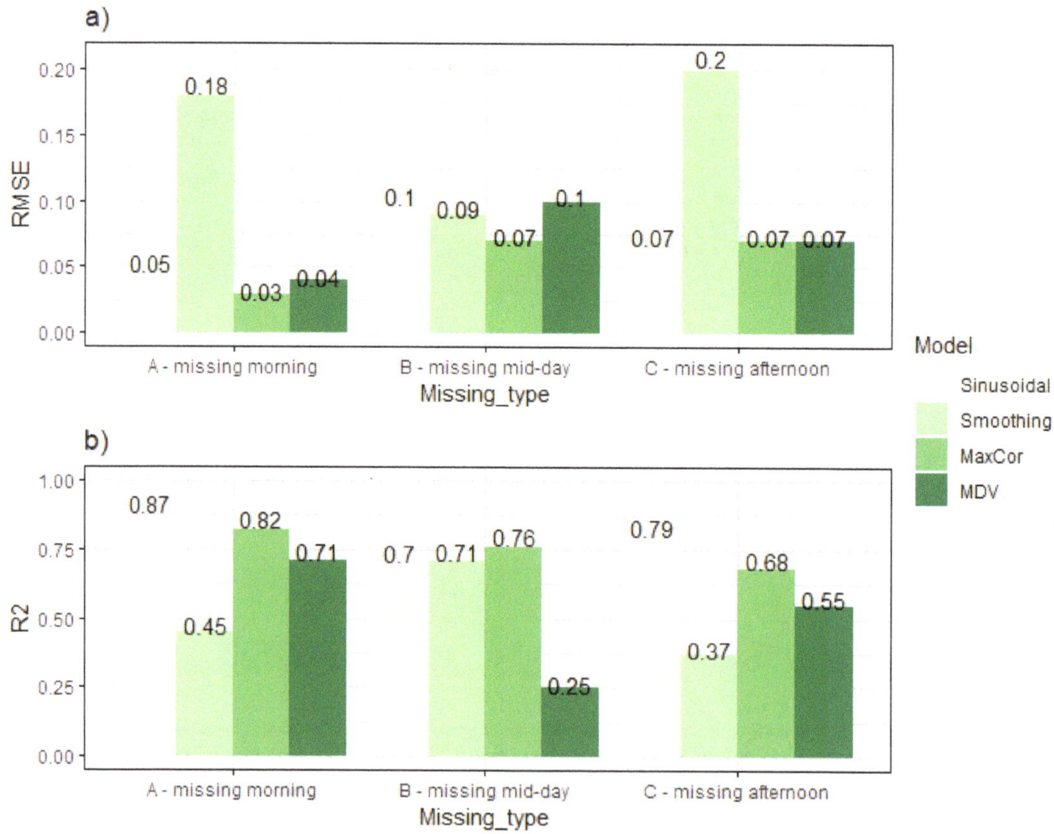

Figure 6. (**a**) RMSE and (**b**) R^2 of the 30-min ET_a (in mm) for the infilled gaps, obtained from the four infilling models for each evaluation dataset that represents typical patterns of missing data: A—missing morning; B—missing mid-day; and C—missing afternoon.

With the above summary of performance of the four infilling models for the 30-min ET_a data, we further aggregate infilling performance for days with missing data to understand the expected accuracy at the scale of daily ET_a. Figure 7 summarizes the daily RMSE of four infilling models for the three types of missing data. The best-performing model, MaxCor, has mean errors of 0.26–0.62 mm in daily total ET_a.

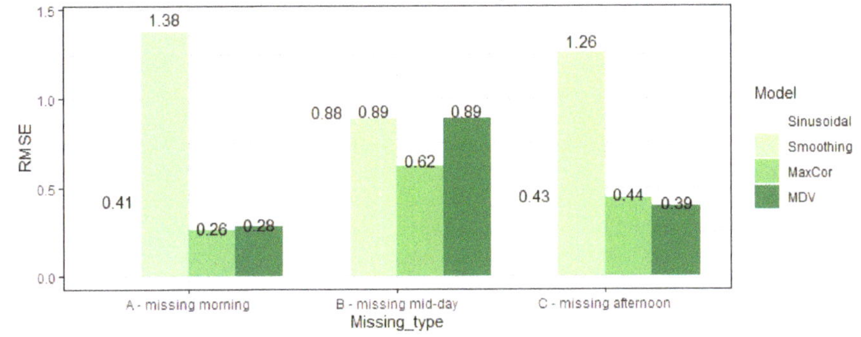

Figure 7. RMSE of daily ET_a (in mm) for the infilled days with gaps, obtained from four infilling models for each evaluation dataset that represents typical patterns of missing data: A—missing morning; B—missing mid-day; and C—missing afternoon.

4. Discussion

4.1. Performances of Gap-Filling Models

Within the four infilling models that we evaluated, we see better performance for two models, which both perform gap filling for a day based on other 'similar' days and another model based on fitting functions to diurnal pattern of each day with gaps. Specifically, the MaxCor model identifies another day that has a complete record, while also having a similar diurnal pattern in the 30-min ET_a to the existing records of the day to infill. The MDV model estimates missing records using the mean value of the specific time step in a day from neighboring days. The Sinusoidal model fits a sinusoidal function to the existing records within each day to infill and then uses the sinusoidal function for gap filling. The performances of all these three models are relatively stable across different missing data types, with less than 0.06 difference in the RMSE of 30-min ET_a across different types of missing data.

In contrast, we see lower and more variable performances for the infilling model that uses smoothing functions to fill each day with gaps (Smoothing), where the maximum difference of RMSE of 30-min ET_a across different missing data types is 0.11. This suggests a potential limitation of infilling performance due to the model structure. Specifically, the second-degree polynomial which is used as the smoothing function introduces more flexibility in the diurnal patterns of the 30-min ET_a compared to either the MaxCor or the Sinusoidal models in which the diurnal pattern for the day to infill is bounded by either a similar temporal pattern in the actual records of the other day or by a sinusoidal function. Consequently, the fitted smoothing functions can be highly sensitive to fluctuations and outliers in the existing data, which can lead to spurious diurnal patterns and thus large errors in the infilled records.

There is no systematic pattern of how model performance varies across different types of missing data, suggesting that these variations are likely a result of individual model structures. For examples, the Sinusoidal and the MaxCor models show the worst performances for missing data Type B (missing mid-day), which may indicate the critical role of mid-day records. For the Sinusoidal model, these mid-day records generally consist of higher absolute values of ET_a and thus have large impact on the calibration of the sinusoidal infilling function. For the MaxCor model, this low performance for missing mid-day records could be a result of the relatively high day-to-day variation of the temporal patterns in mid-day ET_a (Figures 2 and 3), which leads to difficulties to infill missing records reliably using the temporal pattern within another day.

This study focuses on parsimonious gap-filling models for ET_a, which do not rely on any input variables other than ET_a itself. Thus, they do not explicitly take into account any impact of weather conditions (e.g., solar radiation, cloudiness, rainfall, temperature), which are often considered important driving variables for ET. As such, all these parsimonious models share a common and natural caveat in maintaining robust performance on days when the influence of weather conditions on ET_a is strong, and when weather conditions change abruptly. The infilling errors would be greatest when the data gap falls across a period of abrupt change in weather conditions. This is a fundamental limitation of all parsimonious gap-filling models that solely rely on ET_a data.

To further understand the impact of this general limitation of parsimonious infilling models, we performed an additional analysis on the performance of the four gap-filling models under various cloud cover and solar conditions. Specifically, we plot the daily RMSE of each gap-filling model under the three types of data gaps against the daily ratio of actual solar radiation to clear-sky solar radiation (Figure S1.4 in the Supplementary Materials). We found that none of the four models are systematically influenced in performance by various cloud cover conditions within our dataset. This is likely due to the relatively limited variation in cloud cover conditions within our dataset to comprehensively characterize the effects of clouds on the accuracy of these gap filling approaches. Another plausible hypothesis is that these parsimonious gap-filling models, by considering the temporal patterns of sub-daily ET_a, have already effectively represented variation due to changes in cloud cover conditions (since solar radiation is an input when estimating ET_a from EC-systems). However, this could only be the case if the cloud cover is relatively stable throughout the day; under situations where the amount of cloud cover is highly variable within a day, the diurnal variations of ET_a would be much more difficult to be predicted from a simple smooth curve and/or averaging values from another day(s). This analysis illustrates the potential limitation of the infilling approaches under highly variable weather conditions, which is a general limitation to all parsimonious infilling models as discussed above. Similarly, we can expect much higher influence of the weather conditions on model performance when the weather is more highly variable throughout a day. Therefore, we strongly recommend individual investigation of this limitation when testing these (and potential other) parsimonious models to new datasets.

4.2. Recommendations for Practical Situations with Different Data Availability

In addition to the above comparison of model performances, we discuss the data requirement of individual infilling models to provide recommendations for different practical situations.

While the MaxCor and the MDV models are the best-performing models, both models also have higher data requirements for both days with complete data and the available data within each day to infill. Higher data availability on each day to infill enables a better understanding and thus a more reliable match of that day to the appropriate day with complete records. Large numbers of days with complete records are also critical for both the MaxCor and the MDV models: for the former, these provide a diverse set of diurnal patterns to match with the data in the day to infill; for the latter, more days with complete records can provide reliable mean estimates for each time step to infill.

Both the Sinusoidal and the Smoothing models use infilling functions that are fitted to the existing 30-min ET_a records within the day with missing data. Therefore, neither require any day with complete data. Considering this together with the model performances in Section 3, the Sinusoidal model becomes the best choice to infill a dataset with limited complete days of record. An example for this situation is where the monitoring location experiences regular unfavorable wind direction that occurs for part of most days, leading to low-quality data (i.e., effective gaps) on most days of the eddy-covariance observations. Such a data quality issue is likely a result of inappropriate selection for the location of the eddy-covariance system, which may be due to practical constraints in many cases (e.g., to avoid conflict with machinery access to cropping fields, sites with naturally unfavorable conditions in certain upwind directions). It is also worth noting that both the Sinusoidal and the Smoothing models do require a reasonable amount of data available in each day to infill, to enable a reliable infilling function to be developed. This data requirement is less strict for the Sinusoidal model, as the sinusoidal functions pose greater constraint on the diurnal patterns of 30-min ET_a. We encourage individual model evaluation in further case studies to obtain a specific and precise understanding of the impacts of data availability on model performance.

5. Conclusions

We adapted three parsimonious data-driven models to infill gaps in sub-daily ET_a observations from eddy-covariance systems and evaluated these models together with another commonly used benchmarking model of similar data requirement. We applied these models to infill gaps in the 30-min ET_a data collected from an eddy-covariance monitoring station installed in a maize field in southeastern Australia, over the 2020–21 summer season. We identified the best gap-filling model as a pattern-matching model to inform the diurnal pattern of the day to infill by another day with complete data (MaxCor). The second-best model is the benchmarking model, mean diurnal variation (MDV), closely followed by another proposed model which performs gap filling with sinusoidal functions fitted to the diurnal pattern of each day with gaps (Sinusoidal).

Further recommendations on model choice were made considering practical data availability. The best-performing MaxCor model relies on high data availability for both days with complete data and the available records within each day to infill. The Sinusoidal model does not rely on days with complete data while also offering reasonable performance, which makes it the best choice in situations where complete days of records are limited. We acknowledge that the performance of individual infilling models assessed may be specific to our study site and monitoring period, and results may differ across different climatic conditions, evaporative surfaces (i.e., crop type) and data availability. Therefore, local evaluation is highly recommended for future studies aiming to apply these infilling techniques. The strategies to allocate available records—to be used for calibration and evaluation of the infilling models and to be excluded from infilling due to data scarcity—should also be tailored for individual case studies, based on specific assessments of data availability (e.g., Figure S1.1). To facilitate further applications, we made the R codes of all four models evaluated in this study publicly available on GitHub https://github.com/DanluGuo/ETinfilling/blob/main/4_ETinfilling_Models_V2.R (accessed on 1 February 2022) along with the data we used to facilitate further applications.

Supplementary Materials: The following are available online at https://www.mdpi.com/article/10.3390/rs14051286/s1, Figure S1.1: Percentage 30-min ETa data availability within each day, sorted from the lowest to highest across the full monitoring dataset. Figure S1.2: Split of the training and evaluation subsets to represent missing data Types B i.e., missing mid-day.. Figure S1.3: Split of the training and evaluation subsets to represent missing data Types C i.e., missing afternoon. Figure S1.4: Daily RMSE of the four gap-filling models under the three typical patterns of missing data (A—missing morning; B—missing mid-day; and C—missing afternoon), plotted against the daily ratio of actual solar radiation to clear-sky solar radiation. Each panel shows one gap-filling model where the three missing data types are differentiated by colours. Table S1.1: Existing approaches to infill gaps in latent heat flux, carbon flux or directly for ETa. Orange cells highlight models that rely on additional input variable other than the variable to infill. Green cells highlight the only two existing parsimonious gap-filling models, the mean diurnal variation (MDV) and the analogue period (AP). Reference [26] is cited in the Supplementary Materials.

Author Contributions: Conceptualization, A.W.W., D.R., Q.J.W. and D.G.; methodology, D.G., A.W.W., D.R.; software, A.P., D.G.; validation, D.G., A.W.W., D.R., Q.J.W., A.P.; formal analysis, D.G.; investigation, D.G.; resources, A.W.W., D.R., Q.J.W.; data curation, A.P., D.G.; writing—original draft preparation, D.G.; writing—review and editing, D.G., A.P., A.W.W., D.R., Q.J.W.; visualization, D.G.; supervision, A.W.W., D.R., Q.J.W.; project administration, D.G.; funding acquisition, A.W.W., D.R., Q.J.W. All authors have read and agreed to the published version of the manuscript.

Funding: This research has been supported by the Australian Research Council via a Linkage Project (grant no. LP170100710), with contributions from our industrial collaborator, Rubicon Water.

Institutional Review Board Statement: Not applicable.

Informed Consent Statement: Not applicable.

Data Availability Statement: The authors have made all data used in this study publicly available at: https://github.com/DanluGuo/ETinfilling/blob/main/4_ETinfilling_Models_V2.R (accessed on 1 March 2022).

Acknowledgments: The authors would also like to thank Kevin Saillard, David Aughton, Emil Somers, Zitian Gao and Rodger Young for their assistance in the field monitoring campaign.

Conflicts of Interest: The authors declare no conflict of interest.

References

1. Dingman, S.L. *Physical Hydrology*, 3rd ed.; Waveland Press: Long Grove, IL, USA, 2015.
2. McMahon, T.A.; Peel, M.C.; Lowe, L.; Srikanthan, R.; McVicar, T.R. Estimating actual, potential, reference crop and pan evaporation using standard meteorological data: A pragmatic synthesis. *Hydrol. Earth Syst. Sci.* **2013**, *17*, 1331–1363. [CrossRef]
3. Grafton, R.Q.; Williams, J.; Perry, C.J.; Molle, F.; Ringler, C.; Steduto, P.; Udall, B.; Wheeler, S.A.; Wang, Y.; Garrick, D.; et al. The paradox of irrigation efficiency. *Science* **2018**, *361*, 748–750. [CrossRef] [PubMed]
4. Boudhina, N.; Zitouna-Chebbi, R.; Mekki, I.; Jacob, F.; Ben Mechlia, N.; Masmoudi, M.; Prévot, L. Evaluating four gap-filling methods for eddy covariance measurements of evapotranspiration over hilly crop fields. *Geosci. Instrum. Methods Data Syst.* **2018**, *7*, 151–167. [CrossRef]
5. Zitouna-Chebbi, R.; Prévot, L.; Chakhar, A.; Abdallah, M.M.-B.; Jacob, F. Observing Actual Evapotranspiration from Flux Tower Eddy Covariance Measurements within a Hilly Watershed: Case Study of the Kamech Site, Cap Bon Peninsula, Tunisia. *Atmosphere* **2018**, *9*, 68. [CrossRef]
6. Pastorello, G.; Trotta, C.; Canfora, E.; Chu, H.; Christianson, D.; Cheah, Y.-W.; Poindexter, C.; Chen, J.; Elbashandy, A.; Humphrey, M.; et al. The FLUXNET2015 dataset and the ONEFlux processing pipeline for eddy covariance data. *Sci. Data* **2020**, *7*, 1–27. [CrossRef]
7. Aubinet, M.; Vesala, T.; Papale, D. *Eddy Covariance: A Practical Guide to Measurement and Data Analysis*; Springer: Dordrecht, The Netherlands, 2012.
8. Wutzler, T.; Lucas-Moffat, A.; Migliavacca, M.; Knauer, J.; Sickel, K.; Šigut, L.; Menzer, O.; Reichstein, M. Basic and extensible post-processing of eddy covariance flux data with REddyProc. *Biogeosciences* **2018**, *15*, 5015–5030. [CrossRef]
9. Alfieri, J.G.; Blanken, P.D.; Yates, D.N.; Steffen, K. Variability in the Environmental Factors Driving Evapotranspiration from a Grazed Rangeland during Severe Drought Conditions. *J. Hydrometeorol.* **2007**, *8*, 207–220. [CrossRef]
10. Falge, E.; Baldocchi, D.; Olson, R.; Anthoni, P.; Aubinet, M.; Bernhofer, C.; Burba, G.; Ceulemans, R.J.; Clement, R.; Dolman, A.; et al. Gap filling strategies for defensible annual sums of net ecosystem exchange. *Agric. For. Meteorol.* **2001**, *107*, 43–69. [CrossRef]
11. Chen, Y.-Y.; Chu, C.-R.; Li, M.-H. A gap-filling model for eddy covariance latent heat flux: Estimating evapotranspiration of a subtropical seasonal evergreen broad-leaved forest as an example. *J. Hydrol.* **2012**, *468–469*, 101–110. [CrossRef]
12. Goodrich, J.; Wall, A.; Campbell, D.; Fletcher, D.; Wecking, A.; Schipper, L. Improved gap filling approach and uncertainty estimation for eddy covariance N2O fluxes. *Agric. For. Meteorol.* **2020**, *297*, 108280. [CrossRef]
13. Cleverly, J.; Dahm, C.N.; Thibault, J.R.; Gilroy, D.J.; Coonrod, J.E.A. Seasonal estimates of actual evapo-transpiration from Tamarix ramosissima stands using three-dimensional eddy covariance. *J. Arid Environ.* **2002**, *52*, 181–197. [CrossRef]
14. Alavi, N.; Warland, J.S.; Berg, A. Filling gaps in evapotranspiration measurements for water budget studies: Evaluation of a Kalman filtering approach. *Agric. For. Meteorol.* **2006**, *141*, 57–66. [CrossRef]
15. Abudu, S.; Bawazir, A.S.; King, J.P. Infilling Missing Daily Evapotranspiration Data Using Neural Networks. *J. Irrig. Drain. Eng.* **2010**, *136*, 317–325. [CrossRef]
16. Hoeltgebaum, L.E.B.; Dias, N.L.; Costa, M.A. An analog period method for gap-filling of latent heat flux measurements. *Hydrol. Process.* **2021**, *35*, e14105. [CrossRef]
17. Moffat, A.M.; Papale, D.; Reichstein, M.; Hollinger, D.Y.; Richardson, A.D.; Barr, A.G.; Beckstein, C.; Braswell, B.; Churkina, G.; Desai, A.R.; et al. Comprehensive comparison of gap-filling techniques for eddy covariance net carbon fluxes. *Agric. For. Meteorol.* **2007**, *147*, 209–232. [CrossRef]
18. Peel, M.C.; Finlayson, B.L.; McMahon, T.A. Updated world map of the Köppen-Geiger climate classification. *Hydrol. Earth Syst. Sci.* **2007**, *11*, 1633–1644. [CrossRef]
19. LI-COR Biosciences. *EddyPro® Software (Version 7.0.8) [Computer Software]*; LI-COR Biosciences: Lincoln, NE, USA, 2021. Available online: https://www.licor.com/env/support/EddyPro/software.html (accessed on 1 December 2021).
20. LI-COR Biosciences. *EddyPro® Software Version 7.0 Instruction Manual*; LI-COR Biosciences: Lincoln, NE, USA, 2021. Available online: https://www.licor.com/documents/1ium2zmwm6hl36yz9bu4 (accessed on 1 December 2021).

21. Allen, R.G.; Pereira, L.S.; Raes, D.; Smith, M. *FAO Irrigation and Drainage Paper No. 56*; Food and Agriculture Organization of the United Nations: Rome, Italy, 1998; Volume 56, p. e156.
22. Hocke, K.; Kämpfer, N. Gap filling and noise reduction of unevenly sampled data by means of the Lomb-Scargle periodogram. *Atmos. Chem. Phys.* **2009**, *9*, 4197–4206. [CrossRef]
23. Santanello, J.A.; Friedl, M.A. Diurnal Covariation in Soil Heat Flux and Net Radiation. *J. Appl. Meteorol.* **2003**, *42*, 851–862. [CrossRef]
24. Cha, M.; Li, M.; Wang, X. Estimation of Seasonal Evapotranspiration for Crops in Arid Regions Using Multisource Remote Sensing Images. *Remote Sens.* **2020**, *12*, 2398. [CrossRef]
25. Pappas, C.; Papalexiou, S.M.; Koutsoyiannis, D. A quick gap filling of missing hydrometeorological data. *J. Geophys. Res. Atmos.* **2014**, *119*, 9290–9300. [CrossRef]
26. Eamus, D.; Cleverly, J.; Boulain, N.; Grant, N.; Faux, R.; Villalobos-Vega, R. Carbon and water fluxes in an arid-zone Acacia savanna woodland: An analyses of seasonal patterns and responses to rainfall events. *Agric. For. Meteorol.* **2013**, *182*, 225–238. [CrossRef]

remote sensing MDPI

Article

SkySat Data Quality Assessment within the EDAP Framework

Sebastien Saunier [1], Gizem Karakas [2], Ilyas Yalcin [2,3], Fay Done [4], Rubinder Mannan [4], Clement Albinet [5], Philippe Goryl [5] and Sultan Kocaman [6,7,*]

1 Telespazio France, 26 Avenue Jean François Champollion, 31100 Toulouse, France; sebastien.saunier@telespazio.com
2 Graduate School of Science and Engineering, Hacettepe University, 06800 Beytepe, Ankara, Turkey; gizem.karakas@hacettepe.edu.tr (G.K.); ilyas.yalcin@hacettepe.edu.tr (I.Y.)
3 Baskent OSB Technical Sciences Vocational School, Hacettepe University, 06909 Sincan, Ankara, Turkey
4 Telespazio UK Ltd., 350 Capability Green, Luton, Bedfordshire LU1 3LU, UK; fay.done@telespazio.com (F.D.); rubinder.mannan@telespazio.com (R.M.)
5 European Space Agency, ESRIN, Via Galileo Galilei, 1, 00044 Frascati, RM, Italy; clement.albinet@esa.int (C.A.); philippe.goryl@esa.int (P.G.)
6 Department of Geomatics Engineering, Hacettepe University, 06800 Beytepe, Ankara, Turkey
7 ETH Zurich, Institute of Geodesy and Photogrammetry, 8093 Zurich, Switzerland
* Correspondence: sultankocaman@hacettepe.edu.tr

Abstract: Cal/Val activities within the Earthnet Data Assessment Pilot (EDAP) Project of the European Space Agency (ESA) cover several Earth Observation (EO) satellite sensors, including Third-Party Missions (TPMs). As part of the validation studies of very-high-resolution (VHR) sensor data, the geometric and radiometric quality of the images and the mission compliance of the SkySat satellites owned by Planet were evaluated in this study. The SkySat constellation provides optical images with a nominal spatial resolution of 50 cm, and has the capacity for multiple visits of any place on Earth each day. The evaluations performed over several test sites for the purpose of the EDAP Maturity Matrix generation show that the high resolution requirement is fulfilled with high geometric accuracy, although various systematic and random errors could be observed. The 2D and 3D information extracted from SkySat data conform to the quality expectations for the given resolution, although improvements to the vendor-provided rational polynomial coefficients (RPCs) are essential. The results show that the SkySat constellation is compliant with the specifications and the accuracy results are within the ranges claimed by the vendor. The signal-to-noise ratio assessments revealed that the quality is high, but variations occur between the different sensors.

Keywords: Earth Observation; data quality; SkySat; geometric validation; radiometric validation; maturity matrix

1. Introduction

The increasing number and diversity of Earth Observation (EO) satellite missions operated by various agencies and commercial bodies has led to the requirement of a platform to enable communication between providers. In order to ensure the coherence of the missions and their integration into the overall EO strategy, a coordination framework was defined by European Space Agency (ESA) for Third-Party Missions (TPM) within the Earthnet Programme, which has contributed to the Global Earth Observation System of Systems (GEOSS) for over 40 years [1]. The Earthnet Data Assessment Pilot (EDAP) project aimed to validate the data quality for various missions [2]. The calibration and validation (Cal/Val) tasks in the EDAP were defined in a sensor-specific manner [3]. One of the main outcomes of the project was the Maturity Matrix (MM) framework, which was initially developed within the EDAP and advanced jointly with National Aeronautics and Space Administration (NASA), Washington, DC, USA. The MM provides clear guidelines on the Cal/Val activities of the product and outputs reports on the quality and the suitability of satellite missions on the basis of their mission specifications and potential applications [4].

The SkySat constellation from Planet, CA, USA, is composed of 21 satellite missions operating in orbit at the time of writing. After the launch of the SkySat A and B generations in 2013 and 2014, respectively, the first SkySat C-generation satellite was launched in 2016. SkySat was the first commercial microsatellite with sub-meter accuracy [5]. All SkySat C-generation sensors have similar electro-optics, with a focal length of 3.6 m and 5.5 megapixel Complementary Metal Oxide Semiconductor (CMOS) detectors [6] that facilitate the provision of continuous EO data with a small temporal interval and high consistency. Besides the 4 band multispectral (MS) imagery, the pushframe sensor structure enable the High-Definition (HD) full-frame rate panchromatic video acquisition with a 1 m ground sample distance (GSD) [6]. The main objectives of the mission include the acquisition of very high spatial resolution (i.e., enhanced 50 cm spatial resolution) MS images and video data with rapid revisit time (multiple revisits per day).

The SkySat Cal/Val tasks defined within the EDAP aimed at the fulfillment of the MM analysis requirements. The EDAP Best Practice Guidelines [7] and EDAP Optical Mission Quality Assessment Guidelines [8] were in principle followed for the requirement analysis and MM preparation. For this purpose, besides a review of the documentation (e.g., the review in [6] and the Planet L1 Data Quality Report Q3 [9]), extensive data (including metadata) investigations were carried out with the aim of assessing the image radiometric and geometric quality over various test sites. In addition to the Cal/Val reports provided by the vendor [9] and the collaborators (e.g., [10–12]), the results reported in the literature [13–17] were analyzed as well. On the other hand, mission specific assessments, e.g., stereoscopic capability in the case of SkySat data, were also performed using state-of-the-art processing methods.

Consequently, the main aims of this study are (a) to investigate the geometric and radiometric quality of SkySat products acquired over several EDAP test sites and at different product levels; (b) to discuss the issues related to the data and metadata with respect to the instrument specifications and the literature; and (c) to present the EDAP MM assessment results and potential improvements to the sensor data. Although several studies in the literature (i.e., [13–17]) have analyzed the radiometric and geometric quality of different SkySat products for various purposes, such as surface reconstruction from still and video images, resolution improvement methods, etc., using data acquired over one or two sites, this is the first comprehensive assessment of the SkySat constellation covering various Cal/Val aspects (i.e., radiometry, geometry, visual interpretation) at the same time using data acquired from multiple satellites and over multiple sites. In addition, the concept of MM developed as part of the EDAP is explained in detail and the SkySat MM is presented.

The image quality assessment efforts included the determination of the General Image Quality Equation (GIQE) parameters, which were developed for the quantitative measurement of the quality of raw or processed aerial imagery [18]. A scale consisting of ten levels has been developed by the National Image Interpretability Rating Scale (NIIRS) for detecting and interpreting objects contained in visible, radar, infrared, and multispectral imagery [19]. The GIQE parameters allow these scale values to be calculated quantitatively. The Signal-to-Noise Ratio (SNR), Signal-Difference-to-Noise Ratio (SDNR), Relative Edge Response (RER), Ground Sample Distance (GSD), and Modulation Transfer Function (MTF) are among the main GIQE parameters [20]. In this study, the SNR, RER and MTF values were analyzed over an artificial target. In addition, visual checks were performed over multiple test sites.

The geometric quality assessments include absolute and temporal geolocation accuracy, band-to-band registration (BBR) accuracy, and the quality of digital surface model (DSM) generation. The geometric quality aspects evaluated here were selected on the basis of their importance to users and their significance in the literature (e.g., see [21,22]). The methods were determined on the basis of the data format, data availability, and product levels (e.g., geometrically raw or rectified, provision of sensor model parameters or rational functional model coefficients, etc.).

The absolute radiometric calibration quality was also assessed using one of the Pseudo-Invariant Calibration Sites (PICS), i.e., the Libya-4 site, that were employed by the Committee on Earth Observation Satellite (CEOS) for radiometric Cal/Val activities. The site was first utilized for calibration of the Satellite Pour l'Observation de la Terre (SPOT) [23], and has demonstrated proven potential for acquiring optical visible and near-infrared (NIR) images with various resolutions [24–27]. The average SNR and the SDNR values were also computed over this site and evaluated in comparison with the results of the vendor.

The article is organized as follows. The detailed sensor specifications, the test site and data characteristics, and the methods are given in the next section. The results are presented in Section 3. The discussion and conclusion are provided in the final section.

2. Materials and Methods

2.1. The SkySat Constellation

Among the satellite constellations of Planet, SkySat provides very-high-resolution (VHR) imagery. The satellites belonging to the SKS-C family (60 × 60 × 95 cm) acquire VHR data with the capacity for data collection over large areas, e.g., thousands of square kilometers. The satellites are stabilized in three axes and have the agility to observe different regions of interest (ROI) [6]. In addition, they are equipped with systems for orbital and attitude control [28] and Cassegrain telescopes [6]. In accordance with the principles of push broom imaging, the frame images (scenes) are acquired continuously in five bands, namely blue, green, red, NIR, and panchromatic (Table 1). The SkySat constellation has three cameras per satellite with identical focal plane arrangements and acquires images as overlapping strips. Half of the detector takes images in the panchromatic (Pan) channel, while the other half is split into the four multispectral (MS) bands. In accordance with the principles of pushframe imaging, every pixel is composed of up to 20 acquisitions, and a virtual time-delayed integration (TDI) method is implemented for image registration and fusion [15].

Table 1. The main characteristics of the SkySat constellation and the sensors [5,29].

Parameter	Value
Number of satellites	21
Orbit/Altitude	Polar sun-synchronous/400–600 km
Sensor/Size	Frame CMOS with TDI/5.5 megapixel
Number of sensor	3
Sensor type	Pushframe
Bands	Panchromatic: 0.450–0.900 μm Blue: 0.450–0.515 μm Green: 0.515–0.595 μm Red: 0.605–0.695 μm NIR: 0.740–0.900 μm
Focal length	3.6 m
Image capacity	400 km^2/per day
Swath width	5.5–5.9–8 km at nadir
Image strip length (max.)	200 km
Data collection GSD	0.6–1 m
Radiometric resolution	11 bit [5]
Products	Basic Analytic Scene (BAS), Ortho Scene (OS), Ortho Collect (OC), Video

The GSD of the images at data collection depend on the spectral band and on the acquisition configuration. The recent SkySat products have been acquired in Push-broom High Dynamic Range (PBHDR) mode, which enabled higher GSDs. The PBHDR mode alters the acquisition configuration of the platform and the optical system, and is not a post-processing approach for resolution enhancement. The approach artificially reduces the scan rate with camera vibration according to the platform velocity and the frame rate. The method also improves the SNR of the images. Thus, the resolution can be increased to 50 cm for ortho products (both MS and Pan).

The SkySat images are provided at different processing levels, such as basic analytic scene (BAS), various ortho scene (OS) products (e.g., panchromatic, visual, pan-sharpened, analytic), and ortho collect (OC) products [6]. The latter product type is a mosaic formed from individual OSs. The BAS products are suitable for DSM generation, and rational polynomial coefficient (RPC) files are provided together with the images. An orthorectification process is performed by Planet to obtain OS products by using ground control points (GCPs) and digital elevation models (DEMs) with 30 m or 90 m resolution, depending on the area [6].

2.2. Test Site Characterization and Data

In this study, the Level 1A (BAS), Level 1C (OS), and Level 1C_C (OC data formed with 20 OS from each of the three cameras) products of different SkySat satellites were evaluated with respect to their radiometric and geometric quality. The GSDs of the employed products varied depending on the satellite altitude. The collected GSDs ranged from 0.6 m and 0.7 m for the BAS, and the OC products had a resolution of 0.5 m. The test sites were selected on the basis of the land use/land cover (LULC) characteristics and the availability of the equipment (i.e., MTF target, GCPs, etc.). The geographical distribution of the test sites, which were located in Turkey, Libya and France, is presented in Figure 1. The main characteristics of the data analyzed here are summarized in Table 2. In the table, acronyms were used as test site identifiers. Multiple acquisitions from different SkySat (SKS) missions over each site were used for the evaluations. The levels of the products analyzed over each site are also given in the table, together with the type of assessment performed for each product. The assessment types include geometric validation (GeoVal), visual checks (Vis), MTF, SNR, RER, and radiometric validation (RadVal), as presented in Figure 1. The Libya-4 PICS test site (called Libya) used here is categorized as a "bright" site characterized by a highly reflective surface [30].

The GeoVal activities were performed over the Ankara test site. The GCP distribution, along with a schematic representation of the SkySat scene footprints on the OC products at this site, is presented in Figure 2. This site was used extensively for GeoVal, including stereoscopic capability assessment. A reference DSM obtained from unmanned aerial vehicle (UAV) images with 3 cm GSD was used for the comparison with the SkySat DSM.

On the other hand, the SkySat image quality was compared with pan-sharpened Pleiades images acquired over the Salon de Provence test site, France. The Pleiades images were acquired on 25 September 2014. Furthermore, Sentinel-2A MS image products acquired over Libya were employed as a reference for the RadVal activities. The Sentinel-2 satellite products were acquired on forty different dates in 2019 and 2020.

Table 2. The main features of the SkySat data used in the study.

Test Site ID	Location	Acquisition Date	Satellite	GSD (m)	Product Level	Assessment Type
ANK	Ankara, Turkey (Lat: 39.160°N, Lon: 33.472°E)	18 Jul 2020 23 Sep 2020 27 Sep 2020	SKS4 SKS7 SKS12	0.71 0.66 0.71	OC & BAS [1] OC & BAS [1] BAS [1]	GeoVal & Vis GeoVal & Vis GeoVal & Vis
SAL	Salon de Provence, France (Lat: 44.010°N, Lon: 4.197°E)	14 Mar 2021 29 Dec 2020	SKS4 SKS17	0.66 0.57	OS pan-sharpened BAS [1]	Vis MTF & SNR & RER

Table 2. *Cont.*

Test Site ID	Location	Acquisition Date	Satellite	GSD (m)	Product Level	Assessment Type
LIB	Libya (Lat: 30.341°N Lon: 22.805°E)	5 Sep 2020	SKS10	0.70	BAS [1]	RadVal
		11 Sep 2020	SKS1	0.73	BAS [1]	RadVal
		17 Sep 2020	SKS7	0.67	BAS [1]	RadVal
		25 Sep 2020	SKS7	0.67	BAS [1]	RadVal
		26 Sep 2020	SKS10	0.72	BAS [1]	RadVal

[1] OC: Ortho Collect; OS: Ortho Scene; BAS: Basic Analytic Scene.

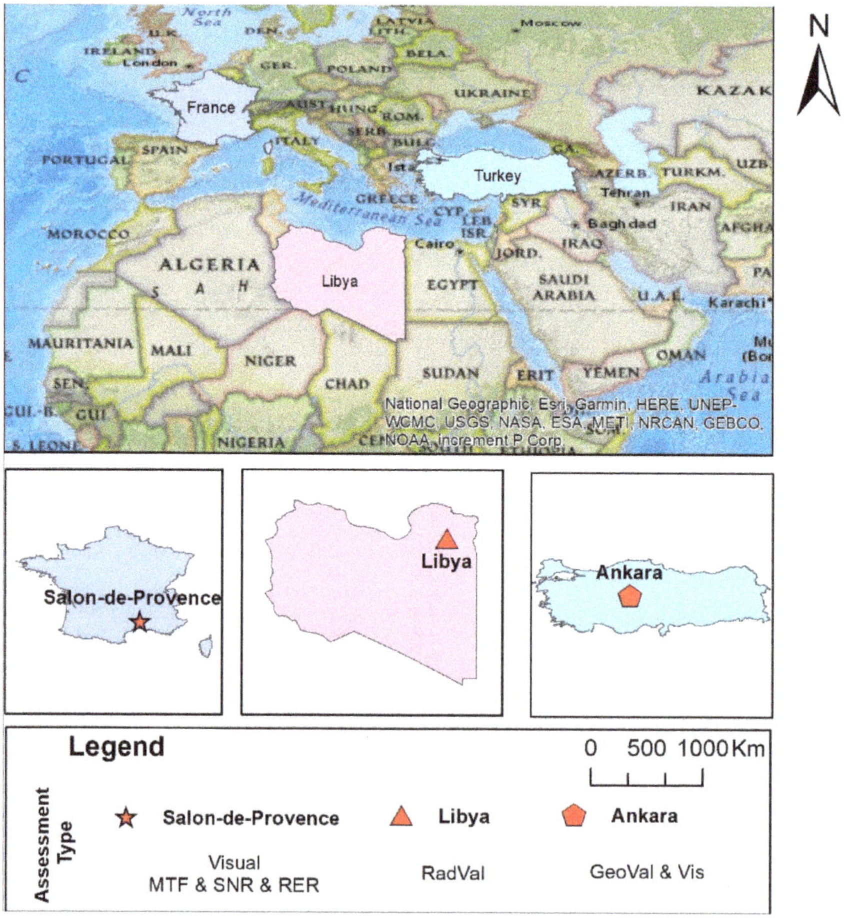

Figure 1. The locations of the EDAP test sites used in this study.

2.3. The EDAP Maturity Matrix

The overall EDAP framework covers optical, synthetic aperture radar (SAR), and atmospheric instruments [7]. While the Cal/Val MMs produced within the framework of the EDAP assignments aim at providing a high-level summary, the sub matrices generated through the validation activities present the detailed analysis results to the users for the selected missions and associated tasks, depending on the mission objectives. The Optical Mission Quality Assessment Guidelines [8] are followed for the production of high-level Cal/Val MM, and these include categories related to product and ancillary information, product generation, uncertainty characterization, and validation. Each category is graded depending on the compliance with the FAIR (findable, accessible, interoperable, reusable) principles by means of six different grades: Not Assessed (outside the scope), Not Assessable (information not available), Basic (does not follow the FAIR principles), Intermediate (partially follows the FAIR principles), Good (largely follows the FAIR principles), and Excellent (fully FAIR compliant) [8].

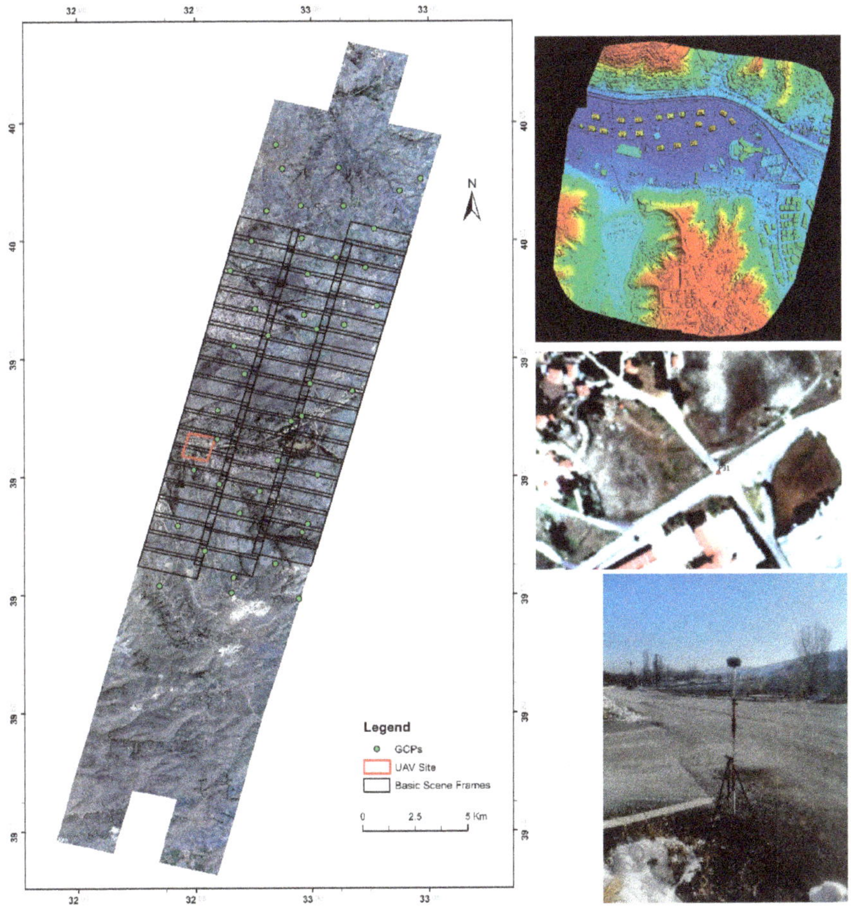

Figure 2. An overview of the SkySat Collect data acquired over the Ankara test site on 18 July 2020 with the GCP distribution (**left**); and the reference UAV DSM and a GCP located on the SkySat Collect image and on the ground (**right**).

Due to the innovative and challenging technology represented by SkySat data, the detailed quality assessments were performed on the basis of a product documentation and format evaluation, image quality, and geometric and radiometric calibration. Evaluations of the product documentation and the format were carried out by means of document and metadata reviews. Detailed assessments of image quality and geometric and radiometric calibration were carried out via visual checks, comparison with external reference data, and by using various radiometric and geometric image processing methods, as explained in the following sections.

2.4. *Image Quality Assessment Approaches*

The approaches for assessing image quality include visual inspection, MTF, SNR, Full Width at Half Maximum (FWHM), and RER analyses. The visual inspection addresses the unusable data mask (UDM) content and the image interpretability components. The UDM provided by the vendor often includes image backfill, cloud, and missing data information. The interpretability was analyzed in selected image sub-areas with respect to an external reference, i.e., Pleiades top of atmosphere (TOA) images. The assessments were carried out for the detection and identification of both objects and textural content. The object delineation and interpretation performances using pan-sharpened SkySat and Pleiades data were anticipated to be similar due to their identical spatial resolution (0.5 m). The selected features (points of interest) included an MTF target and different LULC types, such as transportation elements, large construction sites, forests, airports, crop fields, buildings and other urban structures, etc. Further details on the methods are provided in the following subsections.

2.4.1. SNR Computation

The SNR is an important image quality indicator [20]. Although a high SNR may not affect visual interpretation, it is an important parameter for reducing uncertainty in image measurement. The SNR is a mean value, and two types of SNR, i.e., temporal and spatial, are typically analyzed in the literature. The basic formulation of the SNR used here is given in Equation (1).

$$SNR = \frac{\mu}{\sigma} \tag{1}$$

where

- μ—mean signal; and
- σ—standard deviation of the signal.

Here, the spatial SNR over the Libya site was computed via image patches with a size of 9 × 9 pixels. The average (μ) and the σ of the intensity values within the patches were calculated for each spectral band (radiances). This method was first developed by Zanoni [31], and was adapted by performing patch selection in homogeneous and flat areas. As this method was applied over the Libya site, which high spatial image frequencies (e.g., desert dune summits) exist, a Sobel operator was applied to detect and eliminate the high-frequency content. Thus, patches with low-frequency content could be filtered and used. To fulfill both conditions, the proposed algorithm employed the Sobel-filtered edge image and a DEM. After eliminating unsuitable windows, the frequency distribution (histogram) of the remaining ones was used for histogram analysis. The peak value and the mean radiance value are reported for system SNR, and are presented with graphics as a function.

Although high SNR values are expected from Libya due to its bright surface, a noise model could not be obtained from a single site. On the other hand, the differential SNR (gSNR) [9], also referred to as Signal-Difference-to-Noise Ratio (SDNR) [18], was documented by the vendor. The SDNR is the ratio of the signal difference between two Lambertian surfaces and its noise (Equation (2)). The noise is often computed for the highest reflectance by using high (ρ_2) and low (ρ_1) reflectance values.

$$SDNR = \frac{S(\rho_2) - S(\rho_1)}{N(\rho_2)} \tag{2}$$

where

- $S(\rho)$—signal;
- $N(\rho)$—noise; and
- ρ_1, ρ_2—reflectance values of the Lambertian targets.

The SNR value can also be defined using the approach presented in Equation (3). An approximation between gSNR and SNR is proposed by [18] and presented in Equation (4).

$$SNR = \frac{S(\rho)}{N(\rho)} \tag{3}$$

$$SDNR \approx SNR(\rho) \cdot \frac{0.08}{\sqrt{0.15 \cdot \rho}} \tag{4}$$

2.4.2. MTF Computation

Although the spatial resolution of a sensor can be associated with the GSD or Instantaneous Field of View (IFOV), its determination can be challenging for imaging sensor systems. The MTF and SNR are often assessed together as image quality metrics. As a measure of image sharpness, MTF needs to be analyzed separately for the data taken from each orbit to ensure that the quality is not degraded through vibration, medium changes (air to vacuum), or thermal alterations [32]. Although several different approaches exist, the slanted edge method [32], which analyzes the MTF, the RER and the FWHM values, was used here with a specifically designed MTF target at the Salon de Provence test site. The evaluations were carried out for the MS channels. To obtain the MTF, the Edge Spread Function (ESF) and Line Spread Function (LSF) were computed first in a direction normal to the edge. In the case of a slanted edge (with a slope in the image), the mean values of several sample patches were used, as shown in Figure 3.

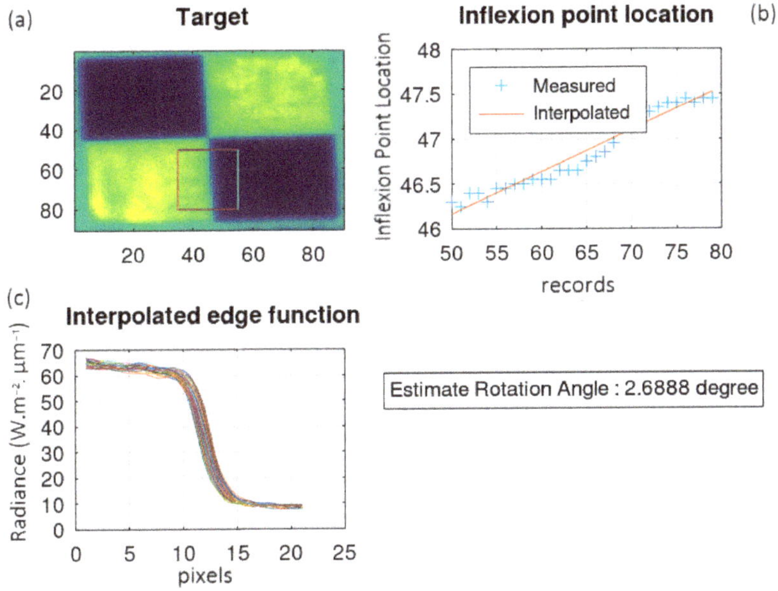

Figure 3. Slant-edge method—edge modeling output: (**a**) the MTF target and the selected image area (red rectangle) for the ESF analysis; (**b**) the line fitting output; and (**c**) the interpolated edges.

Assuming that the point spread function (PSF) is separable, it is possible to characterize the sensor behavior by using a 1D LSF model associated with each detector dimension (i.e., in the along-track direction and across-track direction). Considering a satellite with a sun-synchronous descending orbit, the objectives was to measure the MTF (i) in the along-track direction (AL MTF), assessed using a nearly horizontal (H) edge, and (ii) in the across-track direction (AC MTF), assessed using a nearly vertical (V) edge. Since a checkboard target was used, the values of AL and AC MTF were computed twice depending on the location of the region of interest in the checkboard, and in particular, contrast due to the change from high to low sensor response (i.e., the 'l' configuration) or from low to high sensor response (the 'h' configuration). The purpose was to reflect either the transition from high to low sensor response values or from low to high sensor response values. The along-track (AL)/across-track (AC) directions were also assessed separately with respect to their spatial resolution. The four directions of assessment can briefly be summarized as (i) Vl–AL (vertical) edge from high to low; (ii) Vh–AL (vertical) edge from low to high; (iii) Hl–AC (horizontal) edge from high to low; and (iv) Hh–AC (horizontal) edge from low to high sensor response.

Several parameters drive the selection of the window for performing the MTF calculations. The window parameters are presented in Figure 4 [33]. The parameter descriptions are as follows:

- ΔL—the differential radiance between the dark and bright part of the target.
- L_W—the width of the target in the direction of the MTF profile.
- α—the orientation angle α with respect to the direction of the MTF profile.
- L_H—the height of the target in the orthogonal direction of the MTF profile.

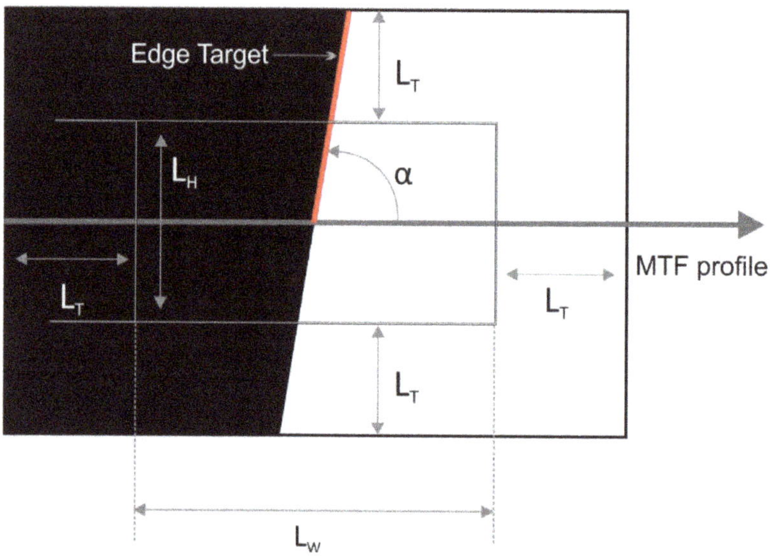

Figure 4. Schematic representation of the parameters for edge target selection (image courtesy [33]).

The ESF calculation proposed by Kohm [34] uses the measurements for both orientations (AL and AC). As an example, edge interpolation was performed for every line for the area shown with a red rectangle in Figure 3a. Afterwards, parametric functions were fitted in order to estimte the bottom of the inflection point, and least squares fitting was used to obtain the rotation (Figure 3b). The edge functions (Figure 3c) were finally analyzed, and noisy measurements were eliminated.

Because the MTF approach can be affected by noise, aliasing, and phase effects, and the edge rotation angle needs to be calculated with high accuracy, to this end, multiple image lines were analyzed. Each sample was projected on a line normal to the edge by using the angle with a bin size of 0.25 pixels (upsampled). To avoid empty bins, the selection of the right rotation angle is important. On the other hand, due to the empty bins, the distribution of the samples for ESF can be irregular. The LOcally WEighted Scatter-plot Smoother (LOESS) curve-fitting algorithm with a second-order polynomial was employed for resampling to obtain a regular distribution. It must be noted that the proposed approach is non-parametric, which makes it possible to obtain results that are closer to system behavior by including the noise.

The LSF was computed from the ESF. To eliminate outliers, a smoothing approach via fourth-order Savitsky Golay filtering with a window size of 11 bins and a Hann window was performed. Sample ESF, LSF, ESF, and MTF graphs obtained from the analysis are presented in Figure 5. Further details on the application of the approach can be found in [32,35,36].

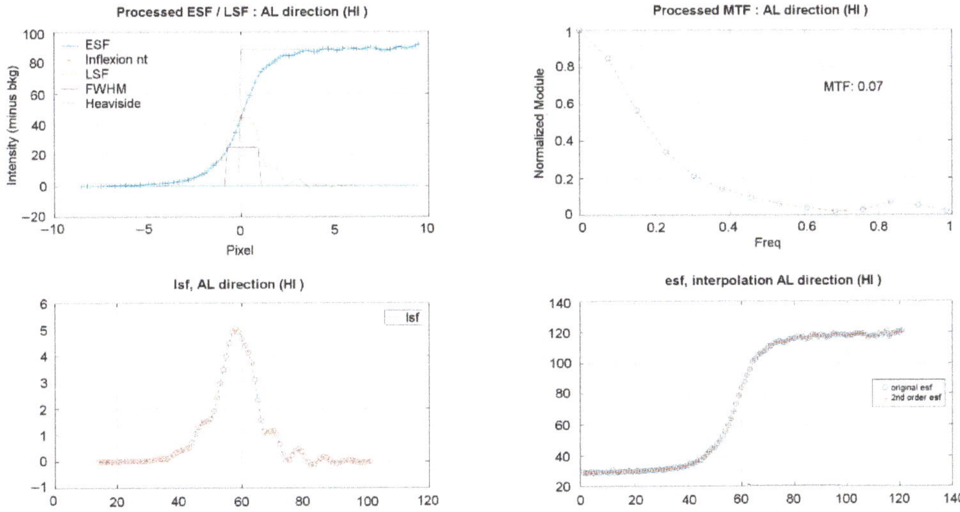

Figure 5. The output graphs of the MTF processing.

On the other hand, the SNR was also computed using the MTF target at the Salon de Provence test site. This approach uses ESF, and the mathematical expression used for this purpose is given in Equation (5).

$$SNR = \frac{\left|\overline{X_{high}} - \overline{X_{low}}\right|}{\frac{\sigma(X_{high})+\sigma(X_{low})}{2}} \tag{5}$$

where:

- X_{high} is the statistical variable of the ESF high-intensity (radiance) values;
- X_{low} is the statistical variable of the ESF low-intensity (radiance) values; and
- $(\overline{X}, \sigma(X))$ are the corresponding statistics.

The high- and low-intensity regions are averaged from three pixels from the inflexion point of the curve, and the cardinality of high- and low-intensity samples is the same.

2.5. Radiometric Calibration Quality Assessment Approach

The radiometric calibration quality of SkySat data was assessed by comparing the reference data obtained from Sentinel-2 satellites, as they possess high accuracy, i.e., a low level of uncertainty (<3%). The TOA measurements simulated from the reference data (Figure 6) were compared for the assessment. The bottom of atmosphere (BOA) data were processed to obtain the TOA values by (i) estimating the Bidirectional Reflectance Distribution Function (BRDF) model; (ii) generating the spectrum over spectral interval; (iii) computing the TOA values by using the Second Simulation of the Satellite Signal in the Solar Spectrum (6S) Radiative Transfer (RT) [37]; and (iv) interpolating the TOA spectrum with SkySat relative spectral response (RSR—spectral convolution). The BOA reference spectra were employed to simulate TOA references in consideration of the image acquisition conditions (observation geometry) and the spectral properties of the sensor. The Sentinel-2 BOA time series data were from a two-year period (2019 and 2020) and the SkySat RSRs were provided freely by Planet. After the visual inspections, the RSR data of SkySat-10 were employed for all missions.

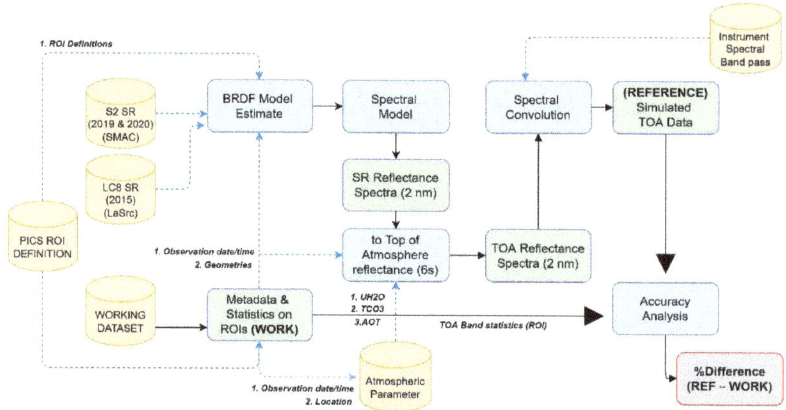

Figure 6. The workflow of absolute radiometric calibration using PICS data.

The radiometric calibration quality assessment procedure presented in Figure 6 includes:

- Surface reflectance estimation from Sentinel-2 data for the ROIs (one per SkySat camera) (Figure 7).
- Estimation of Sentinel-2 BRDF model and correction. Here, the BRDF varies linearly against scattering angle, a linear relationship between the reflectance data and scattering angle is expressed, and this is used to normalize the data.
- BOA values for each Sentinel-2 channel are estimated on the basis of the SkySat image acquisition geometries.
- BOA spectrum computation by interpolating with an interval of 2 nm.
- Atmospheric parameters are obtained at the time of image acquisition from Copernicus Atmospheric Monitoring Service (CAMS).
- Generation the TOA Spectrum using the observation geometry, atmospheric parameters and the BOA spectrum.
- Image merge per observation date, camera, and band.
- Estimation of the MS TOA values from the merged images ($TOA\_Measured$).
- Production of simulated TOA values at SkySat band central wavelength ($TOA\_Simulated$) by convolving the TOA Spectrum using SkySat spectral response.
- Computation of the calibration ratio (Q) and the percent difference between simulated and measured TOA values using Equation (6).

$$\%Difference = \frac{100 * (TOA\_Simulated - TOA\_Measured)}{TOA\_Simulated} \tag{6}$$

where:

- $TOA\_Measure$ is the measurement processed from the SkySat product;
- $TOA\_Simulated$ is the measurement obtained from the Sentinel-2 PICS data.

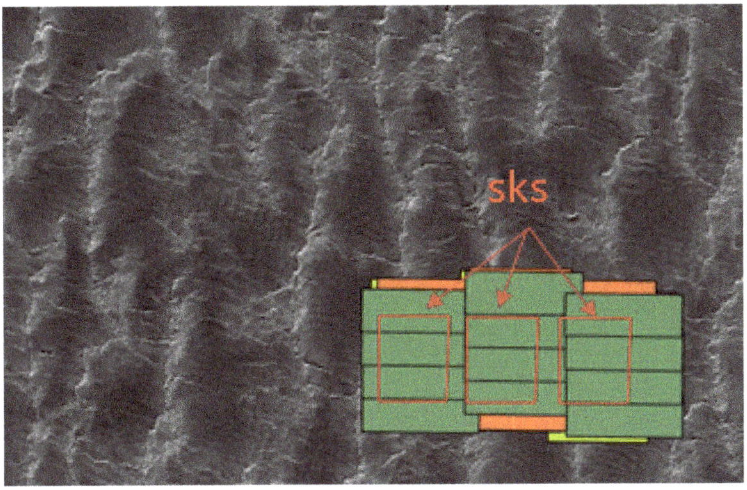

Figure 7. The SkySat scene footprint over the Libya site and the ROIs per SkySat (sks) camera.

Image data numbers (DNs) were transformed into TOA values using the coefficients given in the metadata, and the coefficients were compared with the results by using the DN-to-TOA reflectance equations. The in-band solar irradiance data were also given by the vendor (Thuillier spectrum). The Sentinel-2A (S2A) BOA values with respect to scattering angle are presented in Figure 8. The scattering angle was calculated using Equation (7). The linear correspondence between the scattering angles and the BOA measurements was used to compute the expected S2A BOA measurements for a given SkySat scattering angle. The spectral model used here is shown in Figure 9. One set of ROIs with footprints within the SkySat frame images was selected for this purpose.

$$\cos \xi = \cos \theta \cos \vartheta + \sin \theta \sin \vartheta \cos \phi \quad (7)$$

where:

- θ, ϑ are solar/view zenith angle;
- ξ is the phase or scattering angle related to conventional angles; and
- ϕ is the "view-sun" relative azimuth angle.

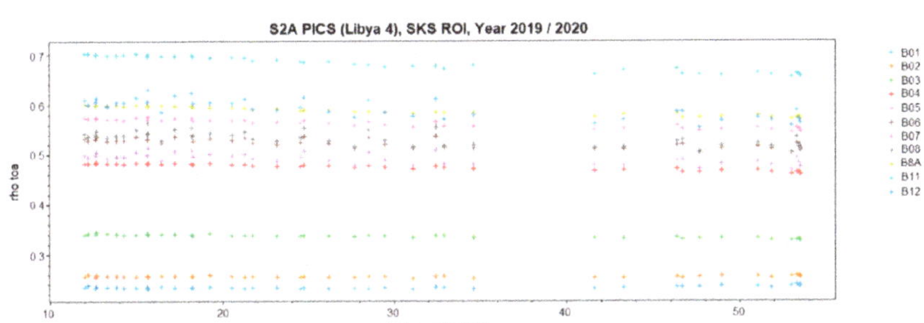

Figure 8. Input data for the BRDF modeling of Sentinel-2A time series.

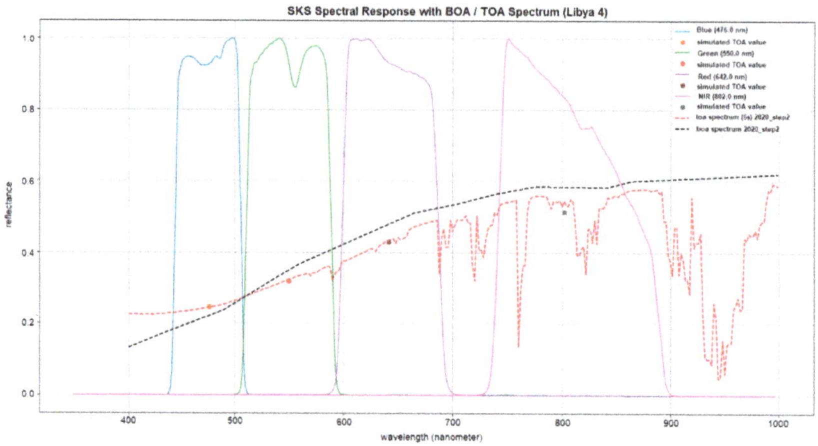

Figure 9. Simulation of TOA reflectance values for SkySat central wavelength.

2.6. Geometric Calibration Quality Assessment Approach

The geometric calibration quality was assessed using SkySat BAS and OC data from the Ankara test site (see Table 2). The BAS and OC data used for the geolocation assessments were acquired on 18 July 2020 and 23 September 2020 (Table 2) with collection GSDs of 73 cm and 66 cm, respectively. The spatial resolution of ortho products was 50 cm. The GCP distribution is given in Figure 2. The assessments were performed in terms of absolute and temporal geolocation accuracy, as well as BBR accuracy. The methods used for assessing the different accuracy assessment types were selected on the basis of the data product level (BAS or OS), the availability of georeferencing data and their format (RPC or GeoTiff), and the availability of the reference information (GCPs and dense DSMs). The methods employed here have also been used in previous scientific publications (e.g., [21,38,39]) and technical reports [22], in which further details can be found.

The absolute geolocation accuracy was assessed using a total of 44 GCPs selected on OC images and surveyed by using Global Navigation Satellite System (GNSS) instruments in the field. The assessment approaches were different for the BAS products, which were provided together with RPCs and the OC products. A ground-to-image backprojection approach was applied to the BAS products (pan only) using the GCP ground coordinates and the provided RPCs [40]. The outputs of the backprojection function were the image coordinates of the GCPs. A comparison was performed between the computed and measured image coordinates. Statistical metrics (i.e., mean, absolute mean, median, σ, and root mean square error (RMSE)) were derived from all GCPs in the x (row) and y (column) directions. The OC products were assessed by comparing the GCP ground coordinates obtained from the GNSS measurements and those obtained from the orthorectified images by manual measurements. The image measurement accuracy was assumed to be 0.5 pixels due to the manual selection of the points. The georeferencing of the OC data was defined in the Universal Transverse Mercator (UTM) Zone 36 projection system on the World Geodetic System 1984 (WGS84) datum. Similar to the BAS products, the statistical metrics were derived in the X (East–West) and Y (North–South) directions.

The temporal geolocation accuracy was obtained on the basis of a comparison of GCP coordinates measured on the OC data over Ankara. Furthermore, a dense image matching approach was applied to the two datasets. This method is based on the Kanade–Lucas–Tomas (KLT) tracker [41], with an expected measurement accuracy of 0.1 pixel, and has been successfully applied in the assessment of lower-resolution satellite images in previous studies (e.g., [39,42,43]). The assessment was performed on the red band images, since this band has good contrast and a higher SNR.

Regarding the BBR accuracy, the KLT tracker was again employed by using the blue band as reference and the other MS bands as the search image. The discrepancies between the image coordinates in the reference and the search images of the matched keypoints were evaluated statistically to assess the BBR errors. Again, the MS BAS images acquired on 18 July 2020 and 23 September 2020 over Ankara were assessed for this purpose.

The stereoscopic capability of SkySat data was assessed by DSM generation, and the output was compared with a reference DSM obtained from the UAV data. The Catalyst software package, by PCI Geomatics, Ontario, Canada, was used for the RPC adjustment, using a total of six GCPs. Three different acquisitions over Ankara, given in Table 2, were used for this purpose. The DSM comparison was performed with the LS3D software from 4Dixplorer AG, Switzerland, and assessed in terms of the height differences between the DSMs. This approach is based on the co-registration of 3D surfaces and computation of the Euclidian distances between them as discrepancy values [44].

3. Results and Discussions

The assessment results for the SkySat maturity matrix, and the radiometric, geometric and image quality are presented and discussed in this section.

3.1. Image QA Results

The image QA assessments were performed on the basis of SNR and MTF evaluations and by means of visual inspection. With respect to the SNR measurements, more than 50 images acquired over the Libya-4 site were processed. The average SNR/SDNR (Equation (4)) results for blue, green, red and NIR bands are listed in Table 3, together with Planet's results (last column). It is worth noting that the minimal SNR requirement disclosed by Planet is 30.

Table 3. SkySat averaged SNR results for products acquired over the Libya-4 test site.

Spectral Bands	Mean SNR (ρ)	SDNR Equation (4)	σ SNR	Reference Radiance	ρ (TOA)	No. of Images	Planet's Results
Blue	134.04	57	7.78	118.33	0.23	52	34
Green	174.32	65	8.73	141.44	0.30	53	43
Red	203.98	68	7.84	158.35	0.38	53	45
NIR	190.55	58	7.40	134.48	0.46	53	41

The high reflectance of the Libya-4 site is not appropriate for assessing image SNR over the full sensor dynamic range. This limitation does not prevent performing a comparison between the EDAP and Planet results. Additionally, for all bands, the results were found to be within the SNR accuracy specification claimed by the data provider. For the same observation date, the accuracy does not vary significantly between the cameras. However, the accuracy varies depending on the satellite involved, as shown in Figure 10. The image quality of the SkySat-1 (ssc1) products is degraded compared to the image quality of the SkySat-7 (ssc7) and SkySat 10 (ssc10) products, which remain consistent.

As shown in Figure 10, a linear relationship between the SNR values and the in-band radiances exists, with a coefficient of determination (linear model estimate) reaching 0.6 for SkySat-1 products and 0.8 for SkySat-7 and SkySat-10 products, as listed in Table 4. The SNR analysis demonstrated that the image quality varies within the constellation. In this case, the accuracy loss might be due to satellite ageing, with SkySat-1 being launched in the year 2014, while SkySat-7 and SkySat-10 were launched in the year 2017.

This latter observation was also confirmed during the visual inspection. The older SkySat-1 images had a higher degree of noise than the other satellite images, and this could not be attributed to the atmosphere, which was stable during the observation period [36].

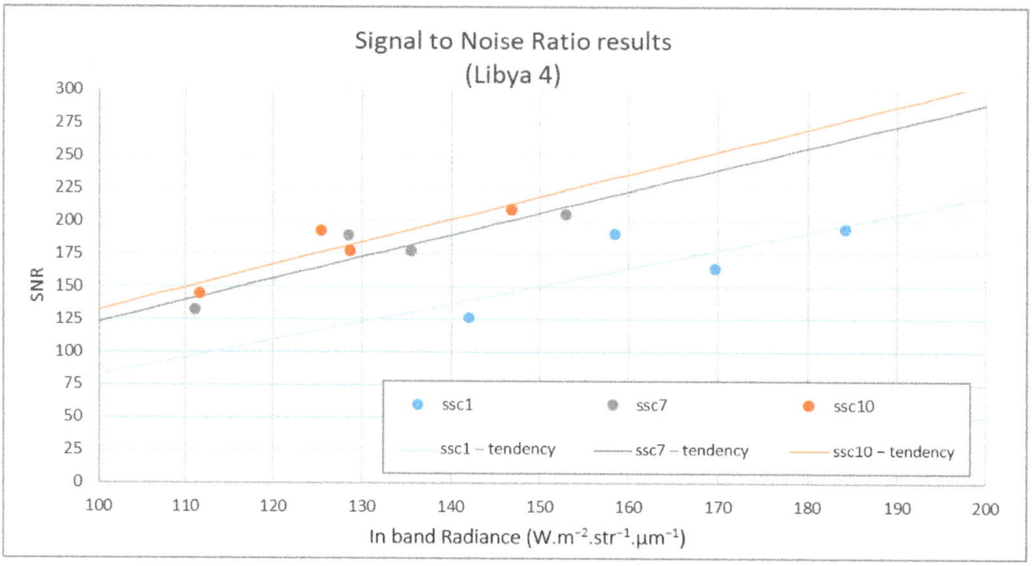

Figure 10. The mean SNR values obtained from three SkySat satellites (ssc1, ssc7, ssc10) as a function of in-band radiance and linear representation.

Table 4. The SNR and in-band radiance linear model estimate: slope (a), offset (b), and coefficients of determination.

Satellite	ssc1	ssc7	ssc10
Number of products	12	27	14
a	1.3655	1.6633	1.7235
b	−54.337	−43.45	−39.865
R^2	0.5985	0.8386	0.8288

The RER was calculated for the MS and Pan bands using edges from 5000 airports worldwide, supplied by the data provider. According to the results, the RER values were (i) blue: 0.28; (ii) green: 0.30; (iii) red: 0.34; and (iv) NIR: 0.26; Pan: 0.30. The other image quality parameters, such as FWHM and MTF value at Nyquist frequency were not shared by the provider. The results obtained here indicate that the image contrast in the MS channels at a resolution of 0.5 m was degraded, and image blurring was observed (e.g., see building rooftops in Figure 11). The pan-sharpening process applied to SkySat imagery is useful for restoring the high-frequency content, especially for urban structures (motorways, bridges, roundabouts, buildings, etc.), and performed well, resulting in the data being comparable with Pleiades images (Figure 12). Yet, as can be seen in Figures 11 and 12, the 0.5 m resolution of SkySat data is not equivalent to Pleiades in terms of image interpretability. Important drawbacks of the pan-sharpening method are that it does not preserve physical quantities in areas with low texture. Thus, these areas are smoother than the original images, resulting in a loss of image content, such as trees, crop fields, etc. In addition, on the basis of visual inspections, artefacts such as edge jitter and saturation in panchromatic images acquired over the Ankara test site were occasionally observed (Figure 13).

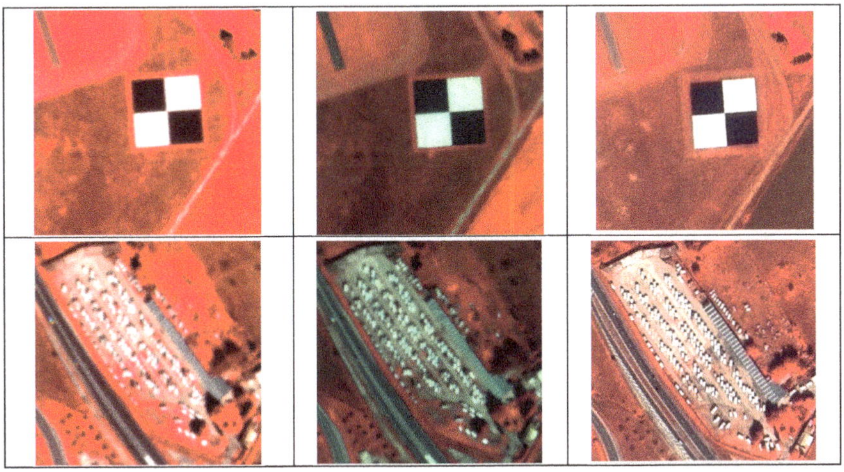

Figure 11. Examples of parts of images from the SkySat green band (**left**), the pan-sharpened green band (**middle**), and Pleiades (**right**) used for visual inspection.

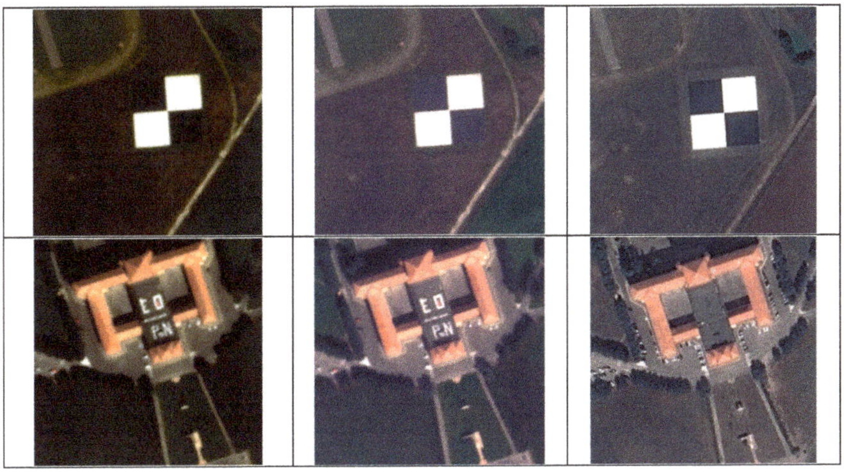

Figure 12. Examples of parts of images from SkySat RGB (**left**), pan-sharpened RGB (**middle**), and Pleiades (**right**), used for visual inspection.

The MTF results obtained from the target in Salon de Provence site (Figure 14) showed that the image sharpness had decreased. The image product was at the BAS level, and was acquired on 29 December 2020. Although haze was observed in the images, this information was not available in the metadata. As can be observed from Figure 14, the dark and bright image regions were not homogeneous, and thus it cannot be considered to be a Lambertian surface, and not all artefacts can be related to the SkySat sensor. The MTF results presented in Table 5 were found to be stable in all directions. The FWHM values were larger than 2 pixels, and confirmed the blurring on the basis of reference criteria [45], which is a similar outcome to the visual inspection results. The RER values were a little worse for red and NIR channels when compared with the above-mentioned vendor results. The image quality of the different channels, in the order of best to worst, was blue, green, red, and NIR. Since the RER remained stable for all bands in different directions, the approach applied here is considered to be consistent. However, the compatibility between the RER values obtained here and those provided by the vendor is low. In addition, the MTF results confirm a loss of contrast at the Nyquist frequency. A comparison with a true 0.5 m GSD image shows that the information content (contrast) of the SkySat image was lower, and the image sharpness decreased with 0.5 m upsampling.

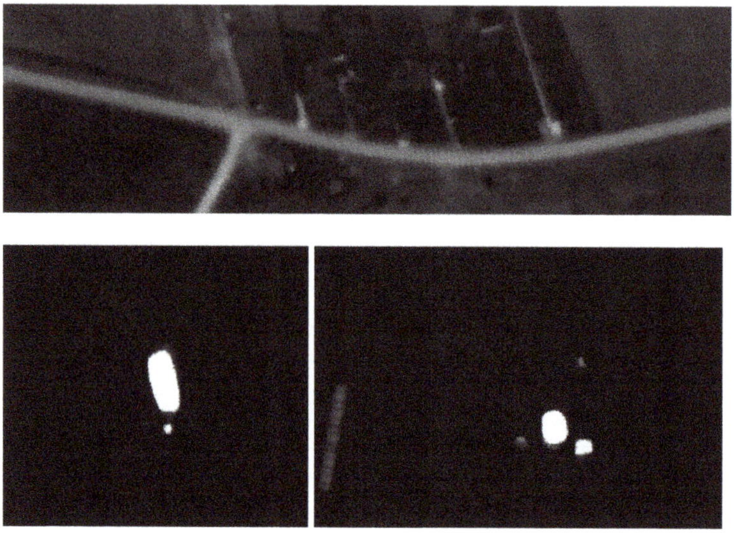

Figure 13. Examples of edge jitter (**above**) and saturation (**below**) artefacts in SkySat data.

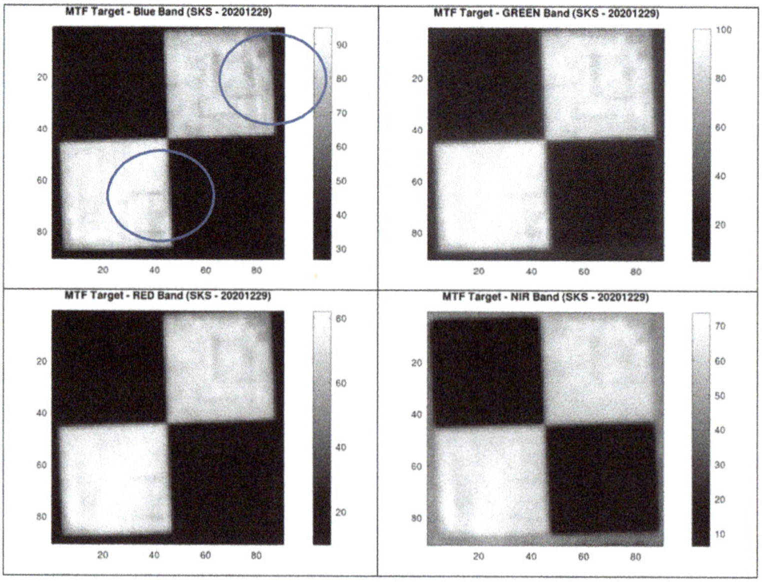

Figure 14. The SkySat image parts (29 December 2021) from the MTF target in Salon de Provence Airport and some visible degradations (blue circles). The axes are given in pixels. The legend is provided in radiance.

Table 5. The MTF results obtained from SkySat-17 BAS data acquired on 29 Dec 2020 over the Salon test site with 57 cm collection GSD and 50 cm pixel spacing. Satellite inclination angle: 53.0°.

Spectral Band and Rotation Angle	Direction of MTF Profile	MTF Along Track		MTF Across Track	
		Hh	Hl	Vh	Vl
Blue 2.46°	SNR	21	15	10	12
	FWHM	2.5	2.25	2	2.5
	RER	0.30	−0.31	0.29	−0.27
	MTF@Nyquist	0.05	0.04	0.06	0.05
	L_w (pixel)	14.00	19.00	16.00	16.00
	L_H (pixel)	12.75	12.75	20.25	22.75
	Delta_L *	59.74	51.49	53.01	56.61
Green 2.60°	SNR	25	15	11	12
	FWHM	2.5	2.5	2.25	2.5
	RER	0.28	−0.29	0.27	−0.26
	MTF@Nyquist	0.035	0.034	0.045	0.035
	L_w (pixel)	14.00	19.00	16.00	16.00
	L_H (pixel)	12.75	12.75	20.25	22.75
	Delta_L *	66.61	59.61	60.36	63.77
Red 2.49°	SNR	21	13	12	10
	FWHM	2.5	2.5	2	2.25
	RER	0.26	−0.26	0.26	−0.25
	MTF@Nyquist	0.04	0.025	0.035	0.031
	L_w (pixel)	14.00	19.00	16.00	17.00
	L_H (pixel)	12.75	12.75	20.25	22.75
	Delta_L *	61.44	57.32	56.71	59.54
NIR 2.68°	SNR	17	12	12	10
	FWHM	2.5	2.75	2.5	2.25
	RER	0.24	−0.23	0.24	−0.23
	MTF@Nyquist	0.03	0.03	0.02	0.04
	L_w (pixel)	12.75	12.75	20.25	22.75
	L_H (pixel)	14.00	19.00	16.00	17.00
	Delta_L *	52.62	52.83	51.05	52.37

* Delta_L = $W.m^{-2}\ str^{-1}\ \mu m^{-1}$.

3.2. Radiometric Calibration QA Results

The radiometric QA results obtained here were found to be within the vendor specifications [6], and varied among the different bands (Table 6). It must be noted that the assessments performed by Planet used RadCalNet data as reference data source. By using RadCalNet measurements, a better uncertainty (2%) was expected compared to the PICS measurements (5%). Furthermore, the RadCalNet measurements were almost suitable for calibration purposes, while PICS measurements required preprocessing (simulated values). The results are the statistics of percentage difference values computed on the basis of Equation (8). The uncertainty, precision and accuracy values in Table 6 were calculated using Equations (9)–(12). It was observed that the Planet results were strongly dependent on the spectral band. The red band results were within 3%, while for the other bands, the accuracy was above 8%. However, the precision values of the Planet results remained above 20%, independently of the selected spectral bands. The mean accuracy obtained in this study was mostly below 10%, which is in accordance with the Planet results. However, it was observed that the σ value was very high. The precision values obtained (below 7%) were better than the results reported by the vendor (<<28%). Regarding the mean accuracy, the Planet and the EDAP results were similar (below 11%), except for the red band, for which the Planet results were significantly better (2.5%). The overall uncertainty reached here was about 12%, which was also significantly better than the Planet results (25%).

Table 6. The absolute radiometric calibration uncertainty results acquired over the Libya site.

Source	Band	Accuracy	Precision	Uncertainty
EDAP	Blue	10.72%	5.69%	12.14%
	Green	8.18%	6.53%	10.46%
	Red	11.23%	4.55%	12.11%
	NIR	9.70%	4.71%	10.78%
Planet	Blue	8.58%	26.77%	28.11%
	Green	11.83%	26.42%	28.95%
	Red	2.55%	22.94%	23.08%
	NIR	8.33%	23.68%	25.10%

$$Percent\_difference = 100 * (Rho_{image} - Rho_{RadCalNet}) / Rho_{RadCalNet} \quad (8)$$

$$\Delta_i = \frac{Scene_{reflectance} - Site_{reflectance}}{Site_{reflectance}} \quad (9)$$

$$Accuracy = \frac{\sum_{i=1}^{n} \Delta_i}{n} \quad (10)$$

$$Precision = \sqrt{\frac{\sum_{i=1}^{n} (\Delta_i - \mu)^2}{n}} \quad (11)$$

$$Uncertainty = \sqrt{\mu^2 + \sigma^2} \quad (12)$$

where

- Δ_i is normalized reflectance difference;
- *Accuracy* is the mean Δ_i;
- *Precision* is the standard deviation of Δ_i; and
- *Uncertainty* is the root of squared sum of accuracy and precision values.

As a result of the small number of products, the test data set is not fully representative of the calibration accuracy of the constellation. Nonetheless, considering the average ROI values for a single date, it was observed that the radiometric calibration varied depending on the satellite involved, as shown in Figure 15. The strong variability between the different dates was mainly due to the satellite involved. When considering a single satellite (e.g., SkySat-7, SkySat-10), date-to-date results were in agreement with each other to within 2%, which is within the uncertainty budget of the method.

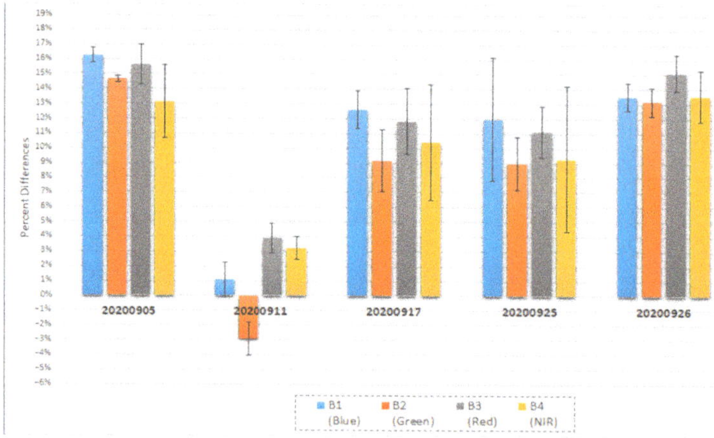

Figure 15. The SkySat radiometric calibration stability (September 2020) given as average results over ROIs.

The detailed results for the absolute radiometric calibration of SkySat given in Table 7 are listed with respect to observation date, depending on the ROI identifier and the detector number. As mentioned previously, the SkySat instrument embeds three independent detectors. An equivalent calibration accuracy among these three detectors is expected. The calibration accuracy of these detectors with respect to the satellite and observation dates is listed in Table 7. In most cases, even when the absolute calibration was degraded (above 10%), there were no significant differences in the results between the three different detectors. Only the SkySat-7 results exhibited large differences between detectors, and this is true for the two dates 17 September 2020 and 20 September 2020, where variability across the detectors exceed 5% in the blue and NIR bands. The calibration was stable among the detectors for SkySat-1, SkySat-10. However, it was not stable among the detectors for SkySat-7, in particular the blue band and the NIR band.

Table 7. Detailed absolute radiometric calibration results over the Libya site.

Observation Date	ROI ID	SkySat Satellite	CCD Number	Percent Difference			
				Blue	Green	Red	NIR
5 September 2020	1	10	d3	16.61%	14.51%	14.67%	11.42%
5 September 2020	2	10	d2	15.91%	14.81%	16.56%	14.88%
11 September 2020	1	1	d3	0.53%	−2.63%	3.05%	3.58%
11 September 2020	2	1	d2	0.34%	−4.18%	3.60%	2.34%
11 September 2020	3	1	d1	2.39%	−2.00%	5.03%	3.76%
17 September 2020	1	7	d1	13.01%	8.85%	10.81%	5.96%
17 September 2020	2	7	d2	13.67%	11.41%	14.41%	13.31%
17 September 2020	3	7	d3	11.19%	7.23%	10.29%	12.02%
25 September 2020	1	7	d1	16.66%	11.01%	11.34%	5.40%
25 September 2020	2	7	d2	8.74%	7.54%	9.32%	7.63%
25 September 2020	3	7	d3	10.51%	8.42%	12.70%	14.83%
26 September 2020	1	10	d3	14.24%	13.42%	14.22%	11.98%
26 September 2020	2	10	d2	12.45%	12.11%	14.64%	13.34%
26 September 2020	3	10	d1	13.86%	13.98%	16.51%	15.39%

It must be noted that the results presented here omit the daily variation, which requires further investigation in order to identify the sources of the differences (e.g., meteorological or sensor behavior). Furthermore, the calibration accuracy of the cameras may be different for different dates/times and within one image. The differences between simulated and actual TOA measurements for every product, every band and every camera show that the radiometric calibration accuracy varies among SkySat missions. The results show that the calibration of SkySat-1 (ssc1) is below 5%, irrespective of the camera. The ssc7 is temporally stable for the first (d1) and the third (d3) cameras, but unstable for the second camera (d2). Variations between the different bands were also observed for ssc7 (e.g., blue was 15% and NIR was 5.5%). The calibration results of ssc10 were temporally stable, and the results of all of the cameras were compatible (i.e., above 12% for all channels). In addition, correlations between the observation geometry properties (sun, viewing angles) and calibration results were also found.

3.3. Geometric Calibration QA Results

Here, the geometric accuracy of the absolute, relative (temporal), BBR and stereoscopic capabilities was validated. According to the vendor [6], the georeferencing accuracy specification is better than 10 m RMSE for OC products and better than 50 m for BAS products. In addition, the accuracy report provided by Planet [9] demonstrated a mean of 3.4 m mean and a σ of 2.6 m, computed from the RMSE values obtained from the absolute geolocation assessments of 50,139 ortho products. Similarly, the RMSE values obtained from the temporal geolocation assessment of 55,496 orthoimages had a mean of 3.1 m and a σ of 5.7 m. The BBR accuracy obtained on the basis of 63,900 products and expressed in mean and σ values between the different bands was also high (i.e., blue–green: 0.11 m/0.06 m;

blue–red: 0.13 m/0.08 m; blue–NIR: 0.22 m/0.14 m; green–red: 0.09 m/0.05 m; green–NIR: 0.20 m/0.12 m; red–NIR: 0.18 m/0.12 m).

The planimetric absolute geolocation accuracy values of BAS products over the Ankara test site were 1.7 m/1.3 m/1.7 m (Mean/σ/RMSE), which were determined on the basis of two BAS products and were found to be within specifications. For the OC products, the absolute geolocation accuracy values were 1.4 m/1.1 m/1.2 m (Mean/σ/RMSE). The temporal geolocation accuracy between the two OC products was 1.8 m/1.4 m/1.5 m (Mean/σ/RMSE). The relative accuracy results obtained on the basis of the image matching of 1193 points exhibited comparable accuracy to the results obtained from the GCPs for the matched scenes within the uncertainty threshold of the techniques, that is, the manual measurements had lower precision compared to the image matching method used here. The BBR results defined as Mean/σ/RMSE values were 0.03 m/0.02 m for blue–green, 0.07 m/0.05 m for blue–red, and 0.12 m/0.05 m for blue–NIR image pairs. The values were computed from six BAS products. All results were within the Planet specifications given above. A summary of all geolocation accuracy assessment results is presented in Table 8.

Table 8. SkySat geolocation uncertainty results. All results are in meters.

	BBR (mean/σ) m	Absolute (mean/σ) m	Temporal (mean/σ) m
EDAP Results	Blue–green: 0.04/0.01 Blue–red: 0.07/0.02 Blue–NIR: 0.11/0.03	1.45/0.18 (BAS) 1.14/0.24 (OC)	OC: 1.51/N.A.
SkySat Q3 quality report [9]	Blue–green: 0.11/0.06 Blue–red: 0.13/0.08 Blue–NIR: 0.22/0.14	3.4/2.6	3.1/5.7

The DSM accuracy obtained from the SkySat triplet indicated a 0.55 mean shift in Z, with a σ of 2.2 m, on the basis of the comparison with the UAV DSM. The discrepancies between the SkySat and UAV DSMs are presented in Figure 16. The results obtained here indicate that the accuracy of DSM generation is affected by the temporal differences between the acquisitions, image geometry (stereo or multi-view configuration), and RPC quality. RPC improvement prior to the DSM generation is essential due to the epipolar image requirement of most dense image matching methods, e.g., semi-global matching. In this study, six GCPs were used for the selected area in order to apply a six-parameter affine correction to the RPCs of a total of eight BAS images. The global shift (0.55 m) between the SkySat and the UAV DSMs remained within the boundaries of RPC adjustment accuracy. The larger errors in the discrepancy map (Figure 16) were the result of image artefacts, cloud coverage, shadows of larger buildings, and surface changes between the acquisition dates of the SkySat dataset as well as the UAV data.

3.4. SkySat Maturity Matrix

The maturity matrix obtained from the analyses in the EDAP study is presented in Figure 17. The validation task in Figure 17 can be analyzed in terms of image quality and radiometric and geometric Cal/Val. Regarding the product information, it was observed that although the mission, products, and product format were well documented and the data were easily accessible; the algorithms used for data processing were not provided. In addition, the measurement data quality was not provided in the product format, and the traceability chain was not given. The documentation of the product generation stages included limited details. Although the documentation on the pre-flight activities was not provided, a number of documents showed that appropriate community infrastructure was used to undertake these activities. It was expected that these documents would be updated more regularly. With respect to ancillary information, we observed that useful ancillary information—including product flags with UDM set per pixel—was available. However, the product includes little information in general on ancillary data origin, ancillary data type, uncertainties, etc. In terms of uncertainty characterization, the quality of the products is regularly monitored, and the results are reported along with a comprehensive assessment of the performances [9]. Moreover, there is a quarterly report prioritizing the documentation of image artefacts, such as parallax effects, over-sharpening, saturation, and problems in the processing chain.

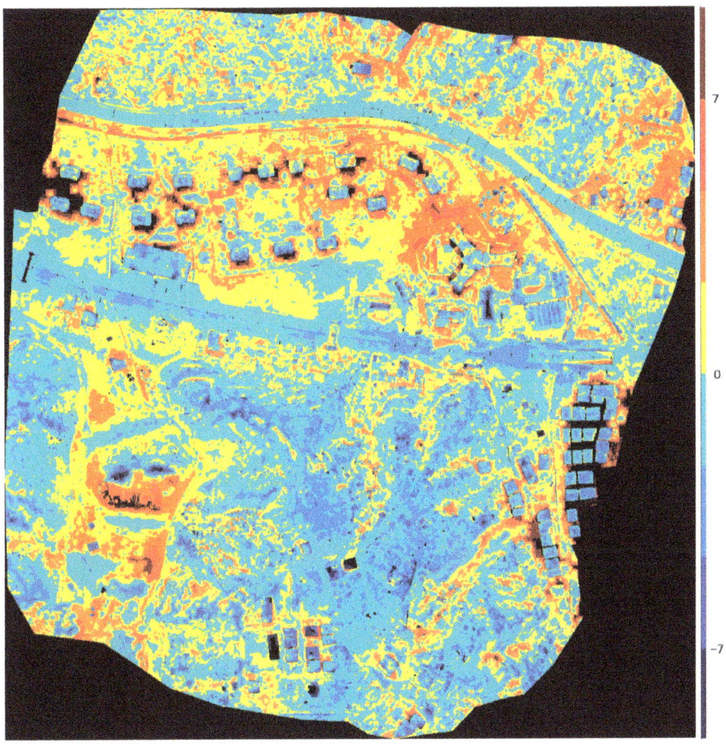

Figure 16. Discrepancy map between the SkySat and the UAV DSMs over the Ankara test site (in meters).

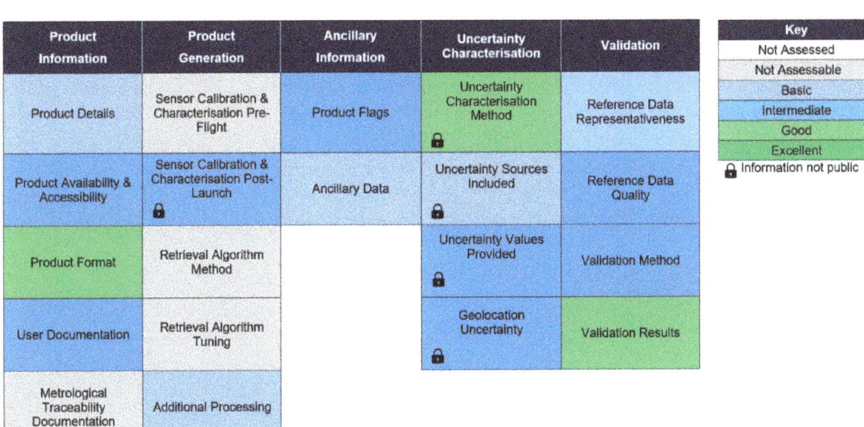

Figure 17. SkySat maturity matrix produced in EDAP activities.

The visual inspections yielded evidence of image artefacts or anomalies that could be detected via visual checks. The SNR evaluations over the bright uniform site indicated that this parameter was stable and compliant with the Planet specification. With the MTF evaluations, the spatial resolution of the 0.5 m data was confirmed, as the processing did not alter the physical measurements of the image, and did not introduce edge overshooting, as classically observed with pan-sharpening/MTF compensation. The geometric Cal/Val assessments under the categories of absolute, temporal, BBR and stereoscopic capability showed that the results were in line with the Planet specifications [6]. The radiometric Cal/Val results obtained from the absolute calibration methodology applied to the Libya data showed that the results were heterogeneous, but remained within the specifications [6].

4. Conclusions

The analyses carried out as part of the EDAP and the results presented here show that the maturity matrix concept is valuable for the development of quality assessment standards for satellite missions and the presentation of the outcomes to users. The SkySat maturity matrix shows that the products have strengths in terms of providing product information and the characterization of uncertainty. The validation results within the EDAP also show that the products are in line with the mission specifications.

The image quality assessments were performed through radiometric calibration assessments, selected GIQE parameters, and visual inspections. The radiometric calibration assessments showed that the accuracy of the SkySat products was compliant with the specifications, and the results were within the range (< 10%) given by Planet. However, the calibration accuracy was inconsistent within the constellation, and further assessments are required using more data and possibly different external references (e.g., RadCalNet data).

The visual checks showed that the feature interpretability in the original and the pan-sharpened SkySat imagery was slightly deteriorated (blurring and smoothing were observed) in comparison to Pleiades data with similar spatial resolution. The SNR results were confirmed here for all bands and the SkySat sensors, indicating that the radiometric equalization between the cameras had been correctly performed for all configurations. On the other hand, the variation between the different sensors of the constellation can be interpreted as the aging of the platform, and must be monitored. Regarding the active PBHDR mode, the spatial resolution and the SNR and the gSNR results were upgraded using this method. It must be noted that improving SNR is an essential task for upscaling the GSD to 0.5 m. The MTF results were in accordance with the visual checks of the original MS images (without pan-sharpening), and the upsampling procedure degrades the image sharpness.

The geometric validation activities performed here involved the BBR, temporal, and absolute georeferencing accuracy, as well as stereoscopic capability (i.e., DSM generation). The BBR accuracy specification given by Planet was described in terms of band pairs. When considering the assessed bands, they were found to be within specifications [6]. The georeferencing accuracy of the BAS and OC products from two different dates was considered as well as the temporal accuracy between them, and they were also compliant with the specifications (1A: <50 m, 1C: <10 m). On the other hand, the stereoscopic capability is usually not a specification provided by vendors, and, similar to many other potential validation subjects (e.g., suitability for various application fields, minimum detectable object size, etc.), only custom assessments can be made. As future work, further validation metrics could be developed and integrated to the current scheme.

The DSM generation exercise performed here showed that the product accuracy is related to the temporal differences between the SkySat images, the image off-nadir configuration, and the RPC quality (or possibility of RPC corrections). The DSM produced from eight SkySat BAS images acquired at three different dates and off-nadir angles was compared with a UAV DSM. The global mean shift between the two DSMs was 0.55 m, with a σ of 2.4 m. The accuracy results obtained by Aati and Avouac [13] using SkySat triplets were 4 m (σ) and 5.3 m (normalized median absolute deviation), when compared with light detection and ranging (LiDAR) data. The larger errors could be explained by the long time interval between the acquisitions of the SkySat data and the reference DSM. d'Angelo et al. (2014) obtained 1–2 m relative and 2–3 m height accuracy from a triplet using data from SkySat 1&2. The results obtained here were also in line with Bhushan et al. [14]. The DSM generation can be improved by eliminating the invalid pixels given in UDM based on the assumption that it is correct.

The main challenge in DSM validation is the correction of RPCs using GCPs, since DSMs cannot be produced without applying this procedure. A total of six GCPs were used for the dataset composed of eight BAS data with a commercial software package (PCI Catalyst). As the footprint of each BAS image is much smaller than that of push broom sensors, and one set of RPCs is provided with each, a large number GCPs is required to cover the same area as that which can be covered using Pleiades triplets. Thus, new methods and tools may be required to reduce the GCP requirement for the same area size. The SkySat image quality and the artefacts that are also affected the DSM quality, such as saturation and repetitive patterns caused by compression or noise, lead to false or no matches. Thus, larger errors were observed in such regions. As future work, radiometric preprocessing methods and image filters can be applied to improve the quality of DSM generation.

It is also considered that employing different satellite data for the EDAP would guide the development of maturity matrix standards, leading to new conclusions and future work.

Author Contributions: Conceptualization, S.S., C.A., P.G. and S.K.; methodology, S.S. and S.K.; software, S.S., I.Y. and G.K.; validation, S.S. and S.K.; formal analysis, S.S., F.D. and S.K.; investigation, S.S., F.D., I.Y., G.K. and S.K.; resources, C.A. and P.G.; data curation, S.S., C.A., F.D., R.M. and S.K.; writing—original draft preparation, S.K.; writing—review and editing, S.S., C.A., S.K. and P.G.; supervision, S.S., C.A., P.G. and S.K.; project administration, R.M. and C.A. All authors have read and agreed to the published version of the manuscript.

Funding: These activities have been funded within the framework of the ESA Earthnet Data Assessment Pilot (EDAP) project (Optical Cluster) (https://earth.esa.int/eogateway/activities/edap accessed on 1 February 2022). SkySat Dataset provision through the ESA EOPI Project—63602, "Quality assessment of SkySat Planet data for the Earthnet Data Assessment Pilot (EDAP)".

Data Availability Statement: Restrictions apply to the availability of these data.

Acknowledgments: The authors are grateful to Planet for their support and the European Space Agency for provision of the SkySat data. The authors also thank to 4DiXplorer, Switzerland and Atay Muhendislik, Turkey for their support. We also would like to thank the anonymous reviewers for their valuable time and contributions.

Conflicts of Interest: The authors declare no conflict of interest.

References

1. Mannan, R.; Halsall, K.; Albinet, C.; Ottavianelli, G.; Goryl, P.; Boccia, V.; Melchiorre, A.; Piro, A.; Giudici, D.; Fox, N.; et al. ESA's Earthnet data assessment pilot: Paving the way for new space players. In Proceedings of the International Society for Optics and Photonics, Sensors, Systems, and Next-Generation Satellites XXIII, Strasbourg, France, 10 October 2019; p. 11151. [CrossRef]
2. ESA EDAP, Earthnet Data Assessment Pilot Project. 2022. Available online: https://earth.esa.int/eogateway/activities/edap (accessed on 3 February 2022).
3. Saunier, S.; Melchiorre, A.; Halsall, K.; Mannan, R.; Albinet, C.; Ottavianelli, G.; Goryl, P.; Boccia, V. EDAP Framework for the geometric validation of high resolution optical data. In Proceedings of the VH-RODA Workshop, Frascati, Italy, 18–22 November 2019. [CrossRef]
4. Hunt, S.E.; Albinet, C.; Nickeson, J.; Hall, A.; Fox, N.; Boccia, V.; Goryl, P. A Quality Assurance Framework for Satellite Earth Observation Missions. In Proceedings of the IEEE International Geoscience and Remote Sensing Symposium, Brussels, Belgium, 11–16 July 2021; pp. 608–611. [CrossRef]
5. Murthy, K.; Shearn, M.; Smiley, B.D.; Chau, A.H.; Levine, J.; Robinson, M.D. Sensors, Systems, and Next-Generation Satellites XVIII. In *SKS-1: Very High-Resolution Imagery from a Small Satellite*; Meynard, R., Neeck, S.P., Shimoda, H., Eds.; International Society for Optics and Photonics: Bellingham, WA, USA, 2014; Volume 9241, p. 92411E. [CrossRef]
6. Planet Imagery Product Specifications, Planet. 2020. Available online: https://assets.planet.com/docs/Planet_Combined_Imagery_Product_Specs_letter_screen.pdf (accessed on 3 February 2022).

7. EDAP.REP.001 Generic EDAP Best Practice Guidelines, 1.1, 23 May 2019. Technical Report. Available online: https://earth.esa.int/eogateway/activities/edap/edap-best-practice-guidelines (accessed on 1 February 2022).
8. EDAP.REP.002 Optical Mission Quality Assessment Guidelines, 1.0, 16 October 2019. Technical Report. Available online: https://earth.esa.int/eogateway/documents/20142/37627/EDAP.REP.002_1.0-Optical-Mission-Quality-Assessment-Guidelines.pdf (accessed on 1 February 2022).
9. *Planet L1 Data Quality Report Q3 2020, Status of calibration and Data Quality for the SkySat Constellation*; Technical Report Planet Labs: San Francisco, CA, USA, 2020.
10. Brunn, A.; Gonzalez, A.; Lenhard, K.; Bahloul, S.; Grennberg, J.; Smith, N.; Kinney-Spano, E.; Zuleta, I. SkySat Initial Radiometric Correction and Radiometric Calibration. In Proceedings of the JACIE Workshop, College Park, MD, USA, 17–19 September 2018. Available online: https://calval.cr.usgs.gov/apps/sites/default/files/jacie/abrunnJACIESkySatAbsCalfin.pdf (accessed on 3 February 2022).
11. Gonzalez, A. Absolute Calibration and Validation of SkySat Constellation. In Proceedings of the VH-RODA Workshop, Frascati, Italy, 18–22 November 2019.
12. Smiley, B. Long Term Geometric Stability of the SkySat Constellation. In Proceedings of the JACIE Workshop, College Park, MD, USA, 17–19 September 2018. Available online: https://calval.cr.usgs.gov/apps/sites/default/files/jacie/bsmileyJACIE2018approveddraft.pdf (accessed on 3 February 2022).
13. Aati, S.; Avouac, J.P. Optimization of Optical Image Geometric Modeling, Application to Topography Extraction and Topographic Change Measurements Using PlanetScope and SkySat Imagery. *Remote Sens.* **2020**, *12*, 3418. [CrossRef]
14. Bhushan, S.; Shean, D.; Alexandrov, O.; Henderson, S. Automated digital elevation model (DEM) generation from very-high-resolution Planet SkySat triplet stereo and video imagery. *ISPRS J. Photogramm. Remote Sens.* **2021**, *173*, 151–165. [CrossRef]
15. D'Angelo, P.; Kuschk, G.; Reinartz, P. Evaluation of skybox video and still image products. *Int. Arch. Photogramm. Remote Sens. Spat. Inf. Sci.* **2014**, *XL-1*, 95–99. [CrossRef]
16. D'Angelo, P.; Máttyus, G.; Reinartz, P. Skybox image and video product evaluation. *Int. J. Image Data Fusion* **2016**, *7*, 3–18. [CrossRef]
17. Anger, J.; Ehret, T.; Facciolo, G. Parallax estimation for push-frame satellite imagery: Application to super-resolution and 3D surface modeling from SkySat products. In Proceedings of the IEEE International Geoscience and Remote Sensing Symposium, Brussels, Belgium, 11–16 July 2021; pp. 2679–2682.
18. Valenzuela, A.Q.; Reyes, J.C.G. Comparative Study of the different versions of the General Image Quality Equation. In Proceedings of the ISPRS Annals of the Photogrammetry, Remote Sensing and Spatial Information Sciences, Enschede, The Netherlands, 10–14 June 2019; pp. 493–500. [CrossRef]
19. The National Image Interpretability Rating Scales. Available online: https://irp.fas.org/imint/niirs.htm (accessed on 5 March 2022).
20. Leachtenauer, J.C.; Malila, W.; Irvine, J.; Colburn, L.; Salvaggio, N. General image-quality equation: GIQE. *Appl. Opt.* **1997**, *36*, 8322–8328. [CrossRef] [PubMed]
21. Baltsavias, E.; Kocaman, S.; Wolff, K. Analysis of Cartosat-1 images regarding image quality, 3D point measurement and DSM generation. *Photogramm. Rec.* **2008**, *23*, 305–322. [CrossRef]
22. Gruen, A.; Kocaman, S. *Optical Sensors High Resolution: Geometry Validation Methodology*; Technical Report Submitted to ESA/ESRIN, RFQ/3-11780/06/I-OL; ETH Zurich, Institute of Geodesy and Photogrammetry: Zurich, Switzerland, 2008; 224p.
23. Cosnefroy, H.; Leroy, M.; Briottet, X. Selection and characterization of Saharan and Arabian desert sites for the calibration of optical satellite sensors. *Remote Sens. Environ.* **1996**, *58*, 101–114. [CrossRef]
24. Saunier, S.; Goryl, P.; Chander, G.; Santer, R.; Bouvet, M.; Collet, B.; Mambimba, A.; Aksakal, S.K. Radiometric, geometric, and image quality assessment of ALOS AVNIR-2 and PRISM sensors. *IEEE Trans. Geosci. Remote Sens.* **2010**, *48*, 3855–3866. [CrossRef]
25. Bouvet, M. Intercomparison of multispectral imagers over natural targets. In Proceedings of the IEEE International Geoscience and Remote Sensing Symposium, Barcelona, Spain, 23–28 July 2007; pp. 2653–2664. [CrossRef]
26. Murakami, H.; Tadono, T.; Imai, H.; Nieke, J.; Shimada, M. Improvement of AVNIR-2 Radiometric Calibration by Comparison of Cross-Calibration and Onboard Lamp Calibration. *IEEE Trans. Geosci. Remote Sens.* **2009**, *47*, 4051–4059. [CrossRef]
27. Chander, G.; Meyer, D.J.; Helder, D.L. Cross calibration of the Landsat-7 ETM+ and EO-1 ALI sensor. *IEEE Trans. Geosci. Remote Sens.* **2004**, *42*, 2821–2831. [CrossRef]
28. Robinson, D. High-resolution imagery and video from SkySat-1. In Proceedings of the Joint Agency Commercial Imagery Evaluation (JACIE), Louisville, KY, USA, 26–28 March 2014.
29. SkySat Instruments. Available online: https://earth.esa.int/eogateway/missions/skysat (accessed on 6 March 2022).
30. Thome, K.J. Post-Launch Calibration of Satellite Sensors. In *In-Flight Intersensor Radiometric Calibration Using Vicarious Approaches*, 1st ed.; Morain, S.A., Budge, A.M., Eds.; Balkema Publishers: London, UK, 2004; pp. 93–102. ISBN 90 5809 693 9.
31. Zanoni, V.; Ryan, R.; Holekamp, K.; Pagnutti, M. IKONOS Signal-to-Noise Ratio Estimation. In Proceedings of the JACIE Workshop, Reston, VA, USA, 25–27 March 2002. Available online: https://ntrs.nasa.gov/api/citations/20040004380/downloads/20040004380.pdf (accessed on 21 November 2021).
32. Viallefont-Robinet, F.; Helder, D.; Fraisse, R.; Newbury, A.; van den Bergh, F.; Lee, D.; Saunier, S. Comparison of MTF measurements using edge method: Towards reference data set. *Opt. Express* **2018**, *26*, 33625–33648. [CrossRef] [PubMed]

33. Blanc, P.H. Calibration Test Sites Selection and Characterisation–WP210. *Technical Report TN-WP210-001-ARMINES submitted to ESA/ESRIN*; VEGA Technologies SAS, 4 March 2008. Available online: http://calvalportal.ceos.org/c/document_library/get_file?uuid=03407bb4-89e4-4271-be26-898298780ee7&groupId=10136 (accessed on 1 February 2022).
34. Kohm, K. Modulation transfer function measurement method and results for the Orbview-3 high resolution imaging satellite. In Proceedings of the International Archives of the Photogrammetry, Remote Sensing and Spatial Information Sciences (ISPRS Arch), Istanbul, Turkey, 12–23 July 2004; Volume 35, pp. 7–12.
35. Yalcin, I.; Kocaman, S.; Saunier, S.; Albinet, C. Radiometric Quality Assessment for Maxar HD Imagery. In Proceedings of the International Archives of Photogrammetry, Remote Sensing and Spatial Information Sciences, Niece, France, 5–9 July 2021; pp. 797–804. [CrossRef]
36. Saunier, S.; Kocaman, S. EDAP Technical Note on Quality Assessment for SkySat. Issue 1.0. 6 September 2021. Available online: https://earth.esa.int/eogateway/documents/20142/37627/EDAP.REP.015+TN+on+Quality+Assessment+for+SkySat_v1.0.pdf/59a2a91d-eecd-20f1-4a13-e670dad8eed3 (accessed on 28 January 2022).
37. Artmo. 2021. Available online: https://artmotoolbox.com/radiative-transfer-models/89-atmospheric-rtms/27-6sv.html (accessed on 3 February 2021).
38. Aksakal, S.K.; Neuhaus, C.; Baltsavias, E.; Schindler, K. Geometric quality analysis of AVHRR orthoimages. *Remote Sens.* **2015**, *7*, 3293–3319. [CrossRef]
39. Kocaman, S.; Debaecker, V.; Bas, S.; Saunier, S.; Garcia, K.; Just, D. A Comprehensive Geometric Quality Assessment Approach for MSG SEVIRI Imagery. *Adv. Space Res.* **2021**, *69*, 1462–1480. [CrossRef]
40. Fraser, C.S.; Dial, G.; Grodecki, J. Sensor orientation via RPCs. *ISPRS J. Photogramm. Remote Sens.* **2006**, *60*, 182–194. [CrossRef]
41. Lucas, B.D.; Kanade, T. An iterative image registration technique with an application to stereo vision. In Proceedings of the 7th International Joint Conference on Artificial Intelligence, Vancouver, BC, Canada, 24–28 August 1981; pp. 674–679.
42. Bas, S.; Debaecker, V.; Kocaman, S.; Saunier, S.; Garcia, K.; Just, D. Investigations on the Geometric Quality of AVHRR Level 1B Imagery Aboard MetOp-A. *PFG–J. Photogramm. Remote Sens. Geoinf. Sci.* **2021**, *89*, 519–534. [CrossRef]
43. Debaecker, V.; Kocaman, S.; Saunier, S.; Garcia, K.; Bas, S.; Just, D. On the Geometric Accuracy and Stability of MSG SEVIRI Images. *Atmos. Environ.* **2021**, *262*, 118645. [CrossRef]
44. Gruen, A.; Akca, D. Least squares 3D surface and curve matching. *ISPRS J. Photogramm. Remote Sens.* **2005**, *59*, 151–174. [CrossRef]
45. USGS Spatial Resolution Digital Imagery Guideline. Available online: https://www.usgs.gov/media/images/spatial-resolution-digital-imagery-guideline (accessed on 1 February 2022).

remote sensing

MDPI

Article

A Quadrifocal Tensor SFM Photogrammetry Positioning and Calibration Technique for HOFS Aerial Sensors

Tao Wang, Yan Zhang *, Yongsheng Zhang, Ying Yu, Lei Li, Shaocong Liu, Xiang Zhao, Zhenchao Zhang and Longhui Wang

Geospatial Information Institute, Information Engineering University, Zhengzhou 450001, China; wangtaoynl@163.com (T.W.); yszhang2001@vip.163.com (Y.Z.); yuying5559104@163.com (Y.Y.); 3110100798@zju.edu.cn (L.L.); scholar_liusc@163.com (S.L.); zx15538019261@163.com (X.Z.); zhzhc_1@163.com (Z.Z.); wlhonline@163.com (L.W.)

* Correspondence: zhangyanxz7806@163.com; Tel.: +86-186-9580-9656

Abstract: Nowadays, the integration between photogrammetry and structure from motion (SFM) has become much closer, and many attempts have been made to combine the two approaches to realize the positioning, calibration, and 3D reconstruction of a large number of images. For the positioning and calibration of high oblique frame sweep (HOFS) aerial cameras, a quadrifocal tensor SFM photogrammetry technique is proposed to resolve the positioning and calibration task of such cameras. It adopts the quadrifocal tensor idea into the OpenMVG SFM pipeline to solve the complexity problem caused by the small single-viewing imaging area and the high image overlapping ratio. It also integrates the photogrammetry iteration idea into the OpenMVG SFM pipeline to enhance the positioning and calibration accuracy, which includes a coarse to fine three-stage Bundle Adjustment (BA) processing approach. In this paper, the overall workflow of the proposed technique was first introduced in detail, from feature extraction and image matching, relative rotation and translation estimation, global rotation and translation estimation, and the quadrifocal tensor model construction to the three-stage BA process and calibration. Then, experiments were carried out in the Zhengzhou area, implementing four types of adjustment methods. The results suggest that the proposed quadrifocal tensor SFM photogrammetry is suitable for large tilt frame sweep camera positioning and calibration without prior information on detailed camera intrinsic parameters and structure. The modifications made to the OpenMVG SFM pipeline enhanced the precision of image positioning and calibration and provided the precision level of professional photogrammetry software.

Keywords: a quadrifocal tensor; SFM photogrammetry; oblique photogrammetry; frame sweep aerial sensor; VisionMap A3 edge sensor

1. Introduction

Given its high spatial resolution and great flexibility, digital frame cameras are widely used in aerial photogrammetry and remote sensing for topographic mapping, archaeological discoveries, forestry and agricultural assessment, natural disaster monitoring, and ecosystem restoration [1–6]. To obtain broader ground coverage, frame digital cameras usually adopt the following technologies: super-wide frame cameras (e.g., Z/I DMC system [4,7]), multi-splicing frame cameras (e.g., Microsoft Ultracam system [4,8], Leica Citymapper system [9], and SWDC-4 cameras [10]), and frame sweep cameras. The representative aerial frame sweep cameras include the EOLOROP [11], DB-110 [12], and VisionMap A3 [13–16].

Most of these cameras are equipped with Position Orientation System (POS) equipment (including Global Navigation Satellite Systems (GNSS) and Inertial Measurement Units (IMU)) to provide initial camera poses and orientation. Due to significant imaging principles and camera inner structures, the specific camera imagery is always processed through corresponding professional software, such as Ultramap software for Microsoft

Ultracam cameras, HxMap software for Leica Citymapper cameras, and LightSpeed for VisionMap A3 cameras [16].

For super-wide and multi-splicing frame cameras, the geometric imaging relationship is relatively simple in nearly vertical photography. Even though the Leica Citymapper camera adopts an oblique photography approach, the angular relationship between different views is fixed. Thus, an accurate geometric imaging model can be established, and high-precision open-source positioning and calibration software can be developed, breaking the limitations of professional software.

In contrast, the frame sweep camera swings along the vertical direction of the flight swiftly for imaging, such as VisionMap A3. It achieves true high-tilt photogrammetry with a long focal length and an ultra-wide sweep angle up to 104 degrees. It can obtain a wide range of clear images at high altitudes, which is particularly useful for surveying, monitoring, and reconnaissance. The obtained multi-angle images are excellent for 3D reconstruction compared to high-resolution satellite images. However, the camera tilt angle varies considerably in dynamic imaging, causing complex geometric imaging relationships and inconsistent imaging scales within one sweep cycle. Moreover, the VisionMap A3 camera is not equipped with an IMU installment, providing no camera orientation information. These increase the difficulty in image positioning processing.

To realize the geometric processing for the VisionMap A3 camera in a more general and simplified way, the research project was implemented, and the performance of the proposed technique was comprehensively analyzed. The rest of the paper is organized as follows. The related works are discussed in the next section, and the VisionMap A3 system is introduced in Section 3. Section 4 presents the methods, including the ideas and workflow. The experimental results and the discussion are provided in Section 5, and a brief conclusion is presented in the last section.

2. Related Work

Photogrammetric techniques have been used to achieve open-source positioning and calibration of the VisionMap A3 camera, but the results have largely been unstable. The main goal of photogrammetry is to generate accurate camera poses (i.e., positions and orientations), intrinsic camera parameters, and 3D points. The processing approach includes matching, aero-triangulation (also called bundle adjustment), orthoimage generation, and digital elevation model (DEM) generation [1–5,17]. Complex geometric imaging relationships and inconsistent imaging scales make it difficult to set up accurate geometric imaging models; thus, many problems occur in processing A3 images by the photogrammetry scheme.

Aside from photogrammetry, SFM has become a common approach in aerial imagery processing. Photogrammetry and SFM approaches have many similarities in feature and bundle adjustment. SFM also computes camera poses, intrinsic parameters, and 3D points cloud [18–20]. SFM methods can be categorized into five types: global SFM, incremental SFM, hierarchical SFM, hybrid SFM, and semantic SFM. Global SFM processes all images simultaneously, first solving global camera rotation using the rotation consistency, then calculating camera displacement, and finally applying BA optimization. While the global strategy significantly improves processing efficiency, it is extremely sensitive to external points. Its performance is unstable in many applications, and its reconstruction accuracy is largely unsatisfactory [21–28].

Incremental SFM starts with two or three view reconstructions, gradually adds new views, and then applies BA operation after each addition. This processing method is robust with high reconstruction accuracy, and most popular SFM pipelines employ incremental approaches. However, incremental SFM has drift risk due to the accumulation of errors [29–37]. Hierarchical SFM is the revision of the traditional incremental SFM. It divides the large-scale datasets into N interrelated sub-datasets, and then parallel incremental SFM processing is performed on the sub-datasets. Finally, the sub-datasets are merged [38–41]. Hybrid SFM combines the advantages of global SFM and incremental

SFM. It solves camera rotation using the global approach and then calculates camera displacement incrementally [42–44]. Semantic SFM uses the semantic label to detect the corresponding models and facilitate image matching and camera calibration. Scene segmentation can be realized in SFM reconstruction [45–47].

The well-known SFM pipelines include Bundler [31,32], VisualSFM [35], COLMAP [34,37], and OpenMVG [29,36]. The properties of these pipelines are summarized in Table 1.

Table 1. Popular SFM Pipelines Analysis.

Pipeline	Properties
Bundler	An early incremental SFM pipeline originated from Photo Tourism Project. The time complexity is $O(n^4)$, the output is sparse point clouds, and the operation is complex.
VisualSFM	A revision on Bundler and the time complexity reaches $O(n^1)$, the output is dense point clouds with higher accuracy, and the operation is much simpler.
COLMAP	COLMAP improves scene graph verification, next best view selection, sampling-based triangulation, and bundle adjustment; it performs better in terms of completeness and robustness. Its efficiency outperforms OpenMVG but does not support POS data input.
OpenMVG	OpenMVG adds an efficient contrario trifocal tensor estimation method and translation registration technique into the original SFM pipeline. It outperforms many incremental and global pipelines in terms of accuracy and running times. The reprojection error and the drift are smaller than COLMAP, and it supports POS data input.

Notes: n is the number of images.

At present, the integration between photogrammetry and SFM has become more pronounced. SFM photogrammetry combines the two approaches to realize the positioning, calibration, and 3D reconstruction of a large number of ordered and disordered images [48].

Photogrammetry often relies on initial POS data and intrinsic camera parameters and can reach higher accuracy through iterative least-squares adjustment, condition adjustment, or indirect adjustment. SFM takes homogeneous coordinates in expression and does not require specific initial values of POS data and intrinsic camera parameters. In image pose recovery and 3D reconstruction, SFM uses the R matrix to model the attitude angle system, which does not involve selecting a specific angle rotation system; thus, it can overcome challenges caused by the swift change of the sweep angle in the ultra-wide imaging range. It also avoids conversion among different coordinate systems, resulting in much simpler operations and faster convergence. Thus, the SFM pipeline was chosen to solve the positioning and calibration task of the VisionMap A3 camera.

By analyzing Table 1 it can be determined that the OpenMVG pipeline has smaller reprojection and drift errors and supports POS data input, which is especially advantageous for remote sensing imagery. Thus, the OpenMVG pipeline was chosen for the geometric processing of VisionMap A3 images, addressing the limitations of the professional LightSpeed software.

Wu et al. [35] introduced a novel BA strategy that provides a good balance between SFM reconstruction speed and accuracy and maintains high accuracy by regularly re-triangulating feature matches that fail to triangulate. Wu's feature detection and description algorithm was adopted into the OpenMVG pipeline. Moulon et al. [36] proposed a new global calibration approach based on the fusion of relative motions between image pairs, presenting an efficient contrario trifocal tensor estimation method and translation registration technique for accurate camera positions recovery in OpenMVG. We followed the tensor cluster idea and designed the quadrifocal tensor-based BA method.

Two main improvements were introduced to the original OpenMVG pipeline. First, the VisionMap A3 frame sweep camera has a small single-viewing imaging area. The high image overlapping ratio and the high number of images enhance the processing complexity. To reduce the complexity and strengthen the robustness, the quadrifocal tensor was added to the OpenMVG pipeline, which deals with the images in the quadrifocal tensor model cluster. The quadrifocal tensor model performs high oblique frame sweep camera positioning and calibration processing as an independent unit. Second, the photogrammetry iteration idea was introduced into the OpenMVG pipeline to enhance the positioning and calibration accuracy. A coarse to fine three-stage BA processing approach was proposed to deal with the pose parameters and the intrinsic parameters.

The quadrifocal tensor-based positioning and calibration method first carries out feature extraction and image matching. Relative rotation and translation estimation and global rotation and translation are performed before establishing the quadrifocal tensor model. The three-dimensional coordinates of matching points in each quadrifocal tensor model are calculated, and the BA cost function is finally set up with the quadrifocal tensor model as the independent unit. Real VisionMap A3 data is then used to evaluate whether the proposed approach can solve the high oblique frame sweep camera position and calibration problem in a more general and simplified way.

3. VisionMap A3 System Introduction

The VisionMap A3 system is a fully automated mapping system established in 2004. The system consists of an airborne digital step-framing double lens metric camera and a ground processing system. The airborne system consists of dual CCDs with two 300 mm lenses (Figures 1 and 2), a fast compression and storage unit, and a dual-frequency GPS. The long focal length yields comparatively high ground resolution when flying at high altitudes, enabling efficient photography of large areas in high resolution.

Figure 1. VisionMap A3.

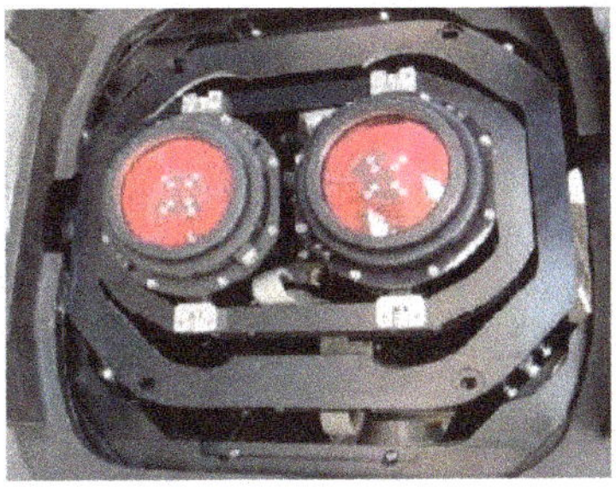

Figure 2. Dual Lenses of VisionMap A3.

During the flight, a sequence of frames is exposed in a cross-track direction at a very high speed to provide a very wide angular coverage of the ground. The two lenses of the camera simultaneously sweep across the flight direction from one side to the opposite side, with each CCD capturing about 27 frames (54 frames for two CCDs) and having a maximum sweep angle of 104 degrees. After completing the first sweep, the lenses return to the start position in preparation for the next sweep. The sweep back time is 0.5 s. Each CCD captures seven frames per second (one frame in 0.142 s); therefore, a single sweep is completed in approximately 3.6 s. The time between sweeps depends largely on the aircraft speed, flight altitude, and the required overlap between two consecutive sweeps. For more details on the technical parameters, refer to Pechatnikov et al. [13].

Two adjacent single frames along the flight form a double frame (Figures 3 and 4). The overlap between two adjacent single frames is about 2% (~100 pix), while the overlap between two adjacent double frames across the flight direction is about 15%. The overlap between two consecutive sweeps along the flight direction is typically 56%, but this value may vary depending on the defined specifications of the aerial survey. Between two consecutive flight lines, the overlap is generally 50% to enable stereo photogrammetric mapping. All overlaps are determined during flight planning and may be altered during the flight. VisionMap A3 camera provides orthogonal coverage of the nadir area and oblique coverage of the remainder of the sweep image. As all images participate in all stages of the analytical computations, after performing matching and block adjustment, accurate solutions for all images, including the oblique images, can be obtained. The generation of accurately solved oblique images simultaneously with regular verticals is a unique and highly important feature of the A3 system.

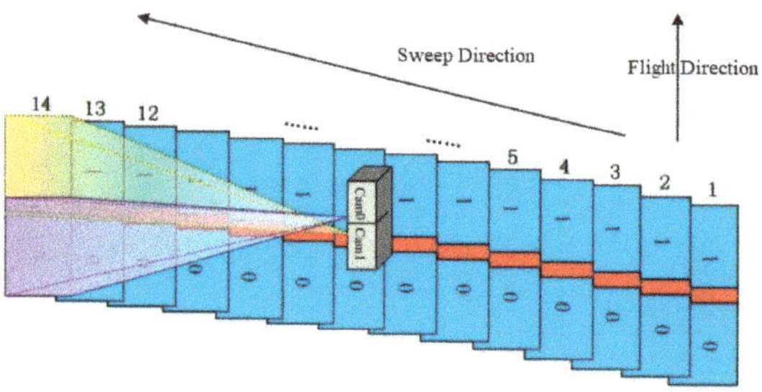

Figure 3. Flights with VisionMap A3. Number 0 represents images captured by lenses 0, number 1 represents images captured by lenses 1.

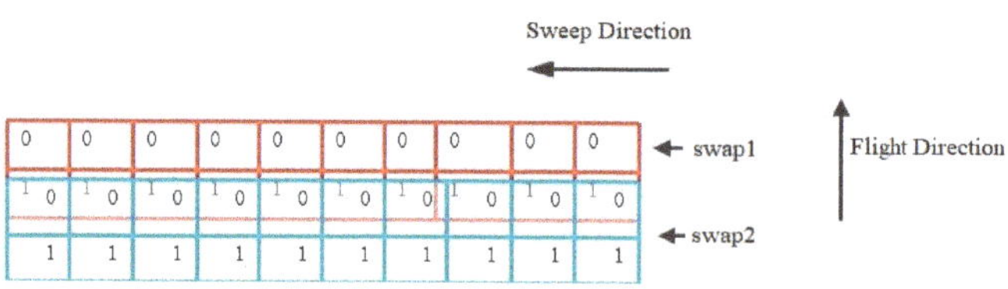

Figure 4. Relationship among Consecutive Swap. Number 0 represents images captured by lenses 0, number 1 represents images captured by lenses 1.

4. The Proposed Quadrifocal Tensor SFM Photogrammetry Positioning and Calibration Technique

The proposed quadrifocal tensor SFM photogrammetry positioning and calibration technique include the following steps. First, all aerial images are sorted by flight route according to imaging time, and the sorting relationship is established. The scale-invariant feature transform (SIFT) algorithm is then used to extract feature points on each view and build corresponding feature descriptors according to the image sorting relationship. The iterative feature matching process is then carried out under the Hamming distance criterion. The RANSAC algorithm is used to eliminate mismatches in the matching results, and the feature point correspondences on the pairwise overlapping images are determined. The pinhole imaging model is then established based on the unknown camera geometric distortion parameters, and the essential matrices between each pairwise overlapping view are calculated using the successful matching point correspondences and redundant constraints. The relative spatial rotation and the translation relationship, i.e., the rotation matrix R and the translation vector t, are derived from the essential matrices.

After performing the Bayesian inferences, the initial global rotations are computed for each view in a global coordinate system, while the initial global translations are generated using the initial POS data or from the translation registration. The quadrifocal tensor model can then be established, and the initial three-dimensional coordinates of the matching point correspondences in each quadrifocal tensor model are calculated. Finally, the BA cost function based on the quadrifocal tensor model is constructed and used in the BA processing. The global rotation matrix and the translation vector for each view are updated, and the intrinsic parameters are calibrated simultaneously. The general workflow is listed in Figure 5.

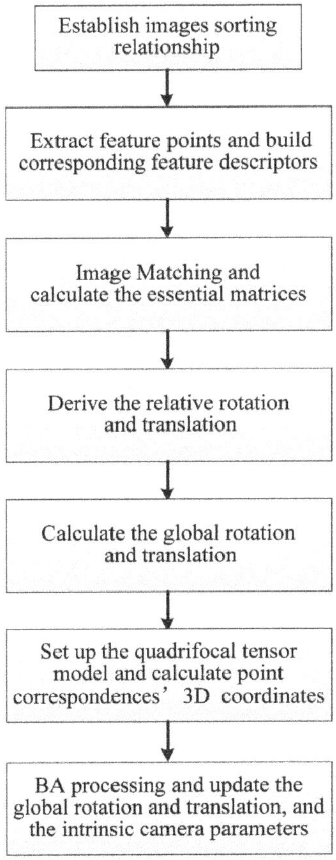

Figure 5. Workflow of the proposed quadrifocal tensor SFM photogrammetry positioning and calibration technique.

4.1. Feature Extraction and Image Matching

The SIFT algorithm is used to extract the feature points on each view and build corresponding feature descriptors simultaneously according to the image sorting relationship [49]. The Hamming distance is taken as the criterion for iterative image matching, and the RANSAC algorithm is adopted to eliminate a large number of mismatches in the matching results.

4.2. Relative Rotation and Translation Estimation

The pinhole camera model is used to describe the frame sweep aerial camera. The three-dimensional point P(X, Y, Z) in the ground object space is projected onto the image plane and forms an image point $p(x, y, -f)$ through the pinhole projection. The geometric relationship between the object point and the image point can be described with the following model:

$$\mathrm{x} = \mathrm{PX} = \mathrm{K}[\mathrm{R}|\,\mathrm{t}]\mathrm{X} \tag{1}$$

where R is the rotation matrix, a 3 × 3 orientation matrix representing the direction of the camera coordinate system; t is the position vector; R and t are the camera exterior parameters; K is the camera intrinsic parameter matrix (also called the camera calibration matrix). The frame sweep camera positioning and calibration determine the R, t, and K matrices; the camera projection matrix is provided by P = K[R I t].

The general form of the calibration matrix for a CCD camera is:

$$\mathrm{K} = \begin{bmatrix} \alpha_x & & x_0 \\ & \alpha_y & y_0 \\ & & 1 \end{bmatrix}. \tag{2}$$

To increase generality, the skew parameter s can be added,

$$\mathrm{K} = \begin{bmatrix} \alpha_x & s & x_0 \\ & \alpha_y & y_0 \\ & & 1 \end{bmatrix}. \tag{3}$$

Based on the successful matching point correspondences obtained in Section 4.1, the relative spatial rotation and translation between any two overlapping images can be calculated. The pairwise point correspondence p_1, p_2 meets the following epipolar constraint:

$$p_2^{\mathrm{T}}\mathrm{K}^{-\mathrm{T}}\mathrm{E}\mathrm{K}^{-1}p_1 = 0 \tag{4}$$

In Equation (4), the points O_1, P, and O_2 are coplanar. E is the essential matrix, which is the outer product of t and R and is perpendicular to t and R. Both translation and rotation are included in the epipolar plane constraint, and the essential matrix E is calculated from the point correspondences. After obtaining the essential matrix E, singular value decomposition (SVD) is performed on the matrix E matrix to obtain R. Thus, the relative rotation and translation of the two overlapping images can be determined [18,29,36].

4.3. Global Rotation and Translation Estimation

Assuming that in the global coordinate system, the global rotation matrix of image i is R_i, that of image j is R_j, and the relative rotation is R_{ij}, which is obtained in Section 4.2. Three rotation matrices meet the consistency Equation (5), which is the basis for solving the global rotation.

$$\mathrm{R}_j = \mathrm{R}_{ij}\mathrm{R}_i. \tag{5}$$

R_j is orthonormal, for $j = 1, \ldots, m$.

While the RANSAC algorithm can remove most mismatches, there are remaining mismatches that cause deviations in the relative rotation estimation. As relative R_{ij} estimates may contain outliers, the global rotation estimation must be robust in identifying the global rotations and the inconsistent/outlier edges (false essential geometry). Moulon's method [36] is used in determining inconsistent relative rotations in Bayesian inference. The iterative use of the Bayesian inference, adjusted with the cycle length weighting, can remove most outliers, check all the triplets of the graph, and reject those with cycle deviations larger than 2°.

When the relative rotations are known, they form a tree graph with the views as vertices connected by an edge. Equation (5) can be solved by least-squares while satisfying the orthonormality conditions, R can be expressed in quaternions [50], and Equation (5) is transformed into

$$\dot{r}^j = \dot{r}^{ij}\dot{r}^i \tag{6}$$

where $\dot{r}^j$ and $\dot{r}^i$ are the unknown quaternions of the *i*th and *j*th view rotation, respectively, and $\dot{r}^{ij}$ is the known relative rotation between views *i* and *j*. Each quaternion can be considered a four-vector. Using known manipulations with quaternions, each equation in (6) can be rewritten as

$$\begin{pmatrix} r_0^j \\ r_x^j \\ r_y^j \\ r_z^j \end{pmatrix} = \begin{bmatrix} r_0 & -r_x & -r_y & -r_z \\ r_x & r_0 & -r_z & r_y \\ r_y & r_z & r_0 & -r_x \\ r_z & -r_y & r_x & r_0 \end{bmatrix} \begin{pmatrix} r_0^i \\ r_x^i \\ r_y^i \\ r_z^i \end{pmatrix} \tag{7}$$

where $\dot{r}^i = r_0^i + ir_x^i + jr_y^i + kr_z^i$, with *i*, *j*, and *k* as imaginary units. There are 4*m* unknowns $r_0^1, r_x^1, r_y^1, r_z^1, \cdots, r_0^m, r_x^m, r_y^m, r_z^m$ with constraints (7) for each view pair *ij* with a known rotation. After the solution, the quaternions can be easily made into units by dividing each by its Euclidean length.

After deciding on the global rotations, the next step is determining the global translations. Given a set of relative motion pairs (R_{ij}, t_{ij}) (rotations and translation directions solved in Section 4.2), the global location $(T_1, \cdots, T_m)$ of all views can be obtained. The different translation directions are reconciled in the global coordinates system.

$$\left\| T_j - R_{ij}T_i - \lambda_{ij}t_{ij} \right\| = 0, \quad \forall i, j \tag{8}$$

To solve *m* global translations and scale factors λ_{ij}, the solution for Equation (7) is optimized using the least-squares method or Moulon's approach [36] under the l_∞ norm. For the VisionMap A3 edge camera, the GPS equipment provides the camera with position information in the world coordinate system, so the processed GPS information can be taken as initial values $(T_1, \cdots, T_m)$.

4.4. Set up the Quadrifocal Tensor Model and Calculate Initial 3D Ground Point

Subsequently, the quadrifocal tensor model would be established, and the global translations and three-dimensional coordinates of the point correspondences would then be determined based on the quadrifocal tensor model. For the VisionMap camera, at a certain sweep angle, the cam0 and cam1 lenses obtain images I and I′, the corresponding projection planes are II and II′, and the projection centers are C and C′. The sweep angle changes swiftly. At the next imaging moment, the cam0 and cam1 lenses obtain images I″ and I‴, with the corresponding projection planes II″ and II‴ and projection centers C″ and C‴. The straight line L in the object space is reflected on four image views (see Figure 6).

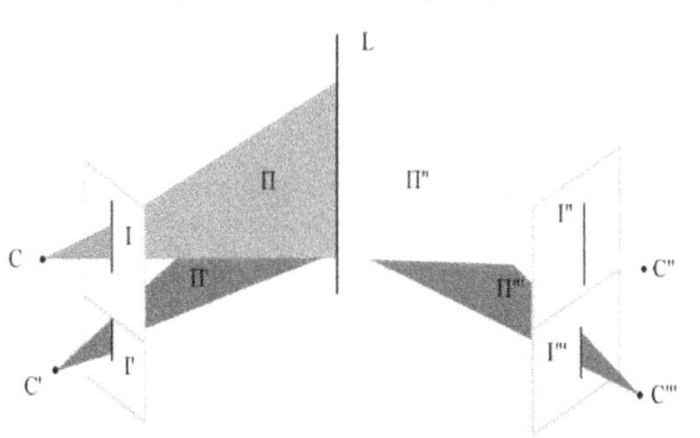

Figure 6. The Quadrifocal Tensor Model.

When the camera projection matrices of the four image views are A, B, C, and D, a three-dimensional ground point X projects onto the four image views, and there is a set of corresponding points $\mathrm{x} \leftrightarrow \mathrm{x}' \leftrightarrow \mathrm{x}'' \leftrightarrow \mathrm{x}'''$ across the four views. The four image views constitute a quadrifocal tensor model, and the projection relationship among the four views is provided by:

$$\begin{bmatrix} \mathrm{A} & \mathrm{x} & & & \\ \mathrm{B} & & \mathrm{x}' & & \\ \mathrm{C} & & & \mathrm{x}'' & \\ \mathrm{D} & & & & \mathrm{x}''' \end{bmatrix} \begin{bmatrix} \mathrm{X} \\ -k \\ -k' \\ -k'' \\ -k''' \end{bmatrix} = 0 \tag{9}$$

where k, k', k'', and k''' represent uncertain scale constants, and a^i, b^i, c^i, and d^i are the row vectors of matrices A, B, C and D, respectively. The rank-4 tensor Q^{pqrs} is defined by

$$\mathrm{Q}^{pqrs} = \det \begin{bmatrix} \mathrm{a}^i \\ \mathrm{b}^q \\ \mathrm{c}^r \\ \mathrm{d}^s \end{bmatrix}. \tag{10}$$

From Equation (9), the quadrilinear relationship can be derived in the following form for the quadrifocal tensor:

$$x^i x'^j x''^k x'''^l \varepsilon_{ipw} \varepsilon_{jqx} \varepsilon_{kry} \varepsilon_{lsz} \mathrm{Q}^{pqrs} = 0_{wxyz} \tag{11}$$

where 0_{wxyz} is a zero tensor with four indices w, x, y and z; $\varepsilon_{ipw}, \varepsilon_{jqx}, \varepsilon_{kry}$ and ε_{lsz} represent the product between the different vectors. Based on the global rotation and translation computed in Section 4.3, the camera projection matrices A, B, C, and D are determined, and the quadrifocal tensor model can then be established.

The next step is realizing 3D reconstruction for point correspondences in the quadrifocal tensor model. The ground point X projects across four views and forms the corresponding image points $\mathrm{x} \leftrightarrow \mathrm{x}' \leftrightarrow \mathrm{x}'' \leftrightarrow \mathrm{x}'''$. The geometric projection relationship between the ground point and the corresponding image point is listed as follows:

$$\begin{cases} \mathrm{x} = \mathrm{AX} \\ \mathrm{x}\prime = \mathrm{BX} \\ \mathrm{x}'' = \mathrm{CX} \\ \mathrm{x}''' = \mathrm{DX} \end{cases} \tag{12}$$

The plane coordinate of the image point x is (x, y). Taking the first expression in Equation (12) as an example, when the cross product is applied, and the homogeneous scalar factor is eliminated, the following equations can be obtained:

$$\begin{cases} x(\mathrm{A}^{3\mathrm{T}}\mathrm{X}) - \left(\mathrm{A}^{1\mathrm{T}}\mathrm{X}\right) = 0 \\ y(\mathrm{A}^{3\mathrm{T}}\mathrm{X}) - \left(\mathrm{A}^{2\mathrm{T}}\mathrm{X}\right) = 0 \\ x(\mathrm{A}^{2\mathrm{T}}\mathrm{X}) - y\left(\mathrm{A}^{1\mathrm{T}}\mathrm{X}\right) = 0 \end{cases}. \tag{13}$$

Taking the first two equations, which are linearly independent, the four image points $\mathrm{x} \leftrightarrow \mathrm{x}' \leftrightarrow \mathrm{x}'' \leftrightarrow \mathrm{x}'''$ develop into eight equations

$$\mathrm{NX} - \mathrm{L} = 0\mathrm{N} = \begin{bmatrix} x\mathrm{A}^{3\mathrm{T}} - \mathrm{A}^{1\mathrm{T}} \\ y\mathrm{A}^{3\mathrm{T}} - \mathrm{A}^{2\mathrm{T}} \\ x'\mathrm{B}^{3\mathrm{T}} - \mathrm{B}^{1\mathrm{T}} \\ y'\mathrm{B}^{3\mathrm{T}} - \mathrm{B}^{2\mathrm{T}} \\ x''\mathrm{C}^{3\mathrm{T}} - \mathrm{C}^{1\mathrm{T}} \\ y''\mathrm{C}^{3\mathrm{T}} - \mathrm{C}^{2\mathrm{T}} \\ x'''\mathrm{D}^{3\mathrm{T}} - \mathrm{D}^{1\mathrm{T}} \\ y'''\mathrm{D}^{3\mathrm{T}} - \mathrm{D}^{2\mathrm{T}} \end{bmatrix}. \tag{14}$$

Solving Equation (14) by the least-squares approach, the 3D coordinates for ground point X are generated.

4.5. Bundle Adjustment Based on the Quadrifocal Tensor Model

The BA processing is performed to update the rotation and translation and simultaneously calibrate the intrinsic matrix. The BA cost function based on the quadrifocal tensor model is defined in the form as

$$f(z) = \left\| \left(x_i - \frac{\mathrm{R}_i^1\mathrm{X} + \mathrm{t}_i^1}{\mathrm{R}_i^3\mathrm{X} + \mathrm{t}_i^3}, y_i - \frac{\mathrm{R}_i^2\mathrm{X} + \mathrm{t}_i^2}{\mathrm{R}_i^3\mathrm{X} + \mathrm{t}_i^3} \right) \right\| \tag{15}$$

where $(x_i, y_i), i \in (1,2,3,4)$ is the observation value corresponding to the four image points $\mathrm{x} \leftrightarrow \mathrm{x}' \leftrightarrow \mathrm{x}'' \leftrightarrow \mathrm{x}'''$; $\mathrm{R}_i, i \in (1,2,3,4)$ is the corresponding rotation matrix to each view; $\mathrm{t}_i, i \in (1,2,3,4)$ is the corresponding translation vector; $\left(\frac{\mathrm{R}_i^1\mathrm{X}+\mathrm{t}_i^1}{\mathrm{R}_i^3\mathrm{X}+\mathrm{t}_i^3}, \frac{\mathrm{R}_i^2\mathrm{X}+\mathrm{t}_i^2}{\mathrm{R}_i^3\mathrm{X}+\mathrm{t}_i^3} \right)$ is the calculation value of the image point; z is the unknown parameter vector comprising the camera pose and calibration parameters (i.e., unknowns in rotation and translation) and the 3D ground point coordinates X.

The differences between the observed values and the calculation values of the image points are the reprojection error of the ground point, representing the errors contained in the rotation, translation, and camera calibration parameters and the ground point coordinates. When the cost function reaches the minimum value, the optimal solution is obtained for the pose parameters, calibration parameters, and 3D coordinates of ground points.

The BA process optimizes the following nonlinear least-squares BA cost function

$$z^* = \operatorname{argmin} \sum_{k=1}^{n} \| f_k(z) \|^2. \tag{16}$$

The BA cost function is optimized by the Levenberg–Marquardt (LM) approach. The LM algorithm decomposes the original nonlinear cost function (Equation (16)) into approximations of a series of regularized linear functions and $J(z)$ is the Jacobian matrix of $f(z)$. In each loop, the LM approach updates the linear least squares problem in the following form:

$$\delta^* = \operatorname*{argmin}_{\delta} \| J(z)\delta + f(z) \|^2 + \lambda \| D(z)\delta \|. \tag{17}$$

If $\|f(z+\delta^*)\| < \|f(z)\|$, then the unknown parameter is updated as $z \to z + \delta^*$. $D(z)$ is the square root matrix of the matrix $J(z)^{\mathrm{T}}J(z)$ and λ is the regularization parameter, which can be adjusted according to the approximation between $J(z)$ and $f(z)$. Solving Equation (17) is equivalent to solving the standard equation:

$$\left(J^{\mathrm{T}}J + \lambda D^{\mathrm{T}}D \right)\delta = -J^{\mathrm{T}}f \tag{18}$$

where $\left(J^{\mathrm{T}}J + \lambda D^{\mathrm{T}}D\right)$ is the extended Hessian matrix. In the BA process, the unknown parameter z includes two parts: the camera pose and calibration parameters (z_c) and the 3D ground point coordinates (z_p). Similarly, D, δ, J can also be divided into two parts. Let

$U = J_c^T J_c$, $V = J_p^T J_p$, $U_\lambda = U + \lambda D_c^T D_c$, $V_\lambda = V + \lambda D_p^T D_p$, $W = J_c^T J_p$; Equation (18) can then be rewritten as the block structure linear system:

$$\begin{bmatrix} U_\lambda & W \\ W^T & V_\lambda \end{bmatrix} \begin{bmatrix} \delta_c \\ \delta_p \end{bmatrix} = -\begin{bmatrix} J_c^T f \\ J_p^T f \end{bmatrix}. \tag{19}$$

Applying the Gaussian elimination method to Equation (19), the 3D coordinates of the ground points can be eliminated, and a linear equation containing just the camera pose and calibration parameters can be obtained:

$$\left(U_\lambda - W V_\lambda^{-1} W^T\right)\delta_c = -J_c^T f + W V_\lambda^{-1} J_p^T f. \tag{20}$$

Therefore, δ_c can be determined by solving Equation (20) and δ_p can then be solved by reverse substitution.

$$\delta_p = -V_\lambda^{-1}\left(J_p^T f + W^T \delta_c\right). \tag{21}$$

The variables δ_c and δ_p are obtained using an iterative solution; after two loops, both can reach high precision. The quadrifocal tensor SFM positioning and calibration process for the frame sweep aerial camera is therefore completed.

To obtain better positioning and calibration precision, the OpenMVG SFM pipeline is improved according to the photogrammetry processing mode. In photogrammetric processing, position deviation is mostly caused by pose errors, while the intrinsic camera parameters generate minor systematic errors for both aerial and satellite-borne cameras [51,52]. So in the photogrammetry BA implementation, BA is always carried out in several stages. Pose parameters are dealt with first, while the intrinsic parameters are processed last. Here, three-stage BA processing is put forward. In the first stage, the intrinsic matrix K is set to E (identity matrix), the rotation R remains unchanged, and only the translation T and the 3D coordinates of ground points X are involved. In the second stage, the intrinsic matrix K remains as E, but the translation T, the rotation T, and the 3D coordinates of ground points X are involved in the BA process. In the last stage, all the unknowns, the intrinsic parameter K, the translation T, the rotation T, and the 3D coordinates of ground points X are adjusted.

5. Experiments and Analysis

5.1. Test Data

The VisionMap Edge A3 camera was used to photograph the Zhengzhou area in January 2020, with a total flight area of about 1000 square kilometers. Two flight missions were carried out at 2300 m flight altitude, generating a ground resolution of about 5 cm. Six routes (L_44, L_45, L_46, L_1, L_2, and L_3), located in the middle region and covering about 120 square kilometers, were used for the experiment (see Figure 7).

The upper half is selected for analysis, comprising 7928 images and covering a total area of 64 square kilometers. The study area (shown in Figure 8) includes plain regions and urban areas.

Figure 7. VisionMap Flight Coverage.

Figure 8. Ground Coverage of Test Field.

Four different calibration approaches were designed to evaluate and compare the proposed quadrifocal tensor SFM photogrammetry method with the original OpenMVG SFM pipeline. The proposed method is tested under different conditions: no Ground Control Point (GCP) support, with GCP support without considering camera intrinsic parameters, and with GCP support considering camera intrinsic parameters.

I. The proposed method with no GCP support

The whole procedure in Figure 5 was implemented using POS. All unknowns, rotation, translation, intrinsic parameters, and 3D ground point coordinates were involved in this process.

II. The proposed method with GCP support, but not considering camera intrinsic parameters

The proposed method with GCP support was evaluated. All the unknowns were employed in the BA procedure except the intrinsic camera parameters. The whole flow was similar to the first test, but the intrinsic parameter matrix remained an identity matrix.

III. The proposed method with GCP support and considering camera intrinsic parameters

The same group of GCPs takes part in the test. All the unknowns, including the intrinsic parameters, are involved in the BA procedure. The intrinsic parameter model adopted Model (4), and the variables α_x and α_y each absorbed two distortion parameters. In the adjustment process, the weight value of the ground control points was 20 times that of the point correspondences.

IV. The original OpenMVG SFM pipeline with GCP support

The same group of GCPs and checkpoints (CHK) was used to evaluate the performance of the original OpenMVG pipeline.

5.2. The Proposed Method with No GCP Support

A total of 7916 images were used in the experiment, and 6,856,633 pairwise point correspondences were extracted from matching. Thirteen ground CHKs were used to measure the discrepancies (i.e., ground point residuals) between the 3D observation values and the calculated estimates. The statistical results of residuals are shown in Table 2, and the root mean squares error (RMSE) is presented in the last row.

Table 2. Statistical Values for CHKs after the First Experiment (Unit: meter).

ID of CHK	X-Axis	Y-Axis	Z-Axis
CHK 1	0.7285	−0.7146	−17.6891
CHK 2	1.3723	−0.5638	−19.4138
CHK 3	0.2392	−0.5449	−16.8526
CHK 4	0.5169	−0.1780	−17.2758
CHK 5	0.2371	−0.5560	−16.6415
CHK 6	−0.0221	−0.3999	−16.5177
CHK 7	1.0317	0.5181	−16.6787
CHK 8	0.6388	0.8845	−16.2159
CHK 9	−1.8655	0.4107	−17.0058
CHK 10	−0.2723	0.1838	−16.7048
CHK 11	0.0234	0.8457	−16.6783
CHK 12	−0.2252	1.3395	−17.2606
CHK 13	−0.6135	0.9441	−17.7322
RMSE	0.7966	0.6950	17.14631

After BA processing, the sparse point cloud of the corresponding area is constructed. The ground point cloud and the corresponding camera projection center are shown in Figure 9. The yellow dashed lines illustrates the position of camera projection centers, and the white form is the ground point cloud.

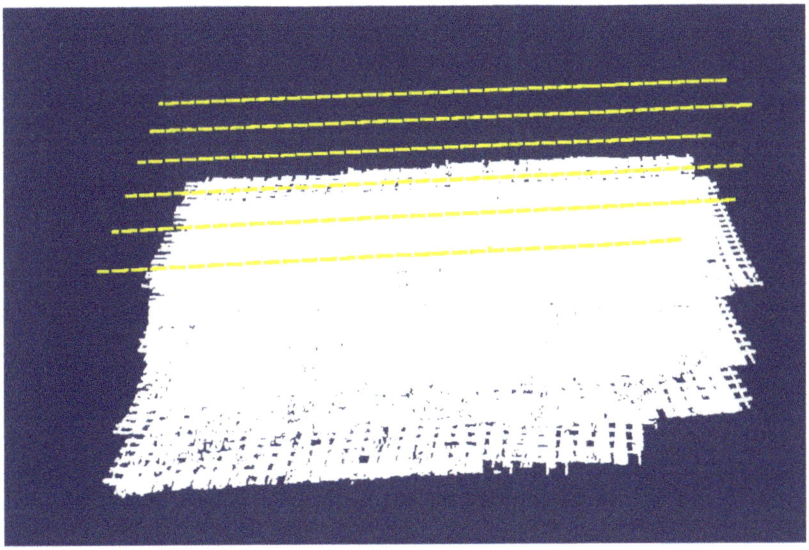

Figure 9. Sparse Point Cloud After the First Experiment.

As presented in Table 1, significant systematic errors were generated when without GCP support, particularly in the Z-axis. From Figure 9, the proposed method can produce a robust sparse, dense cloud.

5.3. The Proposed Method with GCP Support, but Not Considering Camera Intrinsic Parameters

The second experiment was carried out with the same set of images. Four ground control points contributed to the BA process, and the same checkpoints were used to evaluate the adjustment accuracy. A total of 3,091,007 pairwise point correspondences were extracted from matching. The statistical differences and RMSE are shown in Table 3.

Table 3. Statistical Values for CHKS after the Second Experiment (Unit: meter).

ID of CHK	X-Axis	Y-Axis	Z-Axis
CHK 1	—	—	—
CHK 2	—	—	—
CHK 3	—	—	—
CHK 4	—	—	—
CHK 5	0.4117	1.0145	−1.6682
CHK 6	0.0062	0.0014	0.0068
CHK 7	—	—	—
CHK 8	—	—	—
CHK 9	−0.0573	0.0133	0.1149
CHK 10	−1.0118	−0.4301	0.0030
CHK 11	−0.0322	−0.0154	−0.0750
CHK 12	—	—	—
CHK 13	−0.0009	0.0054	−0.0014
RMSE	0.3035	0.3057	0.4642

After BA adjustment with GCPs, the pairwise point correspondences were used to construct the sparse ground point cloud, as presented in Figure 10.

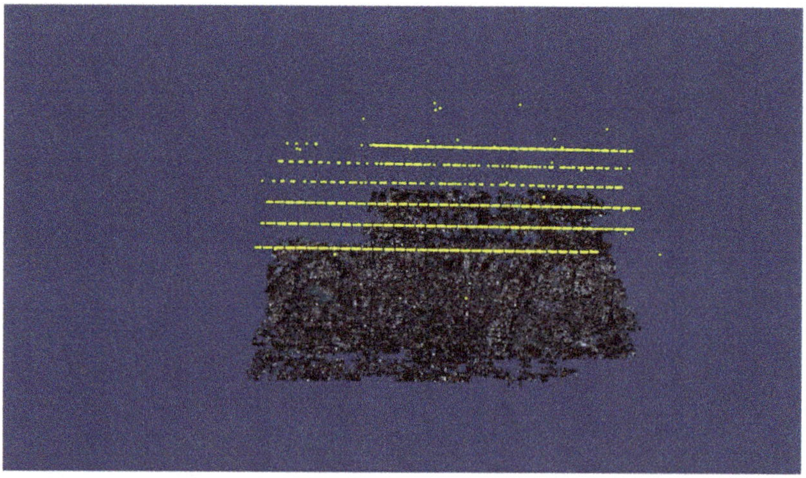

Figure 10. Sparse Point Cloud After the Second Experiment.

Since the intrinsic camera parameters were not adopted in the BA process, not all distortion sources were fully considered, and some images did not meet the consistency constraint. Therefore, image leaks occurred in the BA processing during the adjustment process; only 4567 images passed through the BA process. Since image leaks transpired in the BA process, some CHKs had no statistical residuals (see Table 3), and the sparse point cloud in Figure 10 lacks completeness and robustness.

5.4. The Proposed Method with GCP Support and Considering Camera Intrinsic Parameters

In the third experiment, a total of 6,884,305 pairwise point correspondences were extracted from matching. The residuals between the 3D observation and calculation values are summarized in Table 4. The sparse point cloud is shown in Figure 11. The intrinsic parameters before and after calibration are presented in Table 5.

Table 4. Statistical Values for CHKS after the Third Experiment.

ID of CHK	X-Axis	Y-Axis	Z-Axis
CHK 1	−0.0081	−0.0001	0.0037
CHK 2	0.0035	0.0034	0.0028
CHK 3	−0.0089	−0.0529	−0.0722
CHK 4	0.0401	−0.0213	−0.0754
CHK 5	0.0571	−0.0411	0.0472
CHK 6	−0.0011	−0.0040	0.0107
CHK 7	−0.0364	−0.0023	−0.0755
CHK 8	−0.0157	−0.0030	−0.0250
CHK 9	−0.1061	0.0232	0.2133
CHK 10	−0.0450	0.0122	0.0258
CHK 11	−0.0034	0.0162	0.0015
CHK 12	0.0076	−0.0027	−0.0047
CHK 13	0.0005	−0.0035	−0.0001
RMSE	0.0392	0.0214	0.0711

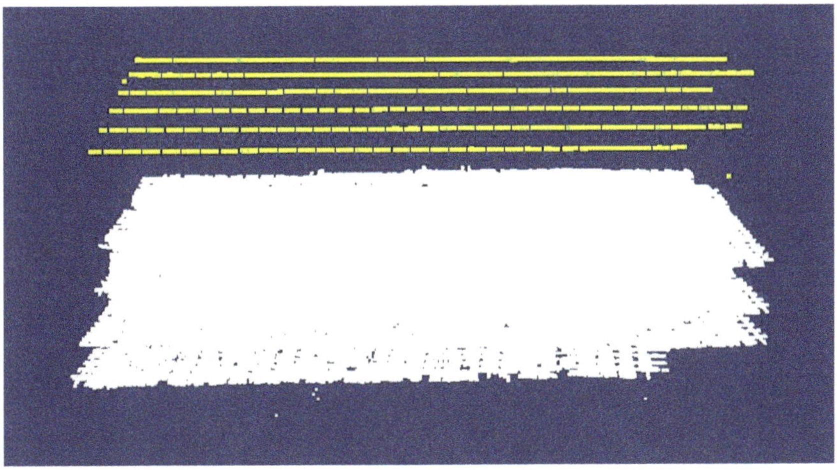

Figure 11. Sparse Point Cloud After the Third Experiment.

Table 5. The intrinsic parameters before and after calibration.

Parameter	Before Calibration	After Calibration
Principal Point x_0 (pixel)	2432.0	2423.1399460647876
Principal Point y_0 (pixel)	1616.0	1603.9860816155979
Calibration parameter 1	0	−0.14123671990384698
Calibration parameter 2	0	−27.234841758167391
Calibration parameter 3	0	3807.0519745535798
Calibration parameter 4	0	−0.00018113363239831057
Calibration parameter 5	0	−0.00069940821735464372

Table 3 show that the RMSE values for the 13 CHKs were below one pixel in the X and Y directions and below two pixels in the Z direction, equivalent to the precision obtained by the professional LightSpeed software [12–15]. The sparse point cloud shown in Figure 11 is much more complete and robust.

5.5. The Original OpenMVG SFM Pipeline with GCP Support

The fourth experiment was carried out using the original OpenMVG pipeline. All unknowns were considered, and the same group of GCPs and CHKs were used in the adjustment. A total of 6,797,129 pairwise point correspondences were extracted from matching. The statistical results are shown in Table 6. The sparse point cloud is displayed in Figure 12.

Table 6. Statistical Values for CHKS after the Fourth Experiment (Unit: meter).

ID of CHK	X-Axis	Y-Axis	Z-Axis
CHK 1	0.2285	−0.2146	0.3560
CHK 2	0.2223	−0.2655	−0.2265
CHK 3	0.2393	−0.2149	−0.2078
CHK 4	0.2362	−0.1163	0.2652
CHK 5	−0.1373	−0.2316	−0.1682
CHK 6	−0.0895	−0.1930	0.3068
CHK 7	0.2187	0.2328	0.1032
CHK 8	0.2167	0.2855	−0.1176
CHK 9	−0.2655	0.2211	0.2149
CHK 10	−0.2353	0.1838	0.2030
CHK 11	0.0215	0.2046	−0.1750
CHK 12	−0.2624	0.2395	0.1582
CHK 13	−0.2297	0.2441	−0.1014
RMSE	0.2998	0.3148	0.3017

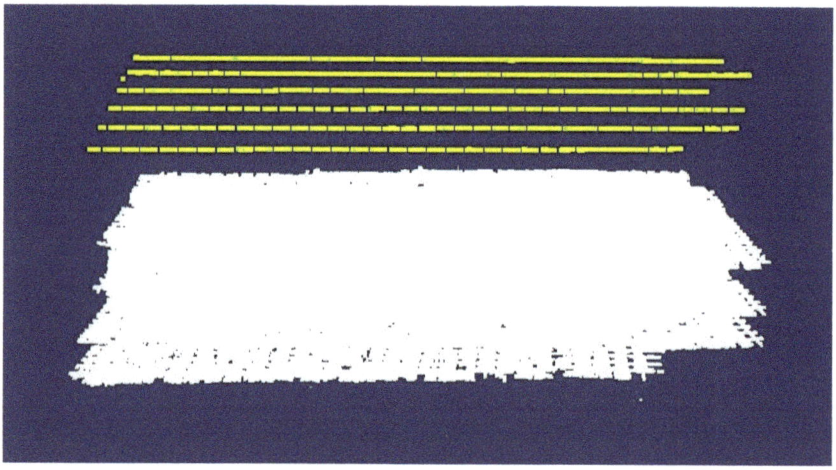

Figure 12. Sparse Point Cloud After the Fourth Experiment.

Table 6 show that the accuracy of the original OpenMVG SFM pipeline is far better than that of the first experiment and that no systematic errors remained in the result. However, the results were still not as good as those of the third experiment.

To better compare and analyze the different test results, the residuals from Tables 2–4 and 6 were plotted. Figure 13 present the residuals at each CHK in the X-direction, Figure 14 show the residuals in the Y-direction, and Figure 15 provide the residuals in the Z-direction. Figure 16 present a more detailed comparison of residuals in the Z-direction.

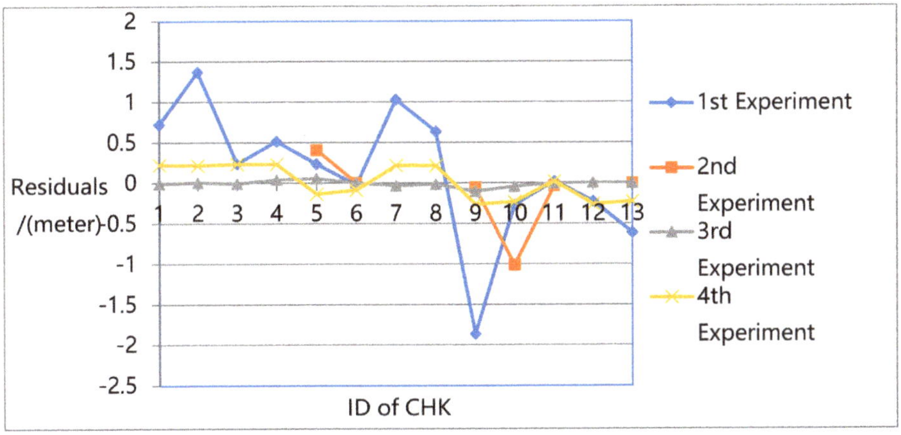

Figure 13. Residuals Plot of CHKs in X-direction.

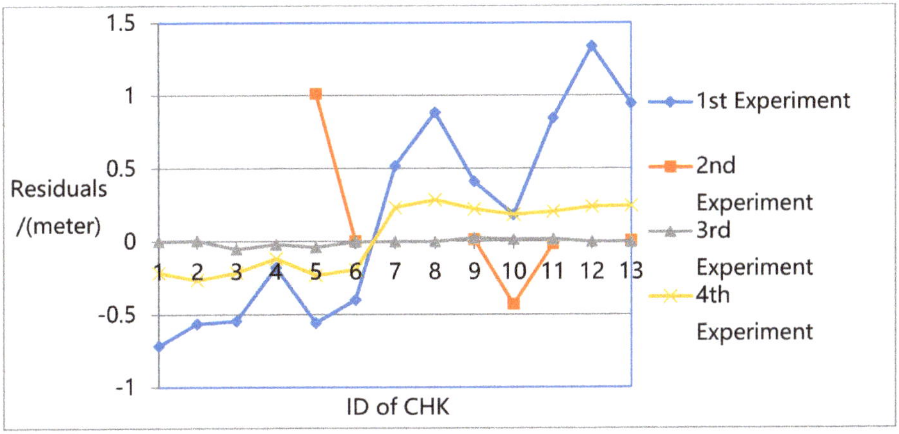

Figure 14. Residuals Plot of CHKs in Y-direction.

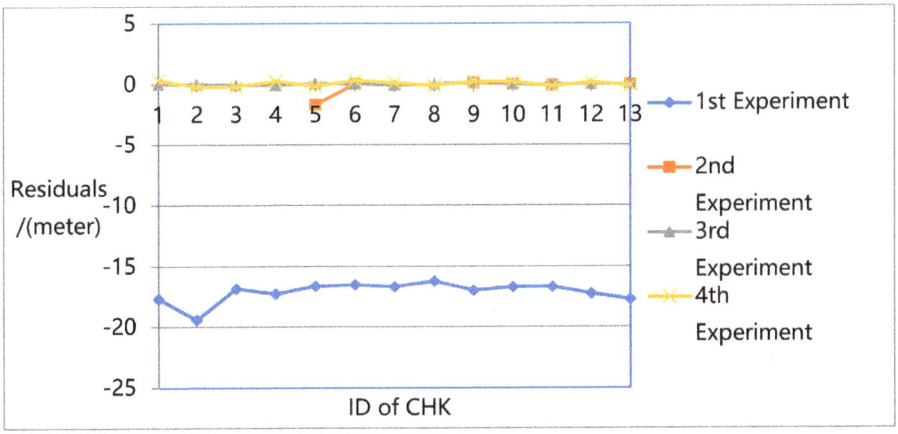

Figure 15. Residuals Plot of CHKs in Z-direction.

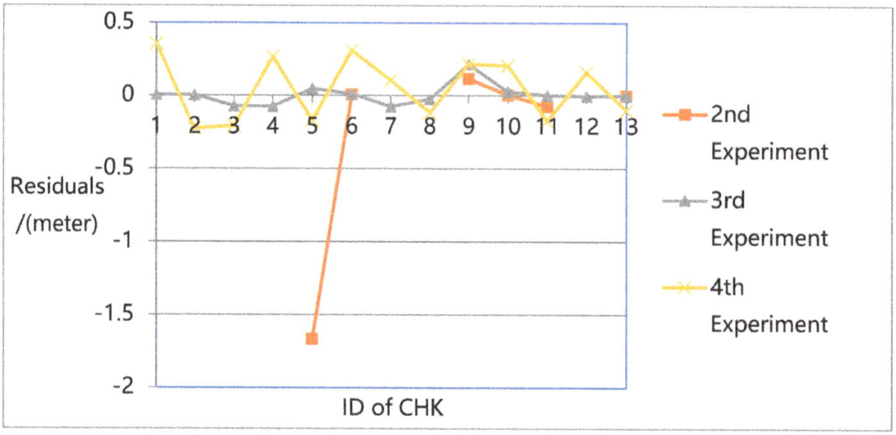

Figure 16. Residuals Plot of CHKs in Z-direction in Detail.

The results show that the largest errors were found in the first experiment. While the first experiment deals with all the unknowns (e.g., rotation, translation, intrinsic parameters, and 3D ground point coordinates) in the overall flow and constructs a robust sparse point cloud, it does not eliminate the systematic errors in the pose data, especially in the Z direction. Due to the lack of ground truth guidance, even with GPS support, significant deviations could be found between the resulting 3D model and the ground truth.

For the second experiment, significant image leaks occurred due to the lack of intrinsic parameters in the adjustment. This means that for high oblique frame sweep images, intrinsic parameters are indispensable in adjustment. For normal vertical photogrammetric images, although the BA process is unable to attain the highest precision when intrinsic parameters are ignored, the robust bundle block can still be obtained.

In the third experiment, the highest precision is reached with a robust sparse point cloud. In the fourth experiment, with GPS and GCP data, the original OpenMVG pipeline can achieve high positioning accuracy in three directions with a robust sparse point cloud. However, its accuracy is not as high as that of the proposed quadrifocal tensor SFM photogrammetry method.

In the third and fourth experiments, while all the unknowns (e.g., rotation, translation, intrinsic parameters, and 3D ground point coordinates) are considered in the BA process, the proposed method achieves much higher accuracy. The proposed approach introduces the iterative photogrammetry idea into the original OpenMVG SFM pipeline following an aerial and satellite photogrammetry processing approach, focusing more on the imaging geometric principle and camera characteristics. Using iterative BA adjustment, the distortion sources are dealt with from a coarse to fine approach, enhancing the BA accuracy step by step.

Figures 17 and 18 show the digital orthophoto map (DOM) and digital elevation model (DEM) products after the BA positioning and calibration process in the third experiment. The ground resolution of the DOM and the DEM was set to 7 cm. Fifteen points were randomly selected from the DOM and DEM products. Measurements were then taken from the two products, and observation values were made in the field test. After comparing the values, the accuracy of the DOM was found to be from 1 to 3 pixels. The largest difference was 2.72 pixels, the smallest difference was 1.01 pixels, and the RMSE was 1.32 pixels. The DEM accuracy ranged from 2 to 5 pixels. The largest difference was 4.23 pixels, the smallest difference was 2.05 pixels, and the RMSE was 2.91 pixels.

Figure 17. DOM Product After Positioning and Calibration.

LightSpeed software was then used for the given dataset, implementing the entire photogrammetry procedure and generating the DOM and DEM products. The photogrammetry results and the DOM and DEM products obtained using the two methods were compared and evaluated. The BA results and the DOM and DEM products obtained by the two approaches were at the same accuracy level, which suggests that our proposed method reaches the precision level of professional photogrammetry software.

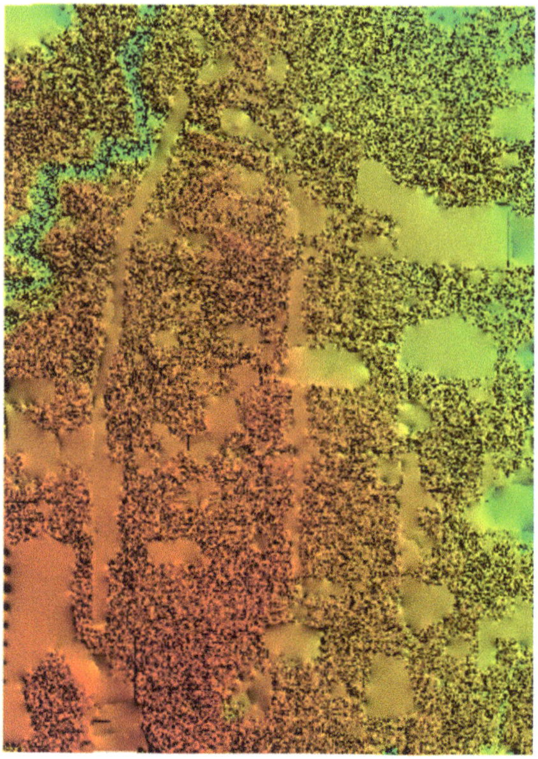

Figure 18. DEM Product After Positioning and Calibration.

6. Conclusions

Sweeping along the flight's vertical direction for imaging, the frame sweep camera is characterized by a large field of view, a wide observation range, and a multi-angle imaging mode. While these characteristics can be useful for 3D reconstruction, they also increase the difficulty of image positioning. For large tilt oblique photogrammetry, the image positioning and calibration process becomes much more complex. Additionally, with the increasing integration between photogrammetry and SFM, more attempts have been made to combine these approaches to realize the positioning, calibration, and 3D reconstruction of large amounts of ordered or disordered images. After an extensive literature review and research, the OpenMVG pipeline was found to be most suitable for VisionMap A3 positioning and calibration.

Using the OpenMVG pipeline, a quadrifocal tensor-based positioning and calibration method was developed for high oblique frame sweep aerial cameras according to the imaging characteristics of VisionMap A3. We comprehensively analyzed the entire processing flow, from feature extraction and image matching, relative rotation and translation estimation, global rotation and translation estimation, and the quadrifocal tensor model construction to the BA process and calibration. Focusing on the imaging character of the VisionMap A3 camera, the quadrifocal tensor was put forward as the basis for BA adjustment. For the BA process and calibration, a coarse to fine three-stage BA processing modification was introduced in the OpenMVG pipeline following photogrammetric processing. Based on the experimental results, our main conclusions are as follows:

First, the SFM photogrammetric approach is suitable for large tilt oblique photogrammetry. When considering the rotation, translation, 3D ground coordinates, and the camera's intrinsic parameters, the SFM pipeline, as OpenMVG, can generate a robust sparse point cloud.

Second, GPS data can only provide an initial value for the BA process. We found significant deviations between the processed bundle block and the ground control points. The results suggest that ground truth is still indispensable in positioning and calibration.

Third, the intrinsic parameters cannot be ignored in the BA process. The lack of intrinsic parameters resulted in image leaks in the bundle block and the sparse point cloud. The results indicate that the intrinsic parameters play an important role in dealing with inner and outer positioning errors.

Fourth, the coarse to fine processing approach in classical photogrammetry is still advisable in SFM photogrammetry. Processing the rotation (R) and translation (T) parameters eliminates most gross positioning errors while processing the intrinsic parameters propels the positioning precision to a higher level.

The experimental results on multiple overlapping routes of VisionMap A3 edge images suggest that our proposed method could achieve a robust bundle block and sparse point cloud and generate accurate DOM and DEM products. However, for the single route, the proposed method generated a bundle block with only contours, and the inner images were mostly considered outliers and excluded from the calculations. Subsequent studies are required to explore how to retain more images and effectively exclude outliers.

Author Contributions: Conceptualization, T.W. and Y.Z. (Yan Zhang); methodology, T.W., Y.Z. (Yan Zhang), and Y.Z. (Yongsheng Zhang); software, T.W. and Y.Z. (Yan Zhang); validation, Y.Y. and L.W.; formal analysis, Z.Z.; writing—original draft preparation, T.W. and Y.Z. (Yan Zhang); writing—review and editing, S.L., L.L. and X.Z.; visualization, Y.Y. and S.L. All authors have read and agreed to the published version of the manuscript.

Funding: This publication has emanated from research conducted with the National Nature Science Foundation of China under grant number 42071340.

Data Availability Statement: The study did not report any data.

Acknowledgments: The authors are grateful to Zhiping Lü and his research team for offering help in processing the GNSS data, which was a preliminary step before we carried out our investigation. Zhiping Lü and his research team have published a free GNSS data process software on www.gnsser.com (accessed on 16 September 2017), and the related paper is https://link.springer.com/article/10.1007/s10291-020-01069-9 (accessed on 19 January 2021).

Conflicts of Interest: The authors declare no conflict of interest. The funders had no role in the design of the study; in the collection, analyses, or interpretation of data; in the writing of the manuscript or in the decision to publish the results.

References

1. Konecny, G. *Geoinformation: Remore Sensing, Photogrammetry and Geographic Information Systems*, 2nd ed.; CRC Press: Boca Raton, FL, USA, 2014.
2. Lindner, W. *Digital Photogrammetry: A Practical Course*; Springer: Berlin, Germany, 2016.
3. McGlone, J.C. (Ed.) *Manual of Photogrammetry*, 6th ed.; American Society of Photogrammetry and Remote Sensing: Bethesda, MD, USA, 2013.
4. Wolf, P.R.; Dewitt, B.A. *Elements of Photogrammetry with Applications in GIS*, 4th ed.; McGraw-Hill Education: New York, NY, USA, 2014.
5. Aber, J.S.; Marzolff, I.; Ries, J. *Small-Format Aerial Photography: Principles, Techniques and Geoscience Applications*; Elsevier: Amsterdam, The Netherlands, 2010.
6. Ikkala, L.; Ronkanen, A.K.; Ilmonen, J.; Similä, M.; Rehell, S.; Kumpula, T.; Päkkilä, L.; Klöve, B.; Marttila, H. Unmanned Aircraft System (UAS) Structure-From-Motion (SfM) for Monitoring the Changed Flow Paths and Wetness in Minerotrophic Peatland Restoration. *Remote Sens.* **2022**, *14*, 3169. [CrossRef]
7. Hinz, A.; Dörstel, C.; Heier, H. *DMC-The digital sensor technology of Z/I-Imaging. Photogrammetric Week'01*; Fritsch, D., Spiller, R., Eds.; Wichmann Verlag: Heidelberg, Germany, 2001.
8. Dhillon, V.S.; Marsh, T.R.; Stevenson, M.J.; Atkinson, D.C.; Kerry, P.; Peacocke, P.T.; Vick, A.J.A.; Beard, S.M.; Ives, D.J.; Lunney, D.W.; et al. ULTRACAM: An ultrafast, triple-beam CCD camera for high-speed astrophysics. *Mon. Not. R. Astron. Soc.* **2007**, *378*, 825–840. [CrossRef]

9. Toschi, I.; Remondino, F.; Rothe, R.; Klimek, K. Combining Airborne Oblique Camera and Lidar Sensors: Investigation and new Perspectives. *Int. Arch. Photogramm. Remote Sens. Spat. Inf. Sci.* **2018**, *42*, 437–444. [CrossRef]
10. Li, J.; Liu, X.; Wan, Y.; Liu, F. Generation of SWDC-4 aerial digital camera virtual images. *Geomat. Inf. Sci. Wuhan Univ.* **2008**, *33*, 450–453.
11. Hoefft, J.R.; Tietz, T. Multispectral EO LOROP camera. In *Infrared Technology and Applications XXIV*; SPIE: Bellingham, WA, USA, 1998; Volume 3436, pp. 402–412.
12. Iyengar, M.; Lange, D. The Goodrich 3rd generation DB-110 system: Operational on tactical and unmanned aircraft. In *Airborne Intelligence, Surveillance, Reconnaissance (ISR) Systems and Applications III*; SPIE: Bellingham, WA, USA, 2006; Volume 6209, pp. 75–88.
13. Pechatnikov, M.; Shor, E.; Raizman, Y. VisionMap A3-super wide angle mapping system basic principles and workflow. In Proceedings of the 21th ISPRS Congress, Beijing, Oral Technical Session SS-8 (2), Beijing, China, 3–11 July 2008.
14. Raizman, Y.; Gozes, A. *High Throughput Aerial Photography, Ortho & 3D Processing. Photogrammetric Week*; Fritsch, D., Spiller, R., Eds.; Wichmann Verlag: Heidelberg, Germany, 2013.
15. Raizman, Y.; Gozes, A. Tel-Aviv. VisionMap A3 Edge—A Single Camera for Multiple Solutions. In Proceedings of the Conference 55th Photogrammetric Week, Stuttgart, Germany, 7–11 September 2015.
16. Pechatnikov, M.; Raizman, Y.; Tel-Aviv, V. Visionmap A3 LightSpeed—A new generation of the ground processing system. In Proceedings of the International Scientific Racurs Conference, Tossa der Mar, Spain, 19–22 September 2011.
17. Triggs, B.; McLauchlan, P.F.; Hartley, R.I.; Fitzgibbon, A.W. Bundle adjustment—A modern synthesis. In *International Workshop on Vision Algorithms*; Springer: Berlin/Heidelberg, Germany, 1999; pp. 298–372.
18. Richard, H.; Andrew, Z. *Multiple View Geometry in Computer Vision*; Cambridge University Press: Cambridge, UK, 2000.
19. Khan, P.W.; Byun, Y.C.; Latif, M.A. Clifford Geometric Algebra-Based Approach for 3D Modeling of Agricultural Images Acquired by UAVs. *IEEE Access* **2020**, *8*, 226297–226308. [CrossRef]
20. Zhang, D.; Jackson, W.; Dobie, G.; West, G.; MacLeod, C. Structure-from-motion based image unwrapping and stitching for small bore pipe inspections. *Comput. Ind.* **2022**, *139*, 103664. [CrossRef]
21. Jiang, N.; Cui, Z.; Tan, P. A global linear method for camera pose registration. In Proceedings of the IEEE International Conference on Computer Vision, Sydney, NSW, Australia, 1–8 December 2013; pp. 481–488.
22. Cui, Z.; Tan, P. Global structure-from-motion by similarity averaging. In Proceedings of the IEEE International Conference on Computer Vision, Santiago, Chile, 7–13 December 2015; pp. 864–872.
23. Cui, Z.; Jiang, N.; Tang, C.; Tan, P. Linear global translation estimation with feature tracks. *arXiv* **2015**, arXiv:1503.01832.
24. Goldstein, T.; Hand, P.; Lee, C.; Voroninski, V.; Soatto, S. Shapefit and shapekick for robust, scalable structure from motion. In Proceedings of the European Conference on Computer Vision; Springer: Cham, Switzerland, 2016; pp. 289–304.
25. Sweeney, C.; Sattler, T.; Hollerer, T.; Turk, M.; Pollefeys, M. Optimizing the viewing graph for structure-from-motion. In Proceedings of the IEEE International Conference on Computer Vision, Santiago, Chile, 7–13 December 2015; pp. 801–809.
26. Ozyesil, O.; Singer, A. Robust camera location estimation by convex programming. In Proceedings of the IEEE Conference on Computer Vision and Pattern Recognition, Boston, MA, USA, 7–12 June 2015; pp. 2674–2683.
27. Wilson, K.; Snavely, N. Robust global translations with 1dsfm. In Proceedings of the European Conference on Computer Vision; Springer: Cham, Switzerland, 2014; pp. 61–75.
28. Zhu, S.; Zhang, R.; Zhou, L.; Shen, T.; Fang, T.; Tan, P.; Quan, L. Very large-scale global sfm by distributed motion averaging. In Proceedings of the IEEE Conference on Computer Vision and Pattern Recognition, Salt Lake City, UT, USA, 18–23 June 2018; pp. 4568–4577.
29. Moulon, P.; Monasse, P.; Marlet, R. Adaptive structure from motion with a contrario model estimation. In Proceedings of the Asian Conference on Computer Vision; Springer: Berlin/Heidelberg, Germany, 2012; pp. 257–270.
30. Cui, H.; Shen, S.; Gao, X.; Hu, Z. Batched incremental structure-from-motion. In Proceedings of the International Conference on 3D Vision (3DV), Qingdao, China, 10–12 October 2017; pp. 205–214.
31. Snavely, N.; Seitz, S.M.; Szeliski, R. Modeling the world from internet photo collections. *Int. J. Comput. Vis.* **2008**, *80*, 189–210. [CrossRef]
32. Snavely, N.; Seitz, S.M.; Szeliski, R. Photo tourism: Exploring photo collections in 3D. In *ACM Siggraph 2006 Papers*; ACM: Boston, MA, USA, 2006; pp. 835–846.
33. Cui, H.; Shen, S.; Gao, W.; Wang, Z. Progressive large-scale structure-from-motion with orthogonal MSTs. In Proceedings of the 2018 International Conference on 3D Vision (3DV), Verona, Italy, 5–8 September 2018; pp. 79–88.
34. Schonberger, J.L.; Frahm, J.M. Structure-from-motion revisited. In Proceedings of the IEEE Conference on Computer Vision and Pattern Recognition, Las Vegas, NV, USA, 27–30 June 2016; pp. 4104–4113.
35. Wu, C. Towards linear-time incremental structure from motion. In Proceedings of the 2013 International Conference on 3D Vision-3DV 2013, Seattle, WA, USA, 29 June–1 July 2013; pp. 127–134.
36. Moulon, P.; Monasse, P.; Marlet, R. Global fusion of relative motions for robust, accurate and scalable structure from motion. In Proceedings of the IEEE International Conference on Computer Vision, Sydney, NSW, Australia, 1–8 December 2013; pp. 3248–3255.

37. Lindenberger, P.; Sarlin, P.E.; Larsson, V.; Pollefeys, M. Pixel-Perfect Structure-from-Motion with Feature metric Refinement. In Proceedings of the IEEE/CVF International Conference on Computer Vision, Montreal, BC, Canada, 11–17 October 2021; pp. 5987–5997.
38. Bhowmick, B.; Patra, S.; Chatterjee, A.; Govindu, V.M.; Banerjee, S. Divide and conquer: Efficient large-scale structure from motion using graph partitioning. In *Asian Conference on Computer Vision*; Springer: Cham, Switzerland, 2014; pp. 273–287.
39. Gherardi, R.; Farenzena, M.; Fusiello, A. Improving the efficiency of hierarchical structure-and-motion. In Proceedings of the 2010 IEEE Computer Society Conference on Computer Vision and Pattern Recognition, San Francisco, CA, USA, 13–18 June 2010; pp. 1594–1600.
40. Farenzena, M.; Fusiello, A.; Gherardi, R. Structure-and-motion pipeline on a hierarchical cluster tree. In Proceedings of the 2009 IEEE 12th International Conference on Computer Vision Workshops, ICCV Workshops, Kyoto, Japan, 27 September–4 October 2009; pp. 1489–1496.
41. Toldo, R.; Gherardi, R.; Farenzena, M.; Fusiello, A. Hierarchical structure-and-motion recovery from uncalibrated images. *Comput. Vis. Image Underst.* **2015**, *140*, 127–143. [CrossRef]
42. Cui, H.; Gao, X.; Shen, S.; Hu, Z. HSfM: Hybrid structure-from-motion. In Proceedings of the IEEE Conference on Computer Vision and Pattern Recognition, Honolulu, HI, USA, 21–26 July 2017; pp. 1212–1221.
43. Zhu, S.; Shen, T.; Zhou, L.; Zhang, R.; Wang, J.; Fang, T.; Quan, L. Parallel structure from motion from local increment to global averaging. *arXiv* **2017**, arXiv:1702.08601.
44. Shah, R.; Deshpande, A.; Narayanan, P.J. Multistage SfM: A coarse-to-fine approach for 3d reconstruction. *arXiv* **2015**, arXiv:1512.06235.
45. Bao, S.Y.; Savarese, S. Semantic structure from motion. In *CVPR 2011*; IEEE: Colorado Springs, CO, USA, 2011; pp. 2025–2032.
46. Bao, S.Y.; Bagra, M.; Chao, Y.W.; Savarese, S. Semantic structure from motion with points, regions, and objects. In Proceedings of the 2012 IEEE Conference on Computer Vision and Pattern Recognition, Providence, RI, USA, 16–21 June 2012; pp. 2703–2710.
47. Chen, Y.; Wang, Y.; Lu, P.; Chen, Y.; Wang, G. Large-scale structure from motion with semantic constraints of aerial images. In *Chinese Conference on Pattern Recognition and Computer Vision (PRCV)*; Springer: Cham, Switzerland, 2018; pp. 347–359.
48. Westoby, M.J.; Brasington, J.; Glasser, N.F.; Hambrey, M.J.; Reynolds, J.M. 'Structure-from-Motion'photogrammetry: A low-cost, effective tool for geoscience applications. *Geomorphology* **2012**, *179*, 300–314. [CrossRef]
49. Lowe, D.G. Distinctive image features from scale-invariant keypoints. *IJCV* **2004**, *60*, 91–110. [CrossRef]
50. Martinec, D.; Pajdla, T. Robust rotation and translation estimation in multi-view reconstruction. In Proceedings of the 2007 IEEE Conference on Computer Vision and Pattern Recognition, Minneapolis, MN, USA, 17–22 June 2007; pp. 1–8.
51. Wang, T.; Zhang, Y.; Zhang, Y.; Zhang, Z.; Xiao, X.; Yu, Y.; Wang, L. A Spliced Satellite Optical Camera Geometric Calibration Method Based on Inter-Chip Geometry Constraints. *Remote Sens.* **2021**, *13*, 2832. [CrossRef]
52. Wang, T.; Zhang, Y.; Zhang, Y.; Jiang, G.; Zhang, Z.; Yu, Y.; Dou, L. Geometric Calibration for the Aerial Line Scanning Camera GFXJ. *Photogramm. Eng. Remote Sens.* **2019**, *85*, 643–658. [CrossRef]

remote sensing

MDPI

Article

Accuracy of the Copernicus High-Resolution Layer Imperviousness Density (HRL IMD) Assessed by Point Sampling within Pixels

Geir-Harald Strand

Department of Survey and Statistics, Norwegian Institute of Bioeconomy Research, 1431 Ås, Norway; ghs@nibio.no

Abstract: The Copernicus high-resolution layer imperviousness density (HRL IMD) for 2018 is a 10 m resolution raster showing the degree of soil sealing across Europe. The imperviousness gradation (0–100%) per pixel is determined by semi-automated classification of remote sensing imagery and based on calibrated NDVI. The product was assessed using a within-pixel point sample of ground truth examined on very high-resolution orthophoto for the section of the product covering Norway. The results show a high overall accuracy, due to the large tracts of natural surfaces correctly portrayed as permeable (0% imperviousness). The total sealed area in Norway is underestimated by approximately 33% by HRL IMD. Point sampling within pixels was found to be suitable for verification of remote sensing products where the measurement is a binomial proportion (e.g., soil sealing or canopy coverage) when high-resolution aerial imagery is available as ground truth. The method is, however, vulnerable to inaccuracies due to geometrical inconsistency, sampling errors and mistaken interpretation of the ground truth. Systematic sampling inside each pixel is easy to work with and is known to produce more accurate estimates than a simple random sample when spatial autocorrelation is present, but this improvement goes unnoticed unless the status and location of each sample point inside the pixel is recorded and an appropriate method is applied to estimate the within-pixel sampling accuracy.

Keywords: land cover; soil sealing; imperviousness; land monitoring; accuracy; verification; systematic sampling; within-pixel sampling

1. Introduction

Population growth and economic development are driving forces behind land use change [1]. A growing population customize their surroundings by replacing natural surfaces with a solid covering consisting of concrete surfaces, buildings, paved roads, and other constructions that create an impervious land cover [2]. This process, known as soil sealing, separates the soils from the atmosphere by one or more impermeable layers [3]. The result is, among other effects, a loss of area available for plant production, reduced biodiversity, and an increase in run-off of surface water [4–6]. Soil sealing is therefore considered as an example of land degradation [7] and imperviousness is frequently used as an indicator of land take in environmental monitoring [8].

There has been an increasing interest in mapping soil sealing and imperviousness over the last 30 years [9]. Around the turn of the century, USGS derived a measurement of soil sealing from Landsat imagery for the 2001 National Land Cover database for the United States, using the degree of imperviousness as their indicator [10,11]. In Europe, statistics of sealed surfaces over larger zones were first generated using the SPOT VEGETATION imagery from 2001, with a spatial resolution of 1 km [12]. Better spatial resolution was achieved in 2005, by mapping soil sealing using the normalized difference vegetation index (NDVI) calculated from QuickBird imagery [13].

A systematic comparison of different approaches to measure soil sealing using air photo interpretation showed an average overall agreement between existing products at 92% [14]. A comprehensive review of techniques using medium resolution (10–100 m) imagery was published in 2012, describing trends in remote sensing of impervious surfaces [15]. These include pixelwise estimation of imperviousness degree, classification of surface material or and mapping land cover or land use features (e.g., buildings or roads) in order to infer imperviousness from land cover/use. A variety of new attempts to develop automated methods for detecting soil sealing and measure the degree of imperviousness have later been published, employing new and improved remote sensing imagery and classification techniques [16–18]. The most recent review of research on soil sealing, land take and impervious surfaces lists eleven different mapping methods currently in use [19].

The European Union's Earth observation program, Copernicus, conducts projects to monitor land cover and land use in order to serve urban and environmental planning and policy making. One outcome from this program is the high-resolution layer imperviousness density (HRL IMD). This is a service provided through the Copernicus land monitoring service (CLMS, https://land.copernicus.eu/, last accessed on 7 June 2022) implemented by the European Environment Agency (EEA). HRL IMD is a raster product estimating the degree of imperviousness within each pixel.

The environmental indicator mapped by HRL IMD is defined as "human-produced surfaces that are essentially impenetrable by rainfall" [20]. This is a binary phenomenon since a location on a surface is either sealed or not sealed. When used as an indicator, imperviousness is reported as the proportion of the land surface covered by a pixel that is impenetrable to water and therefore creates increased surface runoff [11,21], leading to a gradation on the scale 0–100%. The variable is a binomial proportion reported as a percentage.

The HRL IMD product covers the entire European continent and has so far been produced for the reference years 2006, 2009, 2012, 2015 and 2018. The 2018 product has 10 m resolution, while the older products have 20 m resolution. The first updates relied on a semi-automated approach (mainly due to the limited availability of input data). The level of automation of the production has steadily increased due to the availability of better reference data and improved resolution of remote sensing imagery [22].

Data from the Copernicus land monitoring service are used, or are proposed to be used, in studies and reports from Eurostat and the European Environmental Agency [23,24]. It is also a goal to serve urban and environmental planning and policy making at the national and local level. Detailed knowledge of the content and accuracy of the product is, however, a prerequisite for the broader user-uptake of the service. Users need to understand the content of the product to interpret and use the data correctly. Verification has been carried out in several countries, often using a qualitative "look and feel" approach [25] and a technical validation of the 2015 version of HRL IMD is also available [26].

A study of the accuracy of the US National Land Cover Data for 2006 found that the ideal accuracy assessment of imperviousness data would require estimation of the impervious surface for each sample pixel from reference data, but the cost to obtain the required data was considered prohibitive [27]. New technology and high-resolution aerial photography have changed this situation and detailed information about the content of pixels can now be used as reference data. The cost of a complete inventory of the pixel using analytical photogrammetry is, however, still exorbitant. Instead, we may attempt a sampling approach by obtaining a random sample of points inside each control pixel.

The objective of the current study was to examine the content and accuracy of the HRL IMD product for 2018 using a within-pixel sampling approach. Norway was used as a study area due to the availability of high-resolution and geometrically accurate orthophoto acquired through an established and regular national image acquisition program. The goal was to examine the potential and limitations of the within-pixel sampling approach to verification, but also to improve the interpretation and understanding of the HRL IMD product and thus enhance the value of the product for the end-users.

2. Materials and Methods

The high-resolution layer imperviousness density (HRL IMD) 2018 for Norway was downloaded from the Copernicus land monitoring service (CLMS) web site https://land.copernicus.eu, accessed on 6 January 2021. The tiles covering Norway were combined to create a raster with national coverage. The projection of the map was ETRS89-LAEA89 (EPSG: 3035, https://epsg.io, last accessed on 7 June 2022). The dataset is a 10 m raster where each pixel is coded in the range 0–100, representing the degree of imperviousness in percent.

Two web map services (wms) were used to obtain ground truth: the Norwegian national topographic maps at scales ranging between 1:1000 and 1:5000 provided by the National Mapping Authority and the national orthophoto database (*Norge i bilder*) provided by the Norwegian Spatial Data Infrastructure *Norge digitalt*. Data from these sources have national coverage and could be linked directly to the QGIS software used in the study. The imagery and other data that were used as ground truth can be inspected at https://kilden.nibio.no/, last accessed on 7 June 2022.

A two-stage random sample was collected to examine the accuracy of the HRL IMD layer. Stratification was employed in the first stage because most pixels (99.4%) had no reported imperviousness (0%). Only 0.6% of the pixels had imperviousness density in the range 1 to 100%.

The data were divided into twelve strata according to the reported imperviousness (0%, 1–9%, 10–19% { ... } 90–99% and 100%, see also Table 1), providing a basis for stratified random sampling. The two strata "0%" and "100%" represent the extremes. "0%" also constitute a very large part of the population and should therefore be given special attention since a relatively large sample is required to detect small anomalies in large strata. The number and size of the ten remaining strata was a compromise between the need for detail (requiring many strata) and available resources.

Table 1. Imperviousness in Norway as reported by HRL IMD and as estimated by sampling. The total land area (32,380,900 hectares) is divided into strata according to the degree of imperviousness reported by HRL IMD. Mean imperviousness and the corresponding sealed area were calculated for each stratum, as well as for a subtotal of all strata with reported imperviousness (according to HRL IMD) and for the whole country.

Stratum	Area (HRL IMD)	High-Resolution Layer (HRL IMD)		Estimated from Aerial Photographs	
(%)	(Hectare)	Mean (%)	Sealed (Ha)	Mean (%)	Sealed (Ha)
100	24,273	100.00	24,273	86.72	21,050
90–99	7647	94.27	7209	68.41	5231
80–89	8358	84.33	7048	56.77	4745
70–79	10,818	74.53	8063	49.01	5302
60–69	12,685	64.65	8201	47.47	6022
50–59	17,071	53.97	9213	48.33	8251
40–49	31,535	44.29	13,967	35.11	11,072
30–39	38,588	34.34	13,250	35.63	13,749
20–29	33,105	24.78	8204	26.23	8684
10–19	15,204	15.60	2373	12.02	1828
1–9	2430	6.56	159	12.24	298
Subtotal	**201,715**	**50.55**	**101,961**	**42.75**	**86,229**
0%	32,179,185	0.00	0	0.21	67,576
Total	**32,380,900**	**0.31**	**101,961**	**0.47**	**153,805**

The sampling procedure selected (randomly) 1000 points from the stratum where the reported imperviousness was 0% and attempted to select (randomly) 100 points from each of the remaining strata. The stratum with imperviousness in the range 1 to 9% was, however, so rare that we only managed to find 82 sample points in this stratum. The final, stratified random sample consisted of 2082 sample points. This sample size is pragmatic, mainly linked to available funding and acceptable sampling cost. Various aspects of the sampling strategy are discussed in the Section 4 below.

A sampling tool was developed in Python for QGIS. The tool starts from a sample point in the stratified random sample and locates the corresponding 10 m pixel in HRL IMD. A border is drawn around the pixel and a new, systematic random sample of 100 points is placed inside the pixel. The first point is placed 0.5 m north and 0.5 m east of the lower left corner of the pixel. The remaining points are spaced along a regular lattice, one meter apart, throughout the pixel (Figure 1). The observed imperviousness was calculated by counting the number of points (0–100) that fell on impenetrable surfaces as defined in the HRL IMD technical documentation [28].

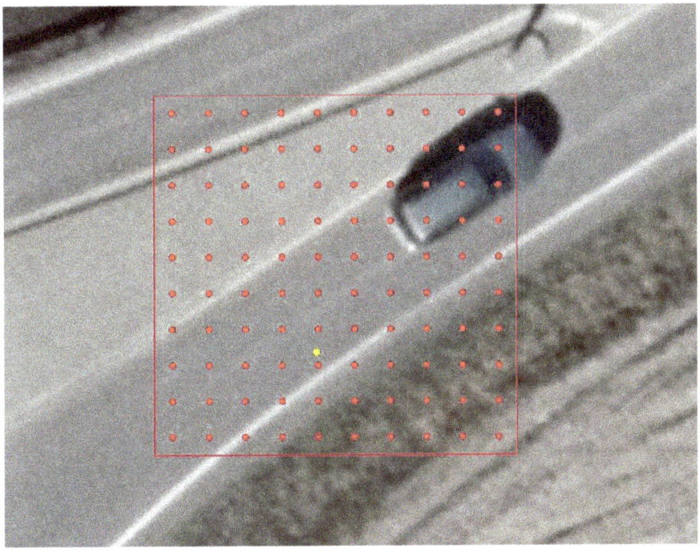

Figure 1. A 10 m pixel in HRL IMD with 100 sample points (red dots). The yellow dot shows the original random sample point used to detect and include this pixel. Photo obtained during the summer of 2017. Data source: © Norge i bilder/Norge digitalt.

Formally, the population examined using this approach consists of all N pixels in the HRL IMD layer divided into M strata with N_i pixels in each stratum. A simple random sample of n_i pixels is drawn from each stratum, resulting in a total sample size of

$$n = \sum_{i=1}^{M} n_i$$

(where n is 2082). Notice that the sample is not proportional, that is $\frac{n_i}{n} \neq \frac{N_i}{N}$ mainly due to the very large number of pixels in the stratum where imperviousness was 0%.

Let x_{ij} represent the imperviousness estimated by HRL IMD and y_{ij} represent the observed imperviousness of pixel j in stratum i. Provided that y_{ij} is an exact measurement, the mean imperviousness of a stratum, a set of strata and the entire sample can be calculated using textbook equations for simple random and stratified sampling [29]. The mean imperviousness of a stratum i is

$$\hat{y}_i = \frac{1}{n_i} \sum_{j=1}^{n_i} y_{ij}$$

with variance

$$s_i^2 = \frac{1}{n_i} \sum_{j=1}^{n_i} (y_{ij} - \hat{y}_i)^2$$

and standard error

$$\sqrt{\frac{s_i^2}{n_i}}$$

while the mean imperviousness across all strata is

$$\hat{y} = \sum_{i=1}^{M} \frac{N_i}{N} \hat{y}_i$$

with

$$Var(\hat{y}) = \sum_{i=1}^{M} \left(\frac{N_i}{N}\right)^2 \frac{s_i^2}{n_i} \left(\frac{N_i - n_i}{N_i}\right)$$

and standard error

$$\sqrt{Var(\hat{y})}$$

Mean difference $(x_{ij} - y_{ij})$ between x_{ij} and y_{ij} in each stratum (d_i) and for the entire population(d) can be calculated using the same approach.

Unfortunately, y_{ij} is an estimate and not an exact measurement. y_{ij} is calculated from the proportion $\hat{p}$ of the 100 systematically distributed sample points inside the pixel. Consequently, $Var(y_{ij})$ is in the range {0–2500} and highest when y_{ij} is 50%. The result is a standard error for y_{ij} itself in the range 0–5% under the assumption that the sample is a simple random sample and the pixelwise population is large compared to the size of the sample inside the pixel.

This is a conservative estimate since the sample inside the pixel is a systematic (and not a simple) random sample, and a systematic sample of this kind is known to provide more accurate estimates than a simple random sample of the same size [30]. An elaboration on this argument is found in the Section 4 below.

With 100 sample points inside a 10 m pixel we acknowledge that the sampling error inside the pixel is present but undetermined. Instead of attempting to estimate this uncertainty, a conservative approach was used for the statistical tests involving the standard error. This was performed by setting the required confidence interval to 99% (instead of the usual 95%).

3. Results

A summary of descriptive statistics based on HRL IMD is found in Table 1. Imperviousness above 0% was found on 0.63% of the pixels in HRL IMD. This is the estimated "built-up land" including a combination of grey and green areas. Weighted with the specific degree of imperviousness for each pixel, the extent of the sealed surfaces (according to HRL IMD) is 101,961 hectares, corresponding to 0.31% of the total land surface. Notice that the term land surface here also includes rivers and lakes (inland water). The mean imperviousness for pixels where imperviousness was present was 50.55% according to HRL IMD.

The areas where HRL IMD reports imperviousness greater than 0% was divided into eleven strata. The mean imperviousness was calculated for each stratum (Table 1). The difference between the reported (HRL IMD) and the observed (from aerial photo) imperviousness is listed in Table 2, together with the standard error and 99% confidence interval of the mean difference in imperviousness (disregarding within-pixel sampling errors). Strata where the difference was statistically significant (different from 0%) at the 99% confidence level are flagged with a double asterisk (**) in the table.

Table 2. Difference (Diff) between the imperviousness reported by HRL IMD (HRL %) and observed on orthophotos (OBS %) calculated by stratum. SErr is the standard error of the difference, and the 99% CI is the confidence interval for the difference. Sign shows the result of the test of the hypothesis Diff $\neq 0$. The null-hypothesis (Diff = 0) is either not rejected (-) or rejected with 99% confidence (**).

					99% CI		
(%)	HRL %	OBS %	Diff	Serr	Lower	Upper	Sign
100	100.00	86.72	13.28	3.06	5.40	21.16	**
90–99	94.27	68.41	25.86	3.92	15.76	35.96	**
80–89	84.33	56.77	27.56	4.07	17.08	38.04	**
70–79	74.53	49.01	25.52	3.93	15.40	35.64	**
60–69	64.65	47.47	17.18	3.27	8.76	25.60	**
50–59	53.97	48.33	5.64	3.51	−3.40	14.68	-
40–49	44.29	35.11	9.18	2.85	1.84	16.52	**
30–39	34.34	35.63	−1.29	2.75	−8.37	5.79	-
20–29	24.78	26.23	−1.45	2.76	−8.56	5.66	-
10–19	15.60	12.02	3.58	2.04	−1.68	8.84	-
1–9	6.56	12.24	−5.68	2.58	−12.33	0.97	-
Subtotal	**50.55**	**42.75**	**7.80**	**0.95**	**5.35**	**10.25**	**
0%	0.00	0.21	−0.21	0.09	−0.44	0.02	-
Total	**0.31**	**0.47**	**−0.16**	**1.09**	**−2.97**	**2.65**	**-**

The reference data obtained by sampling from aerial photographs showed that imperviousness was present in 0.8% of the pixels classified as non-impervious (0%) by HRL IMD. The corresponding area covered by these pixels is approximately 257,433 hectares built-up land neglected by HRL IMD. The mean imperviousness in these pixels was 26%. This corresponds to 67,576 hectares of sealed land omitted by HRL IMD (Table 1).

The results from the interpretation of aerial photographs in the strata found to be impervious by HRL IMD showed generally lower imperviousness than reported by HRL IMD. Weighted by the degree of imperviousness, HRL IMD estimated the sealed surface in these areas to 101,961 hectares, while the estimate from aerial photographs was 86,229 hectares (Table 1).

The observed imperviousness degree in areas mapped as impervious by HRL IMD was 42.75%. This is lower than the imperviousness degree of 50.55% reported by HRL IMD.

There is a degree of reciprocity in the material. HRL IMD is underestimating imperviousness in some areas and overestimating imperviousness in other areas and the errors are to some extent counterbalanced. This can be seen from the fact that there were impervious pixels in the "0%" stratum and non-impervious pixels in the other strata. During the sampling we also noticed that the difference between the HRL values and the observations could be positive as well as negative. The overall result is, however, an underestimation of the total sealed surface for the whole country (0.31% in HRL IMD against 0.47% based on sampling from orthophoto (Table 1). The corresponding absolute figures are 101,961 hectares sealed land based on HRL IMD against 153,805 hectares sealed land based on the orthophoto. The difference is, however, not statistically significant at the 99% CI (Table 2).

Figure 2 is a graphical representation of Table 2, showing the mean difference between imperviousness reported by IMD and imperviousness observed in orthophoto by stratum. The figure also shows the 99% CI of these mean differences. The imperviousness reported by IMD is on average larger than the imperviousness observed in the orthophoto when the mean difference is positive (above the 0% line). The differences are statistically significant (at the 99% confidence level) when the entire bar is above or below the 0% line, as in the strata 0% and 40–49%, and for all strata where IMD reports more than 60% imperviousness.

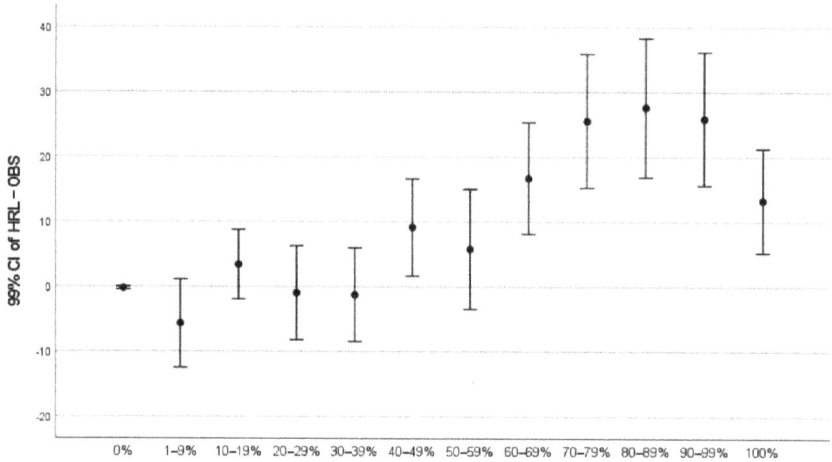

Figure 2. Mean difference between imperviousness reported by IMD and observed in orthophoto together with the 99% CI of these mean differences.

Figure 2 reveals a small but statistically significant neglect of sealed surfaces in areas where HRL IMD reports no imperviousness. The results show good accuracy for land with imperviousness above 0% and up to 40%. The imperviousness on land where the imperviousness is above 40% is often overestimated in HRL IMD, and the error is increasing as the imperviousness increases. The overestimation is high but stable for land where the imperviousness is in the range 70–99%. The estimate is better again, although still high, for land assessed as 100% impervious by HRL IMD.

4. Discussion

This discussion covers four topics. The first topic is the accuracy assessment of HRL IMD. This is followed by a discussion of the within-pixel sampling approach. The third part of the discussion is concerned with alternatives to the within-pixel sampling approach. The final topic is the applicability of the within-pixel sampling approach to other remote sensing products.

4.1. The Accuracy Assessment of HRL IMD

According to Statistics Norway (https://www.ssb.no/statbank/table/10781/, last accessed on 7 June 2022), the physically sealed areas in Norway in 2018 were 167,818 hectares (54,866 hectare covered with buildings or similar constructions and 112,952 hectares road surface). Our survey estimate is around 10% below the figure provided by Statistics Norway. The figure provided by Statistics Norway and the survey estimate both indicate that the figure given by HRL IMD is too low. The estimate provided by HRL IMD is almost 40% below the figure given by Statistics Norway and 33% below the estimate based on aerial photographs.

The data from Statistics Norway are based on official records and are reliable. The two most important sources are the national building register and the roads register. The building register is a complete inventory of buildings in Norway, including information about the ground area of each building. The register is maintained by municipal authorities. The roads register is maintained by the national roads authority and include information about the length, width, and surface material of all roads except logging roads, but logging roads are not paved. The information from these registers is completed with data from municipal authorities documenting parking lots, sports facilities, and other constructions.

Buildings and paved roads in rural areas were frequently classified as non-impervious (0%). There was no obvious pattern behind the omission of these sealed constructions. Part of a road could be marked as impervious, another part as non-impervious. The buildings on one farm could show up as impervious, while the entire farmstead of the next farm was marked as non-impervious. Neither the width of the roads nor the size or substance covering the roof of buildings could explain this variation.

Impervious surfaces below overhanging trees were occasionally mapped as non-impervious in HRL IMD. Examples are smaller buildings (cabins) inside forests and roads running through densely forested areas. This was also sometimes seen in residential neighborhoods with large gardens, where old trees developed an overhanging canopy partially covering constructions underneath.

The inclusion errors could be grouped into (a) permeable areas wrongly classified as impervious, and (b) situations where the size of the area sealed by constructions was overestimated. The latter situation appeared to be an extrapolation of sealed surfaces due to shadows cast by buildings and roadcuts.

Logged timber is stockpiled next to logging roads, awaiting transport to terminals, sawmills and industrial sites where large amounts of logs are stowed while waiting for further transport or processing. These stacks of lumber are frequently registered as impervious land in HRL IMD.

A similar effect is seen in places producing firewood used for heating houses during the cold season. Many farmers produce large quantities of firewood which is sold to house-owners with no private forest of their own. Production of firewood requires that logs transported from the forest are stockpiled (equivalent to the stacks used in ordinary logging operations, but on a smaller scale). The firewood is stored for drying. The process usually takes a year, and the storage can cover substantial areas which occasionally were mapped as impervious in HRL IMD.

Both natural and artificial gravel surfaces were frequently classified as impervious by HRL IMD. This included areas inside gravel pits and quarries as well as natural gravel, e.g., along riverbeds where gravel and boulders are exposed during the dry part of the summer, and along the edge of reservoirs where bedrock is exposed due to low water level part of the year. Artificial gravel surfaces such as parking spaces, playgrounds and sports fields covered with subbus or pebbles of variable grain size were also occasionally mapped as impervious in HRL IMD.

A particular phenomenon were floating docks found in marinas for yachts and smaller boats. They are hard surfaces and registered as impervious in HRL IMD.

According to the technical documentation of HRL IMD [28] railway tracks associated to other impervious surfaces (i.e., inside built-up areas) are impervious while railway tracks not associated to other impervious surfaces (i.e., outside built-up area) are non-impervious. Railway tracks were often represented as impervious in HRL IMD, irrespective of the surrounding environment. Sleepers do represent a perceptible impervious element along the railway track, and the technical documentation should probably be revised accordingly.

Other permeable features frequently reported as impervious are hydropower dams, road cuts, junkyards, parked cars, and silage balls. Dams and road cuts are examples of exposed rock and gravel erroneously classified as impervious. Junkyards, parked cars, and silage are hard surfaces observed by the satellite imagery but should probably not be counted as impervious areas. Junkyards are found in urban as well as rural areas. It can be abandoned industrial sites or farmyards littered with old cars, containers, and other junk. It is probably impossible to classify these areas correctly using satellite imagery.

Parked cars have the same effect as junk. They are interpreted as impervious surfaces. This is not a problem when the cars are parked on an otherwise sealed surface, but parking lots can also be covered by penetrable materials, e.g., gravel.

The silage produced by farmers when meadows are mowed is packed in large balls and wrapped in plastic. The balls are typically assembled in one part of the field and left there until the fodder is needed. The plastic cover around these balls is impenetrable and stacks of silage balls can be seen as impervious areas in HRL IMD.

A final source of error is the deep and dark shadows between or next to buildings and roadcuts. These are sometimes also interpreted as impervious land and inflates the reported imperviousness degree of built-up land.

These results are largely in agreement with results found in other countries. Two studies of HRL IMD products for Poland found that sealed surfaces were omitted in rural areas while imperviousness was overestimated in built-up areas. The omissions were identified as small buildings, narrow roads, etc., which were treated as natural surfaces. Trodden and ridden areas, on the other hand, could appear as impervious. These studies list industrial areas, dirt-roads or dirt-courtyards and areas covered with bare soil as examples of land wrongly assigned an imperviousness above 0% [31,32].

A similar study of the HRL IMD product for Slovakia describes the overall accuracy as acceptable but found that imperviousness tended to be overestimated in areas where soil sealing was regular and underestimated in areas where soil sealing was sporadic. As a result, the proportion of impervious surfaces could be overestimated in urban areas and underestimated or neglected among rural settlements [33].

The interpretation of errors or mistakes in HRL IMD in the current study is based on observations made during the within-pixel sampling. This is an unsystematic approach with respect to errors but may still provide useful information about the data. Systematic studies of the suspected mistakes can be designed based on these observations.

The two most important findings from the accuracy assessment are the apparently unsystematic omission of sealed land in rural areas and the systematic overestimation of imperviousness degree in the most densely built-up areas. Both observations comply with results from other studies, and both indicate that there are challenges regarding the calibration of the model used in the production of HRL IMD. Dedicated studies are needed to quantify these problems and improve the model.

4.2. The Sampling Methodology

This study employed a two-stage sampling strategy to examine HRL IMD. The first stage was a stratified random sample of pixels. The second stage was a systematic point sample within each of the pixels selected at the first stage. Within-pixel sampling will always require two stages, since pixels must be selected for examination before the within-pixel sample can be established, but a simple random sample will usually be sufficient at the first stage.

The decision to use a stratified approach at the first stage was related to the data distribution. The situation with 99.4% of the pixels having 0% imperviousness required special treatment. This skewed distribution of imperviousness data is not unique for Norway, but has also been reported for other countries, among them Poland [31].

The sample of pixels with 0% imperviousness had to be large enough to detect mistakes inside this group. The sample from the remaining pixels (1–100% imperviousness) had to cover the entire range of values and be large enough to represent variation inside each stratum. Finally, the sampling strategy had to be realistic within the available budget. Stratified sampling met these requirements, but only by using an unbalanced sample.

A stratified sample should preferably be a balanced sample. A balanced sample is a sample where the sample size in each stratum is proportional to the size of the stratum, i.e., $n_i/n = \mathrm{N_i}/\mathrm{N}$. This was not possible within the constraints of a realistic sample size.

With 99.4% of the pixels in a single stratum, a balanced sample of 2000 pixels implies that 1988 sample points should be drawn from this stratum alone and only twelve sample points from the pixels with imperviousness > 0%. An unbalanced sample was therefore necessary. A balanced sample would have been preferred if the distribution (of imperviousness values) had been more uniform across the entire range from 0% to 100%. Stratified sampling as well as the unbalanced configuration of strata used in this study are choices justified by the properties of the data being examined and should not be interpreted as general recommendations.

Each pixel selected in the first stage was examined using a systematic sample of 100 sample points inside the pixel (Figure 1). The preconditions for this methodology are the availability of high-resolution orthophoto geometrically aligned with the raster and GIS software that allows for efficient and accurate data collection. The approach was efficient and did provide interesting results, but there were also obvious shortcomings and questions that need to be resolved.

The sample size was set to 100 sample points per pixel. The sample size was chosen due to convenience. The 100 sample points allowed the sample points to be placed one meter apart. This appeared to be a dense sample inside a single 10 m pixel. Proportions could also be calculated without effort. Any sample size could have been used, but smaller samples inside each pixel would weaken the statistical support of the analysis.

The number of pixels examined could have been increased by reducing the sample size inside each pixel. The number of pixels could probably have been increased from 2082 to approximately 8000 by cutting the number of sample points inside pixels from 100 down to 25 (points two meters apart). The effect in terms of accuracy has not been examined. This should be performed to provide better advice about the optimum balance between sample sizes in the first and the second sampling stage.

A systematic random sample was used for sampling inside each pixel. A systematic sample is known to provide more accurate estimates than a simple random sample of the same size when spatial autocorrelation is present in the material [30]. Spatial autocorrelation with respect to imperviousness implies that the sealed area tends to be clustered in certain parts of the pixel rather than distributed randomly over the surface of the pixel. The assumption is justified because soil sealing is caused by constructions.

There is, however, no exact method to calculate the statistical accuracy of a systematic sample. The accuracy must be determined by estimation [34]. A conservative estimate of the uncertainty in a systematic sample can be found by treating the sample as a simple random sample, but the result is that the improved accuracy achieved by the systematic approach goes unnoticed. Trials with forestry data has shown that a systematic sample can reduce uncertainty with as much as 30% when autocorrelation is present [35]. This improvement should be documented. Several estimation methods have been published [36,37] and could be applied to within-pixel sampling. A common requirement is, however, that the (relative) location of each sample point in the systematic sample must be recorded.

The systematic sample within pixels could have been replaced by a simple random sample. The disadvantage, apart from the assumed lower precision, is that a simple random sample may be more difficult to observe when the sample size is large. This assumption has not been tested, but it seems more challenging to work with a completely random distribution of points than counting along a regular lattice with the same number of points.

Finally, by experience, even the resolution of high quality orthophoto is insufficient to allow exact interpretation of impervious vs. permeable parts of the pixel surface. The pixel shown in Figure 1 contains a road, a roadside ditch, and part of an agricultural field. The edge between the solid and the permeable surface is somewhere along the white line painted beside the road, but it is not possible to tell exactly where the edge is. Consequently, the accuracy of the data obtained by sampling within pixels also depend on the quality of the orthophoto and the experience of the analyst [38].

Within-pixel sampling was found to be suitable for accuracy assessment of HRL IMD, but more work is clearly needed to improve the methodology. Geometrical inconsistencies between the raster and the aerial imagery used as ground truth will make the results uncertain or even meaningless, as will interpretation errors with respect to ground truth. The sampling accuracy at stage two was disregarded in this study and replaced by a conservative interpretation of the sampling error at stage one. Better tools are needed to handle the systematic sample and estimate the sampling accuracy inside each pixel.

4.3. Alternative Approaches

A complete inventory of each control pixel is an alternative to the within-pixel sampling approach. Such an inventory could be carried out as a ground survey, but the cost would be excessive. Digitizing from orthophoto or by photogrammetric construction using stereo photography are manageable, but still expensive alternatives. A complete inventory will eliminate the sampling error generated by the second stage of the survey.

Another alternative is a simplified accuracy assessment using a classified version of the dataset. This approach has been used in several unpublished, national verification reports commissioned by EEA. HRL IMD is reclassified into three classes ("0%", "1–29%" and "30–100%"). A stratified sample of control pixels is selected, and a qualitative assessment (based on available imagery) is used to assign the pixels to one of the three simplified classes. The approach allows for calculation of omission and commission errors between the three classes using standard methodology [39,40].

None of these alternatives will remove the observation error and inaccuracies stemming from geometrical misalignment between the data sources, but both alternatives eliminate the sampling error of the within-pixel sampling approach. The complete inventory is probably expensive, even with on-screen digitizing, while the assessment of a simplified version of the product results in drastic reduction in the information obtained from the accuracy assessment.

4.4. Applicability to Other Remote Sensing Products

The within-pixel sampling strategy used in the present study is suitable for assessment of a remote sensing product that represents a binomial proportion. Examples are imperviousness and canopy coverage. Within-pixel sampling may be less suitable for other remote sensing products.

Binary remote sensing products show the presence or absence of a particular feature. An example is the Copernicus high-resolution layer grassland [41] showing the presence or absence of grassland. A binary remote sensing product is essentially a simplified version of a binomial proportion. Sampling within pixels is suitable for assessment of these products provided that the decision rules used in the classification of the pixel are clearly expressed. Examples of decision rules can be that the feature must be present in the pixel, cover a defined proportion of the pixel surface, or dominate the pixel. Correct classification can then be assessed by within-pixel sampling.

Within-pixel sampling is less suitable for assessment of categorical remote sensing products (e.g., land cover maps and crop type maps) unless the classes are scale-independent, pure land cover classes and the decision rules used to classify pixels are known (e.g., crop type determined by the dominant crop type in the pixel). Scale-dependent classifications and classifications involving elements of land use will be difficult to assess by within-pixel sampling.

Within-pixel sampling can, on the other hand, be used for descriptive analysis aiming to improve the explanation of the classes used in a categorical remote sensing product. Descriptive analysis does not assess accuracy but describes the content of classes in terms of more detailed data. An example is the analysis of the content of the European CORINE Land Cover Map by populating the classes with data from more detailed map sources [42]. Within-pixel sampling of pure land cover elements can be used for this purpose.

5. Conclusions

One objective of the current study was to examine the content and accuracy of the high-resolution layer imperviousness density (HRL IMD) product for 2018. Norway was used as the study area. There was a noteworthy neglection of buildings and roads in rural areas. The effect of these omissions was to some extent reduced by overestimation of the imperviousness density in built-up areas. Overall, the amount of sealed surface estimated from HRL IMD was 33% below the amount estimated using high-resolution orthophoto and 40% below the official figure on sealed surface published by Statistics Norway.

The results indicate that the statistics provided by HRL IMD is biased because the omissions of sealed surfaces in rural areas were larger (in absolute terms) than the inclusion errors in built-up areas. It is reasonable to expect that the ratio between built-up and natural land, along with the zoning structure has an impact on the bias and that the errors may be more balanced in regions with more built-up land. Still, bias is present and further work is needed to improve the model used in the production of HRL IMD, aiming to reduce or even eliminate the bias.

The Copernicus land monitoring services (CLMS), including HRL IMD, are standardized products covering all of Europe, but the quality and accuracy are likely to vary across the continent. The products can still be valuable for comparative studies. Time series can provide useful information on changes, provided that the errors are randomly distributed [43]. Authors have also found the products useful for comparing locations, as in the comparative study of suburban patterns in the Barcelona and Milan [44] and as a tool to decompose less detailed data sets, such as CORINE Land Cover [45,46].

The findings from the current study can be used to improve the interpretation and understanding of the product. The user should be aware that bias is present in the material and understand how the errors are distributed between land systems when the results are interpreted. Accuracy assessment and verification studies from various subregions of Europe are therefore needed to establish product credibility when HRL IMD and other CLMS data are incorporated into decision systems, especially at local administrative levels [47]. Interpretative studies of the data provided by the Copernicus land monitoring services should therefore be an integral part of the program.

The within-pixel sampling strategy used in the study was found suitable for assessment of imperviousness. It is probably also appropriate for assessment of other binomial proportions. The study did, however, reveal challenges that require further methodological studies and development. This was linked to the use of systematic random sampling within pixels.

A systematic sample inside each pixel is easy to work with and is known to produce more accurate estimates than a simple random sample when spatial autocorrelation is present. The improvement in accuracy does, however, go unnoticed unless the status and location of each sample point inside the pixel is recorded and an appropriate method is applied to estimate the accuracy. These methods exist but functional tools must be developed. Further research is also needed to help determine the optimal sample sizes at the different stages of a survey.

Funding: The research leading to these results received funding from the Norway Grants 2014–2021 via the Polish National Center for Research and Development [grant no: NOR/POLNOR/InCoNaDa/0050/2019-00] and was partly carried out by the joint Polish-Norwegian project Integration of Copernicus and National Data (InCoNaDa).

Data Availability Statement: HRL IMD can be downloaded from https://land.copernicus.eu and the reference data can be inspected at https://kilden.nibio.no. The sample can be obtained from the author.

Conflicts of Interest: The author declares no conflict of interest. The funders had no role in the design of the study; in the collection, analyses, or interpretation of data; in the writing of the manuscript, or in the decision to publish the results.

References

1. Meyer, W.B.; Turner, B.L. Human population growth and global land-use/cover change. *Annu. Rev. Ecol. Syst.* **1992**, *23*, 39–61. [CrossRef]
2. Jennings, D.B.; Taylor Jarnagin, S. Changes in anthropogenic impervious surfaces, precipitation and daily streamflow discharge: A historical perspective in a mid-Atlantic subwatershed. *Landsc. Ecol.* **2002**, *17*, 471–489. [CrossRef]
3. Burghardt, W. Soil sealing and soil properties related to sealing. *Geol. Soc. Lond. Spec. Publ.* **2006**, *266*, 117–124. [CrossRef]
4. Arnold, C.L., Jr.; Gibbons, C.J. Impervious surface coverage: The emergence of a key environmental indicator. *J. Am. Plan. Assoc.* **1996**, *62*, 243–258. [CrossRef]

5. Shuster, W.D.; Bonta, J.; Thurston, H.; Warnemuende, E.; Smith, D.R. Impacts of impervious surface on watershed hydrology: A review. *Urban Water J.* **2005**, *2*, 263–275. [CrossRef]
6. Jacobson, C.R. Identification and quantification of the hydrological impacts of imperviousness in urban catchments: A review. *J. Environ. Manag.* **2011**, *92*, 1438–1448. [CrossRef] [PubMed]
7. Tobias, S.; Conen, F.; Duss, A.; Wenzel, L.M.; Buser, C.; Alewell, C. Soil sealing and unsealing: State of the art and examples. *Land Degrad. Dev.* **2018**, *29*, 2015–2024. [CrossRef]
8. Li, J.; Pei, Y.; Zhao, S.; Xiao, R.; Sang, X.; Zhang, C. A review of remote sensing for environmental monitoring in China. *Remote Sens.* **2020**, *12*, 1130. [CrossRef]
9. Slonecker, E.T.; Jennings, D.B.; Garofalo, D. Remote sensing of impervious surfaces: A review. *Remote Sens. Rev.* **2001**, *20*, 227–255. [CrossRef]
10. Yang, L.; Huang, C.; Homer, C.; Wylie, B.; Coan, M. An approach for mapping large-area impervious surfaces: Synergistic use of Landsat 7 ETM+ and high spatial resolution imagery. *Can. J. Remote Sens.* **2002**, *29*, 230–240. [CrossRef]
11. Homer, C.; Huang, C.; Yang, L.; Wylie, B.; Coan, M. Development of a 2001 national land-cover database for the United States. *Photogramm. Eng. Remote Sens.* **2004**, *70*, 829–840. [CrossRef]
12. Raymaekers, D.; Bauwens, I.; Van Orshoven, J.; Gulinck, H.; Engel, B.; Dosselaere, N. Spectral unmixing of low resolution images for monitoring soil sealing. In *Urban 2005, Proceedings of the 3rd International Symposium Remote Sensing and Data Fusion Over Urban Areas, Tempe, AZ, USA, 14–16 March 2005*; ISPRS: Hannover, Germany, 14–16 March 2005.
13. Kampouraki, M.; Wood, G.A.; Brewer, T.R. The application of remote sensing to identify and measure sealed areas in urban environments. In *Bridging Remote Sensing and GIS, Proceedings of the 1st International Conference on Object-Based Image Analysis, Salzburg, Austria, 4–6 June 2006*; ISPRS: Hannover, Germany, 2006.
14. Kampouraki, M.; Wood, G.A.; Brewer, T.R. The suitability of object-based image segmentation to replace manual aerial photo interpretation for mapping impermeable land cover. In *Challenges for Earth Observation: Scientific, Technical and Commercial, 426–430, Proceedings of the Remote Sensing and Photogrammetry Society Annual Conference, Newcastle Upon Tyne, UK, 11–14 September 2007*; Heipke, C., Ed.; Remote Sensing and Photogrammetry Society (RSPSoc): Nottingham, Germany, 2007.
15. Weng, Q. Remote sensing of impervious surfaces in the urban areas: Requirements, methods, and trends. *Remote Sens. Environ.* **2012**, *117*, 34–49. [CrossRef]
16. Salvati, L. The spatial pattern of soil sealing along the urban-rural gradient in a Mediterranean region. *J. Environ. Plan. Manag.* **2014**, *57*, 848–861. [CrossRef]
17. Kuc, G.; Chormański, J. Sentinel-2 imagery for mapping and monitoring imperviousness in urban areas. The International Archives of Photogrammetry. *Remote Sens. Spat. Inf. Sci.* **2019**, *42*, 43–47. [CrossRef]
18. Shrestha, B.; Ahmad, S.; Stephen, H. Fusion of Sentinel-1 and Sentinel-2 data in mapping the impervious surfaces at city scale. *Environ. Monit. Assess.* **2021**, *193*, 556. [CrossRef]
19. Peroni, F.; Pappalardo, S.E.; Facchinelli, F.; Crescini, E.; Munafò, M.; Hodgson, M.E.; De Marchi, M. How to map soil sealing, land take and impervious surfaces? A systematic review. *Environ. Res. Lett.* **2022**, *17*, 053005. [CrossRef]
20. Moglen, G.E.; Kim, S. Limiting imperviousness: Are threshold-based policies a good idea? *J. Am. Plan. Assoc.* **2007**, *73*, 161–171. [CrossRef]
21. Parent, J.R.; Lei, Q. Estimating percent impervious cover from Landsat-based land cover with a simple and transferable regression model. *Int. J. Remote Sens.* **2018**, *39*, 3839–3851. [CrossRef]
22. Lefebvre, A.; Sannier, C.; Corpetti, T. Monitoring urban areas with Sentinel-2A data: Application to the update of the Copernicus high resolution layer imperviousness degree. *Remote Sens.* **2016**, *8*, 606. [CrossRef]
23. European Environment Agency. *The European Environment—State and Outlook 2020. Knowledge for Transition to a Sustainable Europe*; Publications office of the European Union: Luxembourg, 2019.
24. Vysna, V.; Maes, J.; Petersen, J.E.; La Notte, A.; Vallecillo, S.; Aizpurua, N.; Ivits, E.; Teller, A. *Accounting for ecosystems and their services in the European Union (INCA). Final Report from Phase II of the INCA Project Aiming to Develop a Pilot for an Integrated System of Ecosystem Accounts for the EU*; Statistical Report; Publications office of the European Union: Luxembourg, 2021.
25. Congedo, L.; Sallustio, L.; Munafò, M.; Ottaviano, M.; Tonti, D.; Marchetti, M. Copernicus high-resolution layers for land cover classification in Italy. *J. Maps* **2016**, *12*, 1195–1205. [CrossRef]
26. Smith, G. GMES Initial Operations/Copernicus Land Monitoring Services—Validation of Products. HRL Imperviousness Degree 2015 Validation Report. GIO HRL IMD Validation Report SC03, Issue D1.3. 2019. Available online: https://land.copernicus.eu/user-corner/technical-library/hrl-imperviousness-2015-validation-report (accessed on 7 June 2022).
27. Wickham, J.D.; Stehman, S.V.; Gass, L.; Dewitz, J.; Fry, J.A.; Wade, T.G. Accuracy assessment of NLCD 2006 land cover and impervious surface. *Remote Sens. Environ.* **2013**, *130*, 294–304. [CrossRef]
28. Steidl, M.; Schleicher, C.; Sannier, C. *Copernicus Land Monitoring Service—High Resolution Layer Imperviousness: Product Specifications Document*; European Environment Agency: Copenhagen, Denmark, 2018.
29. Lohr, S.L. *Sampling: Design and Analysis*, 2nd ed.; CRC Press: Boka Raton, FL, USA, 2019.
30. Cochran, W.G. *Sampling Techniques*, 3rd ed.; John Wiley & Sons: New York, NY, USA, 1977.
31. Krówczyńska, M.; Soszyńska, A.; Pabjanek, P.; Wilk, E.; Hurbanek, P.; Rosina, K. Accuracy of the soil sealing enhancement product for Poland. *Quaest. Geogr.* **2016**, *35*, 89–95. [CrossRef]

32. Pabjanek, P.; Krówczyńsk, M.; Wilk, E.; Miecznikowski1, M. An accuracy assessment of European Soil Sealing Dataset (SSL2009): Stara Miłosna area, Poland—A case study. *Misc. Geogr. Reg. Stud. Dev.* **2016**, *20*, 59–63. [CrossRef]
33. Hurbanek, P.; Atkinson, P.M.; Pazur, R.; Rosina, K.; Chockalingam, J. Accuracy of built-up area mapping in Europe at varying scales and thresholds. In Proceedings of the Accuracy 2010, Ninth International Symposium on Spatial Accuracy Assessment in Natural Resources and Environmental Sciences, Leicester, England, 20–23 July 2010; University of Leicester: Leicester, UK, 2010; Volume 385388.
34. Strand, G.H. A study of variance estimation methods for systematic spatial sampling. *Spat. Stat.* **2017**, *21*, 226–240. [CrossRef]
35. Magnussen, S.; Nord-Larsen, T. Design-consistent model-based variances with systematic sampling: A case study with the Danish national Forest inventory. *Commun. Stat. Simul. Comput.* **2021**, *50*, 38–48. [CrossRef]
36. Aune-Lundberg, L.; Strand, G.H. Comparison of variance estimation methods for use with two-dimensional systematic sampling of land use/land cover data. *Environ. Model. Softw.* **2014**, *61*, 87–97. [CrossRef]
37. Brus, D.J.; Saby, N.P. Approximating the variance of estimated means for systematic random sampling, illustrated with data of the French Soil Monitoring Network. *Geoderma* **2016**, *279*, 77–86. [CrossRef]
38. Strand, G.H.; Dramstad, W.; Engan, G. The effect of field experience on the accuracy of identifying land cover types in aerial photographs. *Int. J. Appl. Earth Obs. Geoinf.* **2002**, *4*, 137–146. [CrossRef]
39. Aronoff, S. The map accuracy report: A user's view. *Photogramm. Eng. Remote Sens.* **1982**, *48*, 1309–1312.
40. Congalton, R.G. A review of assessing the accuracy of classifications of remotely sensed data. *Remote Sens. Environ.* **1991**, *37*, 35–46. [CrossRef]
41. Available online: https://land.copernicus.eu/pan-european/high-resolution-layers/grassland (accessed on 7 June 2022).
42. Aune-Lundberg, L.; Strand, G.H. The content and accuracy of the CORINE Land Cover dataset for Norway. *Int. J. Appl. Earth Obs. Geoinf.* **2021**, *96*, 102266. [CrossRef]
43. Drašković, B.J. Urban expansion of the largest cities in Bosnia and Herzegovina over the period 2000–2018. *Geogr. Pannonica* **2021**, *25*. [CrossRef]
44. Pagliarin, S. Supra-local spatial planning practices and suburban patterns in the Barcelona and Milan urban regions. *Land Use Policy* **2022**, *112*, 105816. [CrossRef]
45. Cole, B.; Smith, G.; Balzter, H. Acceleration and fragmentation of CORINE land cover changes in the United Kingdom from 2006–2012 detected by Copernicus IMAGE2012 satellite data. *Int. J. Appl. Earth Obs. Geoinf.* **2018**, *73*, 107–122. [CrossRef]
46. Rosina, K.; e Silva, F.B.; Vizcaino, P.; Herrera, M.M.; Freire, S.; Schiavina, M. Increasing the detail of European land use/cover data by combining heterogeneous data sets. *Int. J. Digit. Earth* **2020**, *13*. [CrossRef]
47. Manakos, I.; Tomaszewska, M.; Gkinis, I.; Brovkina, O.; Filchev, L.; Genc, L.; Gitas, I.Z.; Halabuk, A.; Inalpulat, M.; Irimescu, A.; et al. Comparison of global and continental land cover products for selected study areas in South Central and Eastern European region. *Remote Sens.* **2018**, *10*, 1967. [CrossRef]

remote sensing

MDPI

Article

An Improved Vicarious Calibration Method Based on Multi-Grayscale Targets

Shiwei Bao [1,2], Hongyao Chen [1,*], Yan Li [3], Liming Zhang [1], Wenxin Huang [1], Xiaolong Si [1], Xianhua Wang [1], Zhou Fang [3], Yuanwei Chen [3], Xinrong Wang [3] and Xiaowen Zhao [3]

1 Key Laboratory of Optical Calibration and Characterization, Hefei Institutes of Physical Science, Chinese Academy of Sciences, Hefei 230031, China
2 Science Island Branch of Graduate School, University of Science and Technology of China, Hefei 230026, China
3 State Key Laboratory of Space-Ground Integrated Information Technology Space Star Technology Company Ltd., Beijing 100095, China
* Correspondence: hychen@aiofm.ac.cn

Abstract: Vicarious calibration is a well-developed method for electro-optical (EO) sensor calibration that has been used since the early 1980s. The radiometric calibration of reflectance solar band is mainly applied to reflection inversion. In this paper, a radiometric calibration-reflectance inversion (RCRII) model is proposed as an improved vicarious calibration method. Taking the reflectance of grayscale targets with constant spectrum, suitable uniformity, and near-Lambertian characteristics as the known information, the grayscale target calibration is realized, and the initial value of calibration coefficient and offset are calculated. Then, the adjacency effect is evaluated and corrected by reflectance inversion, and the results are fed back to the calibration process to realize the iterative process of calibration inversion rescaling. The results indicate that the absolute difference between the reflectance calculated with the RCRII model and measured reflectance is less than 0.01. By comparing with Sentinel-2A images, it is cross-verified that the difference of radiance between them is within 4%, and the absolute reflectance difference is less than 0.01, in the range of 0.1~0.3 reflectance.

Keywords: remote sensing; radiometric calibration; vicarious calibration

1. Introduction

On-orbit calibration has the advantage of being performed under true flight conditions, rather than simulated flight-like conditions. The goals of on-orbit calibration are to measure parameters that cannot be measured on the ground, maintain calibration throughout a sensor's operational lifetime, quantify calibration uncertainty, and update calibration coefficients, if necessary, to meet measurement requirements [1]. It has been proven that the sensor response is affected by long-term attenuation and short-term fluctuation, particularly in the orbital environment during the on-orbit periods; if the calibration results are not continuously updated, it may lead to large errors [2,3]. On-orbit calibration includes onboard calibration, vicarious calibration, and other methods. Vicarious calibration can accomplish radiometric calibration based on actual Earth scenes, in order to realize a series of applications, such as reflectance inversion.

In recent years, scientists have developed a variety of solutions to effectively improve the calibration frequency and accuracy. For example, the Working Group on Calibration and Validation of the Committee on Earth Observation Satellites (CEOS-WGCV) has worked to establish a radiometric calibration network (RadCalNet). RadCalNet is a network of sites that can be used to compare different satellite sensors to each other and a common reference. It is based on the reflectance-based approach, with continuous deployment of automated instrumentation that is calibrated traceably to SI, as well as with known and peer-reviewed uncertainties [4,5].

In addition, pseudo-invariant site calibration realizes the high-frequency calibration of optical satellite sensors through the cross-calibration of MODIS, VIIRS, Landsat 8, and Sentinel 2 equipped with on-board calibration components. Due to the high reflectance of the on-board calibration diffuser, on-board calibration is usually realized at the high end of the sensor response, which is difficult for achieving high-precision calibration in the full dynamic range, and the low end of the response is a lack of verification. For example, although Landsat satellites have good consistency in at-sensor radiance/apparent reflectance, there are great differences in surface reflectance [6]. In addition to the influence of atmospheric aerosols, the sensor radiation response model (possibly with sensor offset) will be another potential reason. Therefore, full dynamic range high-frequency radiometric calibration is particularly urgent and important [7]. Based on the above reasons, full dynamic range and high-frequency calibration method based on grayscale targets are studied in this paper.

Continuous improvement of the spatial resolution of optical remote sensors enables on-orbit radiometric calibration based on grayscale targets. Robert E. Ryan of the NASA Stennis Space Center proposed a simplified approach to radiometric calibration. This new technique uses two neighboring uniform ground target areas with different reflectance values. For each target area, the radiance leaving the ground in the direction of the satellite is measured with a radiometrically calibrated spectroradiometer. Using the radiance measurements from the two targets, atmospheric adjacency and scattering effects can be subtracted. However, offsets, such as the dark current, atmospheric path radiation, and environmental radiation, are deducted as a whole, and each item cannot be separately determined.

In contrast, with remote sensor spatial resolution improvement, the adjacency effect has increasingly become one of the influencing factors that cannot be ignored. The adjacency effect is the image blur phenomenon caused by the entrance pupil of the remote sensor after atmospheric multiple scattering. The higher the remote sensor spatial resolution and larger the aerosol optical thickness are, the more serious the adjacency effect [8]. To date, the common adjacency effect correction methods include the Monte Carlo method and Fourier transform of the modulation transfer function (MTF) method [9], which are mostly used in atmospheric correction models. In addition, there has been much research regarding the adjacency effect based on three-dimensional radiative transfer. However, due to the massive computational efforts required to run three-dimensional radiative transfer codes, sophisticated atmospheric correction algorithms have not been pursued extensively [9]. Additionally, assuming that the surface and atmosphere are uniform in space, as well as only the vertical changes of atmospheric optical properties, the one-dimensional radiative transfer method implicitly explains the modeling of adjacency effect [10].

Based on the above, a radiometric calibration reflectance inversion iteration (RCRII) model is proposed in this paper. Firstly, the calibration coefficient and offset of the sensor can be calculated by using the multi-grayscale target calibration method. Then, the calculated parameters are substituted into the image for reflectance inversion, in order to evaluate the influence of adjacency effect. Finally, the results are fed back to the calibration process, so as to realize the radiometric calibration-reflectance inversion iteration method, which provides a new method for vicarious calibration.

The article is organized as follows: in Section 2, the RCRII model and data processing flow are introduced. Choosing the Pakistan remote sensing satellite (PRSS-1) as an example, the experimental data processing method is introduced in Section 3. In Section 4, PRSS-1 and Sentinel 2A are cross-verified and reflectance inversion verified, respectively, and their differences are compared. Finally, conclusions are provided in Section 5 and 6.

2. Radiometric Calibration-Reflectance Inversion Iterative Model

2.1. Radiometric Calibration Model

According to the atmospheric radiative transfer model, radiation information at the entrance pupil of the satellite sensor is shown in Figure 1, which mainly comprises the intrinsic atmospheric radiance, solar radiation reflected by the surface and directly transmitted from the surface toward the sensor, and contribution of the environment, which reflects the total (direct and diffuse) downward flux. Photons reach the sensor via scattering.

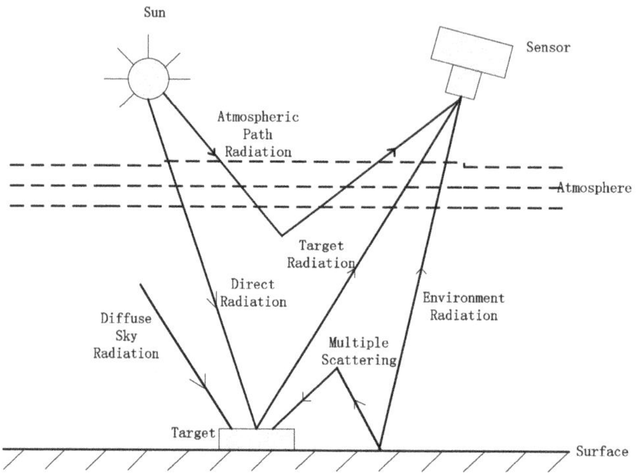

Figure 1. Schematic diagram of sun, target and sensor's reciprocity.

Assuming that the surface is uniform, the apparent reflectance collected by the optical sensor can be expressed as:

$$\rho^* = T_g[\rho_a + \frac{\rho}{1 - s\rho} T(\theta_s)T(\theta_v)] \tag{1}$$

where ρ is the surface reflectance, T_g is the gaseous absorption transmittance, ρ_a is the path reflectance resulting from photons that never interact with the surface, s is the spherical albedo of the atmosphere, $T(\theta_s)$ is the sun-to-ground total transmittance, i.e., $T(\theta_s) = e^{-\tau/\mu_s} + t_d(\theta_s)$, $T(\theta_v)$ is the ground-to-sensor total transmittance, i.e., $T(\theta_v) = e^{-\tau/\mu_v} + t_d(\theta_v)$, $e^{-\tau/\mu_s}$ is the sun-to-ground direct transmittance, $t_d(\theta_s)$ is the sun-to-ground diffuse transmittance, $e^{-\tau/\mu_v}$ is the ground-to-sensor direct transmittance, and $t_d(\theta_v)$ is the ground-to-sensor diffuse transmittance.

When placing grayscale targets with different reflectance values as reference targets, considering that the target reflectance is inconsistent with that of the background, Equation (1) can be rewritten as:

$$\rho^* = T_g\{\rho_a + \frac{T(\theta_s)}{1 - s < \rho >}[e^{-\tau/\mu_v}\rho_t + t_d(\theta_v) < \rho >]\} \tag{2}$$

where ρ_t is the target reflectance, and $< \rho >$ is not only the background reflectance but also the equivalent environmental reflectance affected by atmospheric multiple scattering, related to both the environmental and target reflectance. In the second simulation of the satellite signal in the solar spectrum (6S) radiative transfer model [11], the environmental function $F(r)$ is used to characterize the contribution of the target reflectance to the background, which can be calculated as:

$$< \rho >= F(r)\rho_t + (1 - F(r))\rho_e \tag{3}$$

where ρ_e is the environmental reflectance, r is the target radius.

Combined with the camera radiometric calibration model $DN = A \cdot \rho^* + DN_0$, where DN is the camera digital number, DN_0 is the camera offset, and A is the reflectance calibration coefficient, it can be seen that the equation between the target reflectance and camera digital number is as follows:

$$DN = AT_g \frac{T(\theta_s)}{1 - s < \rho >} e^{-\tau/\mu_v} \rho_t + AT_g [\rho_a + \frac{T(\theta_s)}{1 - s < \rho >} t_d(\theta_v) < \rho >] + DN_0 \quad (4)$$

Then, based on the linear regression between the target reflectance and digital value, the slope and intercept of the regression equation can be expressed as follows:

$$K = AT_g \frac{T(\theta_s)}{1 - s < \rho >} e^{-\tau/\mu_v}. \quad (5)$$

$$b = AT_g [\rho_a + \frac{T(\theta_s)}{1 - s < \rho >} t_d(\theta_v) < \rho >] + DN_0 \quad (6)$$

where K is the slope of the linear regression, and b is the intercept of the linear regression.

Another consideration here is that, as mentioned above, the equivalent environmental reflectance is related to the target reflectance. Then, for different reflectance targets, the equivalent environmental reflectance cannot be regarded as a constant, which may lead to nonlinear error. This phenomenon will be analyzed later.

When comparing the grayscale target calibration method to the traditional vicarious calibration method, the following two points should be considered:

(1) When the calibration experiment is conducted at a uniform site, it is necessary to mark the position of the measurement area to obtain the digital number of the corresponding area in the image [12]. The grayscale target is small, easy to identify, and exhibits a suitable uniformity. This approach reduces the random measurement error in the process of reflectance linear regression, and the measurement accuracy is, thus, generally higher than that of the traditional vicarious calibration method.

(2) When the sensor observes the Earth, there may occur nonlinear, stray light, or dark current issues. The single-point calibration method can hardly eliminate this factor. Assuming that the offset is 0, the site with low reflectance will lead to a large error in the calculated calibration coefficient. Based on grayscale targets, the calibration coefficient and offset can be calculated according to the slope and intercept equation.

2.2. Reflectance Inversion Model

Based on Equations (5) and (6), the initial values of the reflectance calibration coefficient A and offset DN_0 can be obtained. When these values are applied to the image, the apparent reflectance of the whole image can be obtained. The surface reflectance of the image area can be determined through reflectance inversion. Based on the 6S atmospheric correction model and adjacency effect analysis, a reflectance inversion model is proposed in this paper. The inversion steps are as follows:

(1) In regard to atmospheric absorption and atmospheric path radiation correction of the apparent reflectance image, Equation (2) can be rewritten as:

$$\frac{\rho*}{T_g} - \rho_a = \frac{T(\theta_s)}{1 - s < \rho >} e^{-\tau/\mu_v} \rho_t + \frac{T(\theta_s)}{1 - s < \rho >} t_d(\theta_v) < \rho > \quad (7)$$

(2) Choosing any pixel as the target, the reflectance of each pixel can be calculated as follows:

$$\rho_t = (\frac{\rho*}{T_g} - \rho_a) \frac{1}{T(\theta_s) e^{-\tau/\mu_v}} - [(\frac{\rho*}{T_g} - \rho_a) \frac{s}{T(\theta_s) e^{-\tau/\mu_v}} + \frac{t_d(\theta_v)}{e^{-\tau/\mu_v}}] < \rho > \quad (8)$$

where the target reflectance ρ_t and equivalent environmental reflectance $< \rho >$ are unknown parameters.

Assuming that the selected area is a uniform region, the initial reflectance value can be computed according to Equation (1):

$$\rho_{uniform} = \frac{\frac{\rho*}{T_g} - \rho_a}{T(\theta_s)T(\theta_v) + (\frac{\rho*}{T_g} - \rho_a)s} \tag{9}$$

where $\rho_{uniform}$ is the initial environmental reflectance value of each pixel.

Then, the target reflectance of each pixel can be calculated according to Equation (8):

$$\rho_t^{(0)} = \frac{\frac{\rho*}{T_g} - \rho_a}{T(\theta_s)e^{-\tau/\mu_v}} - [\frac{(\frac{\rho*}{T_g} - \rho_a)s}{T(\theta_s)e^{-\tau/\mu_v}} + \frac{t_d(\theta_v)}{e^{-\tau/\mu_v}}]\rho_{uniform} \tag{10}$$

where $\rho_t^{(0)}$ is the initial value of the target reflectance of each pixel.

However, in reality, most of the images are not uniform, that is, $\rho_t \neq \rho_e$. In the above assumption, each pixel is calculated as a uniform scene, that is, $\rho_t = \rho_e$, which will make the initial value of high reflectance higher and initial value of low reflectance lower, providing a basis for the next step.

The equivalent environmental reflectance can be regarded as the mean of all reflectance within a certain range around the target reflectance. Therefore, the equivalent environmental reflectance can be estimated as the average reflectance of the area:

$$< \rho(m_0, n_0) >^{(0)} = \frac{1}{(N+1)^2} \sum_{m=m_0-N/2}^{m_0+N/2} \sum_{n=n_0-N/2}^{n_0+N/2} \rho(m,n)_t^{(0)} \tag{11}$$

where (m_0, n_0) is the target pixel coordinate, $< \rho(m_0, n_0) >^{(0)}$ is the initial value of equivalent environmental reflectance, (m, n) is the coordinate of any pixel within the area of $(N + 1) \times (N + 1)$, and $\rho(m,n)_t^{(0)}$ is the initial value of target reflectance corresponding to (m, n), that is, $\rho_t^{(0)}$ is calculated by Equation (10).

In the previous step, it was mentioned that the assumption of uniform background will lead to the calculation error of reflectance. In this step, the average calculation of equivalent environmental reflectance will effectively eliminate this error and make the calculated equivalent environmental reflectance close to the real value.

Then, we use the equivalent environmental reflectance obtained in the previous step and substitute it into the calculation equation of the target reflectance (Equation (10)), in order to obtain a more accurate target reflectance, and then substitute the target reflectance into Equation (11) to obtain the equivalent environmental reflectance. After g repetitions, the equation can be expressed as follows:

$$\rho(m,n)^{(g)} = \frac{\frac{\rho*}{T_g} - \rho_a}{T(\theta_s)e^{-\tau/\mu_v}} - [\frac{(\frac{\rho*}{T_g} - \rho_a)s}{T(\theta_s)e^{-\tau/\mu_v}} + \frac{t_d(\theta_v)}{e^{-\tau/\mu_v}}] < \rho(m,n) >^{(g-1)} \tag{12}$$

The principle of this step is that all pixels in the region are not only targets, but also components of the environment. Each pixel is involved in the calculation of both the target and equivalent environment reflectance.

When $\left|\rho(m,n)^{(g)} - \rho(m,n)^{(g-1)}\right| \leq 0.001$ is satisfied, the iteration process is stopped to obtain the reflectance of each pixel.

In the above process, there is still a problem. Inversion is for all pixels in the area. When the $(N + 1) \times (N + 1)$ area is selected as the inversion object, as described above, Equation (11) can be applied to calculate the central pixel when calculating the equivalent environmental reflectance. However, when other pixels are used as the target pixels, a large enough environment cannot be selected to calculate the equivalent environmental reflectance. This means that, for the inversion area, it needs to be at least doubled to meet the demand of reflectance calculation accuracy of each pixel in the $(N + 1) \times (N + 1)$ area.

This problem needs to be viewed in combination with the adjacency effect. For most land surfaces, the effective horizontal range of adjacency effects is approximately 100~500 m [13–15]. Based on this range, the effective influence distance of the adjacency effect is selected as N pixels (the distance is N multiplied by the ground sampling distance (GSD)), which is expanded once, namely, the $(2N + 1) \times (2N + 1)$ area centered on the target pixel is selected in the image to calculate the reflectance of the $(N + 1) \times (N + 1)$ area.

(3) Calculation of the equivalent environmental reflectance

After calculating the reflectance of each pixel according to the above steps, it is still necessary to quantitatively analyze the influence of adjacency effect. As described in Section 2.1, the 6S radiative transfer model proposes to characterize the contribution of the target pixel to the environment by the environment function $F(r)$, which means the sum of the environment weighting factors of each pixel in the surrounding area is relative to the target pixel [11]. The equivalent environment reflectance is the sum of the product of the reflectance of all surrounding pixels and their corresponding weighting factors.

If the selected inversion area is large enough to cover all the environmental pixels affected by the target, the sum of the environmental weighting factors should be 1. However, in fact, this condition is often difficult to achieve, so it is necessary to evaluate the total weight of the region, in order to calculate the equivalent environmental reflectance.

Therefore, in this step, the weighting factor corresponding to each pixel in the region is calculated (Equation (13)) and combined with the environmental reflectance obtained in the previous step to calculate the equivalent environmental reflectance (Equation (15)).

The sum of weighting factors $q(m, n)$ in the $(N + 1) \times (N + 1)$ area is:

$$q(m,n) = \sum_{m=m_0-N/2}^{m_0+N/2} \sum_{n=n_0-N/2}^{n_0+N/2} (\Delta F(r)) \tag{13}$$

where $\Delta F(r)$ is the weighting factor of the pixel (m, n) at r km from the central pixel (m_0, n_0), and its functions are derived from the environmental function $F(r)$ in the 6S radiative transfer model (subroutine ENVIRO):

$$\Delta F(r) = \frac{GSD}{2\pi\sqrt{(m-m_0)^2+(n-n_0)^2}} \cdot \frac{\Delta F^R t_d^R + \Delta F^A t_d^A}{t_d^R + t_d^A} \tag{14}$$

where ΔF^R is the Rayleigh weighting factor, t_d^R is the Rayleigh upward diffuse transmittance, ΔF^A is the aerosol weighting factor, and t_d^A is the aerosol upward diffuse transmittance.

In the 6S radiative transfer model, $F(r)$ is integrated by the weighting factor of the circle at the distance from the target r km. Considering the weighting factor corresponding to a pixel at the distance from the target r km, the weighting factor of this pixel is the product of its proportion in the circle (i.e., $\frac{GSD}{2\pi\sqrt{(m-m_0)^2+(n-n_0)^2}}$) and the circle weight factor (i.e., $\frac{\Delta F^R t_d^R + \Delta F^A t_d^A}{t_d^R + t_d^A}$).

Thus, the equivalent environmental reflectance can be expressed as follows:

$$< \rho(m_0,n_0) >_{eq} = \frac{1}{q(m,n)} \cdot \sum_{m=m_0-N/2}^{m_0+N/2} \sum_{n=n_0-N/2}^{n_0+N/2} (\rho(m,n) \cdot \Delta F(r)) \tag{15}$$

Then, the surface reflectance can be obtained as:

$$\rho(m_0,n_0) = \frac{\frac{\rho*}{T_g} - \rho_a}{T(\theta_s)e^{-\tau/\mu_v}} - \left[\frac{(\frac{\rho*}{T_g} - \rho_a)s}{T(\theta_s)e^{-\tau/\mu_v}} - \frac{t_d(\theta_v)}{e^{-\tau/\mu_v}}\right] < \rho(m_0,n_0) >_{eq} \tag{16}$$

Finally, the reflectance of each pixel can be calculated.

2.3. Radiometric Calibration-Reflectance Inversion Iterative Model

In the above process of reflectance inversion, it is worth noting that the equivalent environmental reflectance is obtained by analyzing the adjacency effect of the inversion region. In the calibration process, this is difficult to achieve, because it is impossible to measure a large area point by point. Based on this, this paper proposes a RCRII model that combines radiometric calibration and reflectance inversion in overall iteration. The iteration steps are as follows:

(1) Radiometric calibration is conducted for the grayscale target area, and initial values of the calibration coefficient and dark current are calculated.

(2) The reflectance inversion model described in the above section is applied to the same image to obtain the surface reflectance.

(3) The equivalent environmental reflectance of the target area is determined with the retrieved surface reflectance and substituted into the calibration equation. The calibration coefficient and dark current are again calculated, and reflectance inversion is again performed. The iteration process is repeated a certain number of times until the relative difference between the calibration coefficient and previous iteration result is less than 1‰. Notably, for $\left|A^{(g)} - A^{(g-1)}\right| \leq 0.001 \cdot A^{(g)}$, the iteration process is terminated. A model flow chart is shown in Figure 2.

In the calibration experiment, when the grayscale target placement site is a uniform site or relatively uniform background, the approximate value of the environmental reflectance can be obtained via site measurement. However, when the experimental area is a complex environment, the environmental reflectance cannot be acquired. Only the image count value near the target area can be used to estimate the environmental reflectance; the diffuse-to-global irradiance ratio can be obtained with the irradiance-based method to calculate the calibration parameters. The calibration inversion iterative model adopts the calculated calibration coefficient as the initial value, combined with multiple iterations of the reflectance inversion model, to increase the calibration accuracy.

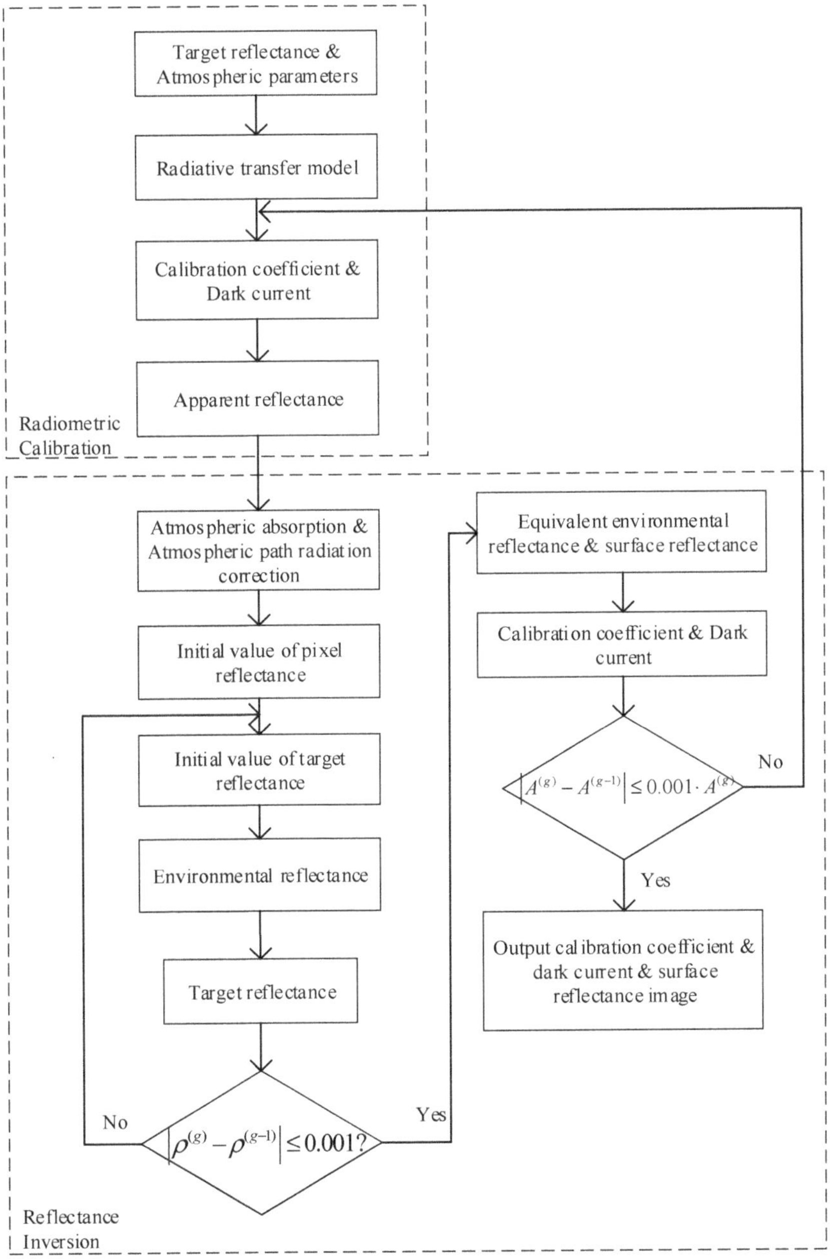

Figure 2. RCRII model flow chart.

3. Experiment and Data Analysis

Pakistan remote sensing satellite (PRSS-1) was successfully launched at China Jiuquan Satellite Launch Center on 9 July 2018. PRSS-1 is equipped with two panchromatic multispectral high-resolution cameras (PMS cameras), with pixel resolutions of 1 m panchromatic and 3 m multispectral, respectively. The camera type is TDICCD push broom type, and the imaging method is linear array push broom imaging, with a single width of about 30 km.

The width of two cameras after splicing is better than 60 km. The design life is 7 years [16]. The spectral response function of the multispectral camera is shown in Figure 3.

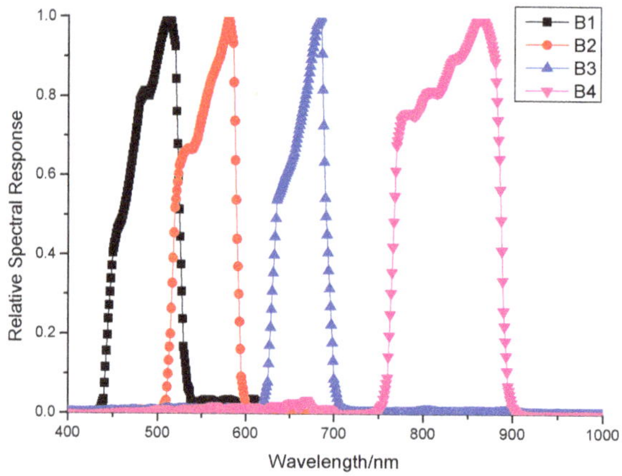

Figure 3. Relative spectral response of PRSS-1 multispectral camera.

3.1. Experimental Data Measurement

The Dunhuang radiometric calibration site is located 15 km west of Dunhuang City, Gansu Province, China. The surface reflectance at this site is stable, and the atmosphere is dry and clean. In August 2018, grayscale targets with reflectance levels of 5%, 20%, 40%, and 60% were placed at the site, and a calibration test was performed in regard to the high-spatial resolution multispectral camera of PRSS-1. The target placement area at the site is greater than 10 × 10 pixels, and an average value of 3 × 3 pixels at the center was chosen for calibration calculation to avoid the influence of the camera modulation transfer function (MTF). Recorded on 18 August 2018, the site test photo, obtained images, and site measurement parameters are as follows (Figures 4–6).

Figure 4. Site test photo.

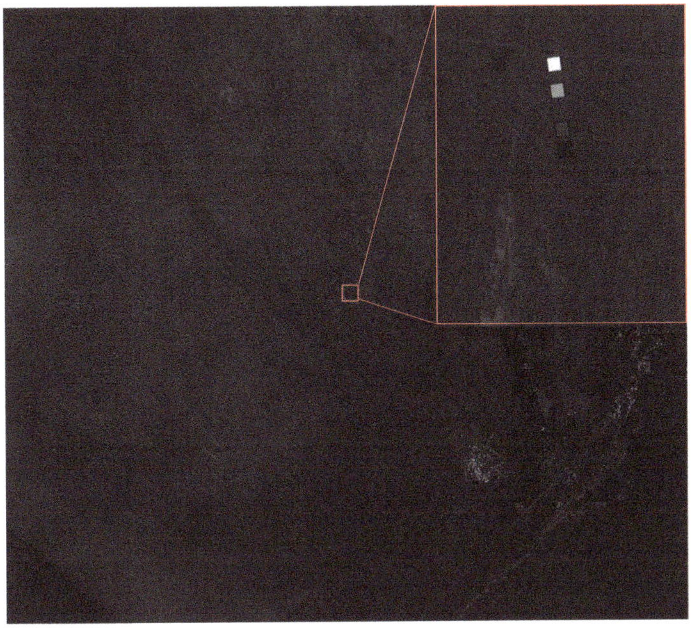

Figure 5. Multispectral image of the target area on 18 August 2018 (band 1).

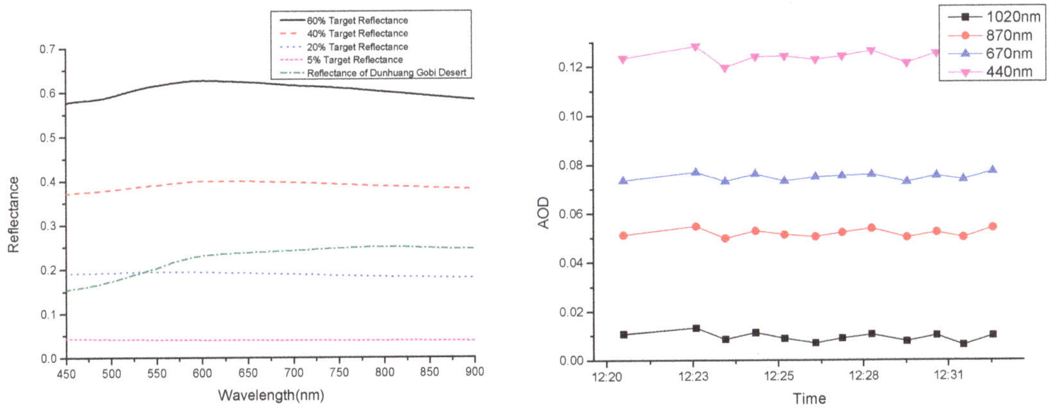

(a) (b)

Figure 6. (**a**) Measured ground reflectance of the multi-grayscale targets on 18 August 2018 (overpass time: 12:28). (**b**) Aerosol optical depth on 18 August 2018.

The acquired image has been relatively calibrated, so the response inconsistency will not be discussed in this paper.

The spectral reflectance of the target and Gobi areas was measured with an ASD spectroradiometer and calibration panel at the time of satellite overpass. With the calibration panel, the bidirectional reflectance factor (BRF) and directional hemispherical reflectance (DHR) were determined in the laboratory, which were used to correct the solar illumination angle of the measured reflectance.

3.2. Calibration Calculation and Reflectance Inversion

Based on the calibration slope and intercept equation mentioned in the previous section, the measured ground reflectance of the emplaced targets was linearly regressed with the corresponding image digital numbers, as shown in Figure 7.

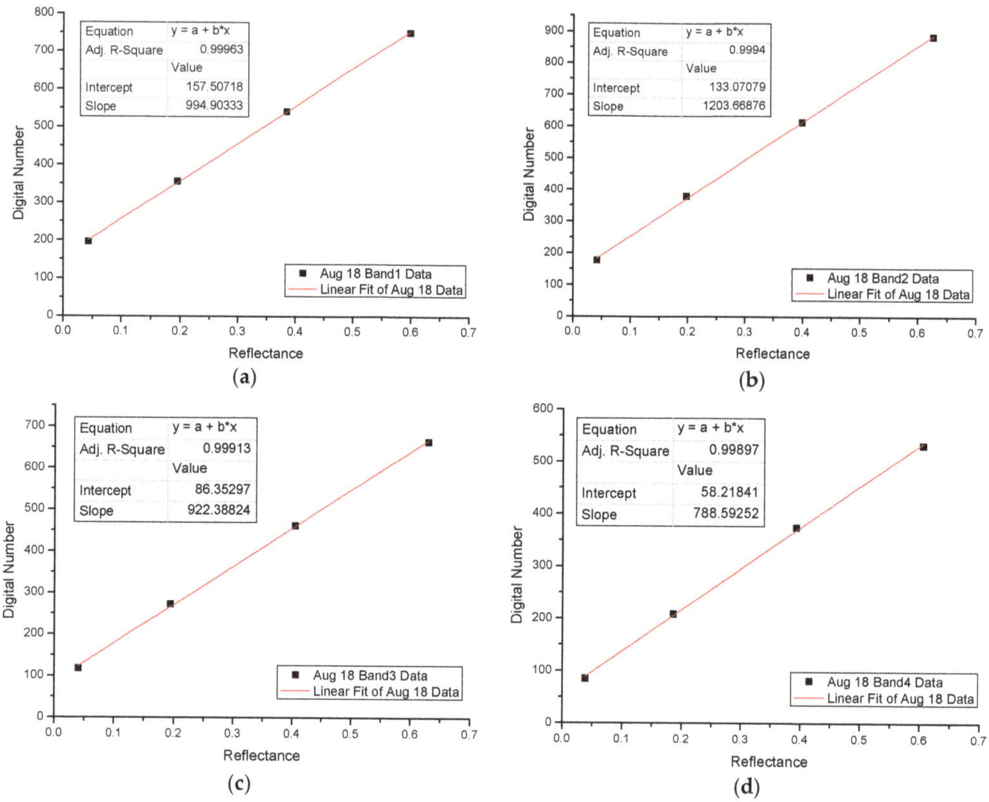

Figure 7. Linear regression curve between the measured ground reflectance of the multi-grayscale targets, and the corresponding image digital numbers on 18 August 2018; (**a**–**d**) show blue, green, red, and NIR bands.

Based on the measured data, the downward direct transmittance calculated with the 6S radiative transfer model was compared to the measured result of the solar photometer, and the transmittance difference remained within 0.015, as shown in Figure 8.

Combined with the above data and Equations (5) and (6), initial values of the calibration coefficient and offset could be obtained. According to Equation (6), the offset in the four bands ranged from 30~50 counts. If this parameter was not deducted and the calibration coefficient directly calculated with the uniform site calculation Equation (2), this may lead to a large difference in the calibration coefficient, which is also related to the low reflectance of the Gobi.

As described in the above section, the RCRII model was used in the test area. Considering the spatial resolution of the camera and influence of adjacency effect, 1023 $\times$ 1023 pixels (corresponding to the $(2N + 1) \times (2N + 1)$ area, mentioned in the previous section, i.e., $N = 511$) area with 60% of the target center as the central pixel is selected. After applying the RCRII model, the calculated equivalent environmental reflectance was substituted into the calibration equation for calculation purposes. After multiple iterations, the final calibration coefficient, offset, and surface reflectance results could be obtained.

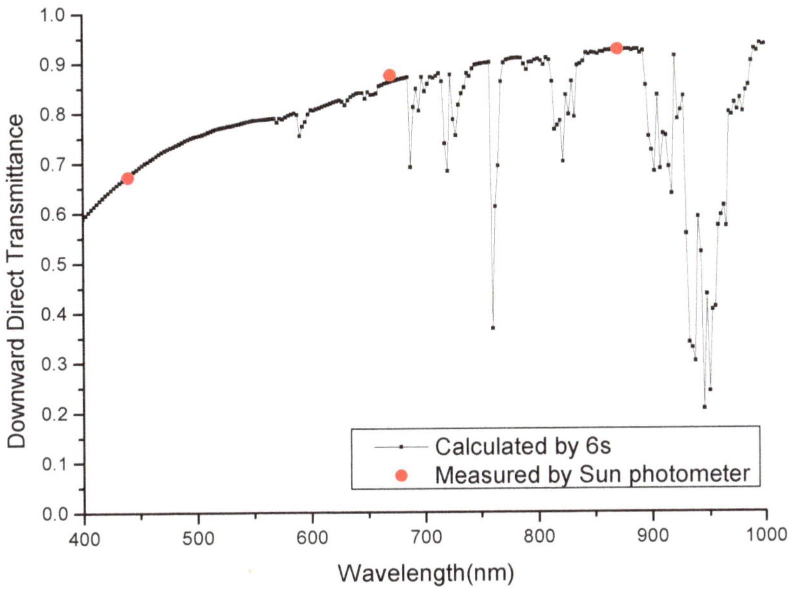

Figure 8. Comparison of transmittance between sun photometer measurement and 6S calculation on 18 August 2018.

The measured reflectance (that is, the reflectance measured with ASD during calibration mentioned in "Experimental Data Measurement" in Section 3.1), reflectance obtained by directly applying the initial value of the calibration coefficient (i.e., the constant calibration method), and reflectance obtained by applying the RCRII model of 60% target are compared, respectively. The results are as follows (Table 1).

Table 1. Comparison of the 60% target reflectance results.

	Measured Reflectance	No Model Used		RCRII Model Used	
		Inversion Reflectance	Absolute Difference	Inversion Reflectance	Absolute Difference
Band 1	0.599	0.591	0.00756	0.597	0.00220
Band 2	0.626	0.619	0.00657	0.623	0.00224
Band 3	0.629	0.623	0.00654	0.626	0.00356
Band 4	0.606	0.598	0.00737	0.600	0.00560

In Table 1, "No modal used" means that the calibration coefficient calculated by linear regression (i.e., the initial value of the calibration coefficient in the model) is directly used in the reflectance inversion model without the radiometric calibration-reflectance inversion iteration.

In summary, the absolute difference between the inversion results and measured data remained within 0.01. Additionally, it is worth emphasizing that the adjacency effect was greater at the short wavelength and smaller at the long wavelength, which is consistent with the inversion results. In band 4, due to the decrease of adjacency effect, there was almost no difference before and after the application of the model.

4. Method Application

In the previous section, by comparing the measured reflectance with the inversion reflectance before and after the application of the model, it can be proven that the calibration coefficient obtained by the application of the model can correct the adjacency effect. The apparent reflectance image can be obtained by applying the calibration coefficient obtained in the previous section to PRSS-1.

In this section, it is planned to cross-verify PRSS-1 and Sentinel-2 satellites, compare the difference of their at-sensor radiance, and compare the difference of ground reflectance through the application of reflectance inversion model, so as to evaluate the calibration accuracy.

Sentinel-2 is an Earth observation mission of the Copernicus Programme that systematically acquires optical imagery at a high spatial resolution over both land and coastal waters [17–19]. To achieve the goal of observation missions and improve the timeliness of observation data, Sentinel-2 comprises the Sentinel-2A and Sentinel-2B satellites. Through networking and collaborative observation, Sentinel-2 can realize high-frequency Earth observation with a revisit cycle of five days. Sentinel-2 provides 13 spectral channels, covering visible near-infrared to shortwave infrared bands. Visible near-infrared mainly includes four bands, namely, B2, B3, B4, and B8, with a spatial resolution of 10 m. The spectral response function is shown in Figure 9.

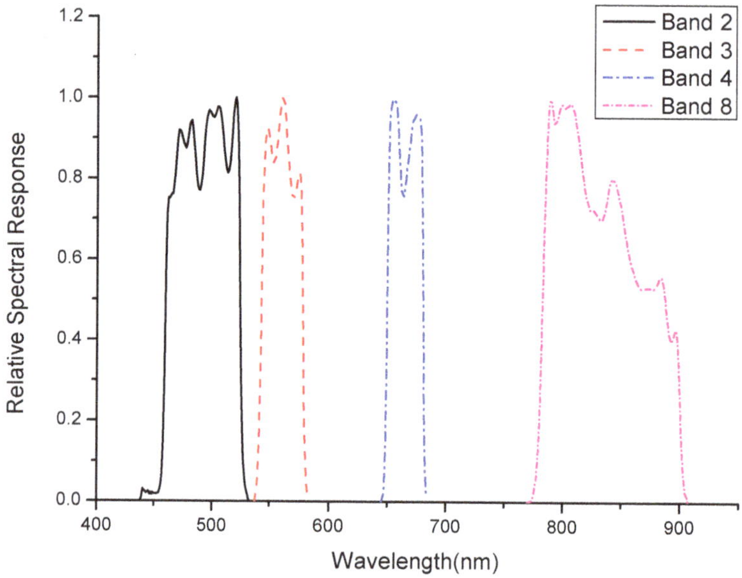

Figure 9. Sentinel-2A relative spectral response.

To verify the reliability of the proposed model, cross-validation and reflectance inversion were conducted with the Sentinel-2A and PRSS-1 image data obtained on 18 August 2018. The imaging time difference between these satellites was less than 10 min, and the weather was good on that day. The aerosol optical depth for 550 nm was 0.0936. The observed zenith angle was lower than 10 degrees.

The central wavelengths of the Sentinel-2A and PRSS-1 visible near-infrared bands are as follows (Table 2).

Table 2. PRSS-1 and Sentinel-2A band center wavelength (nm).

PRSS-1 Band/Sentinel-2A Band	PRSS-1	Sentinel-2A
Band1/Band2	496.165	492.437
Band2/Band3	557.375	559.849
Band3/Band4	664.622	664.622
Band4/Band8	823.127	832.794

4.1. Top-of-Atmosphere Radiance Cross-Validation

The Gobi near the target is selected as the cross-validation object. Since the spectral response functions of the two satellites are inconsistent, it is necessary to calculate the spectral matching factor before cross-validation. The process is as follows:

(1) Using 6S radiative transfer model and field measurement parameters, the at-sensor radiance of the two satellites L_{6S-S2}, $L_{6S-PRSS1}$ is calculated, respectively; then, the spectral matching factor can be obtained as the ratio of the two.

(2) Take the Gobi on the east side of the target as the crossing object; the PRSS-1 calibration coefficient and offset obtained in the above section is used to calculate the at-sensor radiance by $L_{PRSS1} = \frac{\mu_s E_s}{\pi} \frac{DN_{Gobi} - DN_0}{A}$. The at-sensor radiance of Sentinel-2A L_{S2} is directly obtained from the L1C level image.

(3) The at-sensor radiance of PRSS-1 $L_{PRSS1-cross}$ is converted by spectral matching factor and compared with Sentinel-2A.

According to the above process, the results are as follows (Table 3).

Table 3. Cross-validation results of PRSS-1 and Sentinel-2A.

PRSS-1 Band/Sentinel-2A Band	Spectral Matching Factor	$L_{PRSS1-cross}$	L_{S2}	Relative Deviation (%)
Band1/Band2	1.019	101.945	101.040	0.892
Band2/Band3	1.028	103.742	100.241	3.432
Band3/Band4	1.038	97.504	94.385	3.251
Band4/Band8	1.000	67.611	65.427	3.283

The results of cross-validation show that the difference in radiance between PRSS-1 and Sentinel-2A is within 3.5%.

4.2. Reflectance Inversion Verification

It can be seen from Table 2 that the spectrum and bandwidth of these two satellites are inconsistent, but the central wavelengths are similar. Theoretically, it is necessary to reconstruct the reflectance obtained via inversion and then compare the results. However, according to the spectra of the selected three areas (Gobi, grassland, and water areas), the absolute difference in reflectance, due to the inconsistency in spectral response functions between Sentinel-2A and PRSS-1, should occur within 0.005. Therefore, the reflectance inversion results were directly compared here. A water/Gobi area with the Dunhuang Gobi as a uniform background and water/grassland area with Dunhuang City, as a complex environmental background was selected. The selected area images (from PRSS-1) are as follows (Figures 10–13).

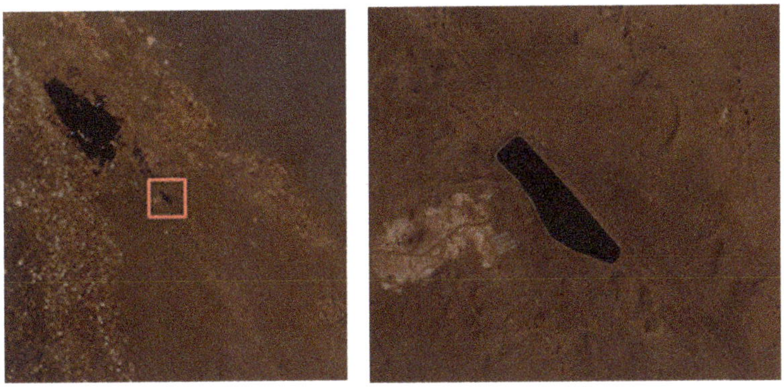

(a) (b)

Figure 10. (**a**) Select Water1 area (in the red box) and its surrounding environment (Gobi background); (**b**) enlarged Water1 area (the area is about (76.48 × 128.06) m^2).

(a) (b)

Figure 11. (**a**) Select Gobi area (in the red box) and its surrounding environment; (**b**) enlarged Gobi area (near the target).

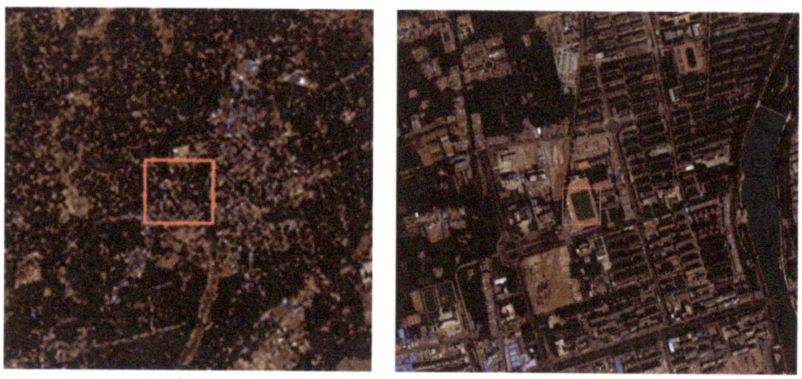

(a) (b)

Figure 12. (**a**) Select Grassland area (in the red box) and its surrounding environment (urban background); (**b**) enlarged Grassland area (the area is about (63.57 × 112.09) m^2).

(a) (b)

Figure 13. (**a**) Select Water2 area (in the red box) and its surrounding environment (urban background); (**b**) enlarged Water2 area.

Using the Sentinel-2 L1C level image, the obtained apparent reflectance was input into the reflectance inversion model proposed in this paper, so as to obtain the ground reflectance results. It was compared with the ground reflectance results of PRSS-1, which was directly applied to the RCRII model. These results are shown in Figures 14 and 15.

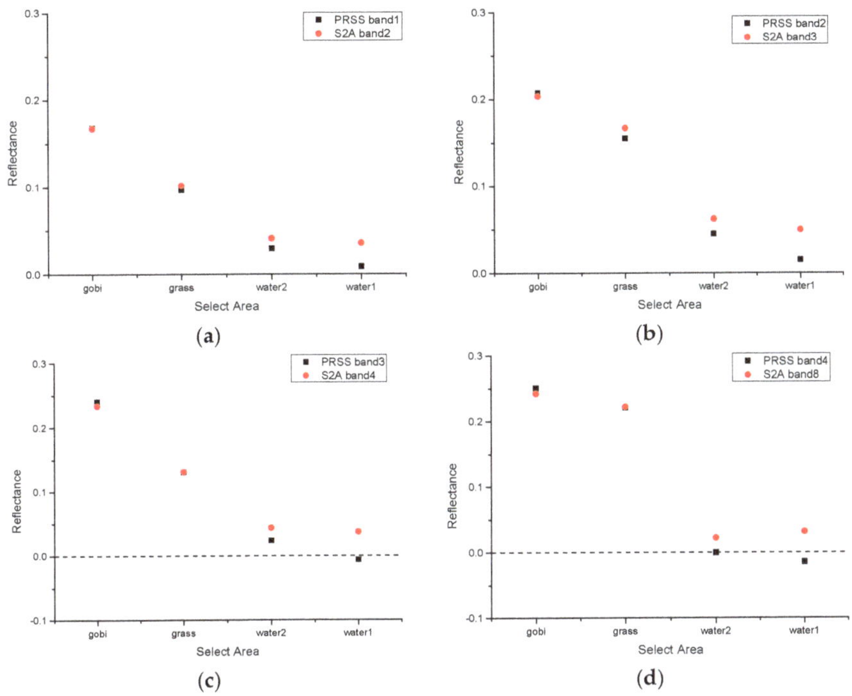

Figure 14. Comparison of ground reflectance in different areas obtained by PRSS-1 using RCRII model and Sentinel-2 using reflectance inversion model; (**a**–**d**) show blue, green, red, and NIR bands.

The inversion results of the Gobi area were compared to the measured values. It can be seen from Figure 16 that the ground Gobi reflectance measured by ASD is compared with the results of PRSS-1 using the RCRII model and reflectance inversion results of Sentinel-2A. The result of PRSS-1 is closer to the measured reflectance. For reflectance inversion, it should be more convincing to take the actual measured reflectance of the site as the standard; however, due to the lack of measured data, only the comparison of Gobi reflectance is given here. Considering that the difference of the inversion results of Sentinel-2A is greater than PRSS-1, it is only for comparison and reference in the follow-up, and quantitative analysis will not be carried out.

According to the above results, the reflectance retrieved by Sentinel-2A and PRSS-1 was close in the Gobi and grassland areas, and the difference remained within 0.01. However, there occurred a large reflectance difference in the two selected water areas. The reflectance retrieved by Sentinel-2A in the water area was obviously high, and a 3~4% reflectance was obtained in the near-infrared band. Inversion of PRSS-1 in the Water1 area yielded a negative value in the near-infrared band. There are many possible reasons for this phenomenon, i.e., the model error of reflectance inversion, estimation error of offset, and so on. Most importantly, too little data makes it difficult to obtain statistical results. Therefore, the specific reason still needs further analysis and the determination of more data.

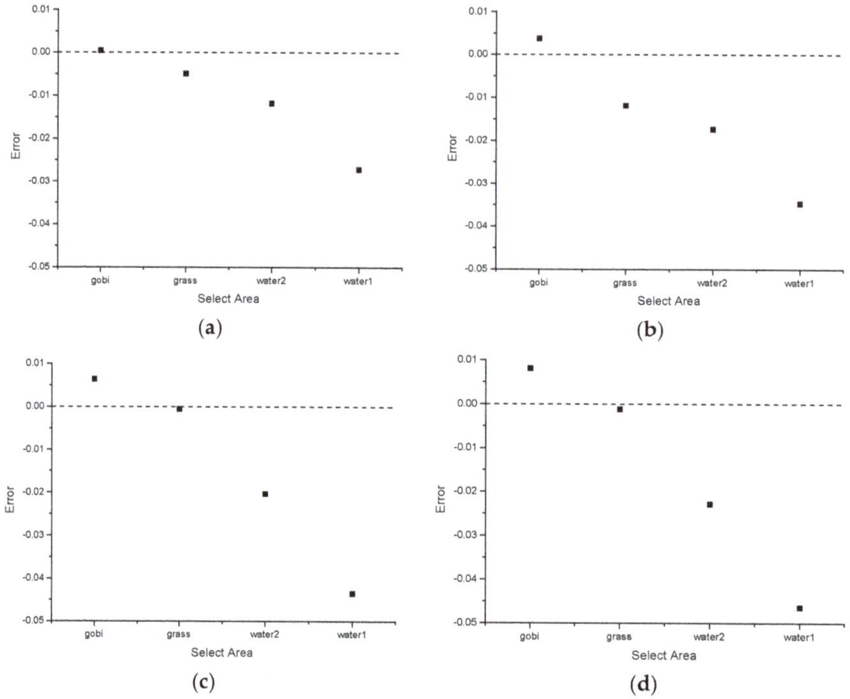

Figure 15. Comparison of absolute difference of ground reflectance in different areas obtained by PRSS-1 using RCRII model and Sentinel-2 using reflectance inversion model; (**a**–**d**) show blue, green, red, and NIR bands.

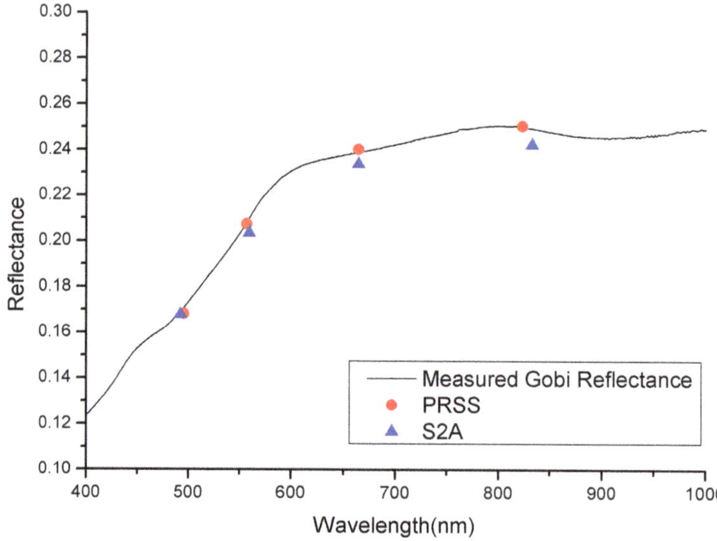

Figure 16. Comparison between the Gobi ground reflectance obtained by PRSS-1 using RCRII model and Sentinel-2 using the reflectance inversion model and measured Gobi reflectance by ASD.

5. Discussion

The RCRII model is mainly composed of two parts: radiometric calibration and reflectance inversion. For the radiometric calibration method, the grayscale target calibration method is selected because the camera may have offset, which will lead to large errors in the calibration of uniform site with low reflectance. The linear regression of multiple reflectance targets can separate the target radiation from the non-target radiation (i.e., atmospheric path and environmental radiation), calculate the calibration coefficient with the target radiation, and calculate the offset with the non-target radiation. However, the multi-grayscale target calibration method needs to consider the influence of adjacency effect. The actual measurement obviously cannot meet the requirements of adjacency effect range, and it is difficult to measure point by point to quantify its influence. Therefore, on this basis, the reflectance inversion model is introduced to form the radiometric calibration-reflectance inversion iterative model in this paper. In addition, there are still several problems to discuss.

In the third section, the selected area is centered on the 60% target, and the inversion results of the 60% target are also discussed. However, for the multi-grayscale target method, the most appropriate method should be to evaluate the combined errors of four targets (60%, 40%, 20%, and 5%). The reason this is not achieved is due to the adjacency effect. The influence of adjacency effect on target method is reflected in the variable of equivalent environmental reflectance, which can be divided into two parts: the influence on the linearity and nonlinearity of the target. This can be seen in Equations (5) and (6) in Section 2. When we use the linear regression method to realize the target calibration, the influence of adjacency effect on nonlinearity has been ignored. However, in fact, for targets with different reflectance, the equivalent environmental reflectance is a variable. Back to the original problem, in the application of the model centered on the 60% target, the evaluation of the equivalent environmental reflectance is completely determined by the 60% target. This assumption may lead to unexpected errors in other reflectance targets. Fortunately, when the aerosol optical thickness is small, the error caused by this factor can be ignored (this can be seen from the inversion error, but the difference can still be observed in the model). However, when the aerosol optical thickness is large and adjacency effect is more obvious, this error is difficult to ignore. Therefore, the model may need to improve this part. However, it should be emphasized that, even if such errors exist in the model, it can still effectively eliminate the adjacency effect.

Another problem is that, in the example provided in this paper, the target was deployed under the background of a uniform site. In fact, in the principles described in Section 2, there are no special requirements for the environment. In other words, this model should be able to be implemented, even in complex environments. The difficulty of complex environment lies in how to obtain the initial value of environmental reflectance to calculate the initial value of calibration coefficient. This problem can be substituted into the model by calculating the initial value of reflectance by selecting the mean value of the area or obtaining the ratio of diffuse-to-global by irradiance-based method. Unfortunately, due to the lack of data, it is difficult to prove this point at present. Try to substitute 0.5 and 0.01 as the initial values of environmental reflectance into the model, and still get good inversion results; however, this fact is still difficult to prove its universality in complex environments.

In regard to the grayscale target calibration method, the uncertainty mainly originates from two sources: (1) the target reflectance-image digital number fitting error, which can be divided into the nonlinear error of the camera sensor, reflectance measurement uncertainty, and uncertainty due to the inconsistent influence of the adjacency effect attributed to varying target reflectance values; (2) the error in the radiative transfer model during atmospheric parameter calculation.

Therefore, in addition to the above two factors and the measurement error of the instrument, the influencing factors can be summarized as the calculation of the total ground irradiance, measurement of the bidirectional reflection distribution function (BRDF) of the target, calculation of the upward atmospheric transmittance, and adjacency effect.

The uncertainty of the total ground irradiance mainly comes from the radiative transfer model, and the uncertainty of the radiative transfer model mainly comes from the assumptions of the aerosol model and calculation of the solar irradiance. Therefore, it is considered that the uncertainty is less than 3%. The grayscale target has the characteristics of constant spectrum and near-Lambertian. Therefore, it is considered that the BRDF measurement uncertainty is less than 2%. The uncertainty of upward atmospheric transmittance mainly comes from aerosol scattering. The parameters in radiation transmission can be calculated by using the Langley method to calculate the atmospheric optical thickness measured by the sun photometer. The uncertainty of atmospheric optical thickness measurement mainly comes from the calibration of the sun photometer, and the measurement uncertainty is less than 2%.

To sum up, compared with the conventional calibration methods, we expect to use the RCRII model to calculate sensor offset, improve the reflectance measurement accuracy, and quantitatively analyze the impact of adjacency effect, so as to improve calibration accuracy. A preliminary assessment of the uncertainty of this method is made, as shown in Table 4. This part will be further analyzed later.

Table 4. Uncertainty analysis for the vicarious calibration using RCRII model.

Uncertainty Factors	Relative Uncertainty (%)
Calculation of total ground irradiance	3.0
Target BRDF measurement	2.0
Calculation of upward transmittance	2.0
Adjacency effect calculation	1.0
Others (geometric factors, etc.)	1.0
Comprehensive uncertainty	4.4

6. Conclusions

Radiometric calibration and reflectance inversion are usually independent processes. Generally, calibration is followed by reflectance inversion. In this paper, a general iterative model for radiometric calibration-reflectance inversion, based on the 6S radiative transfer method, was proposed. The two processes of radiometric calibration and reflectance inversion were combined, and calibration and inversion of the same image were repeated in an iterative process to improve the accuracy. The equivalent environmental reflectance calculated with the inversion model was applied in calibration, and the calibration results were again used for inversion. Via a comparison to the measured values in the calibration experiment, calibration and inversion integration was achieved, and quantitative remote sensing was preliminarily realized. With the use of the proposed model against the background of the Dunhuang uniform site, the results indicated that the difference between the reflectance obtained with the established method and measured reflectance was less than 0.01. By comparing with Sentinel-2a image, it is cross-verified that the radiance difference was within 4%, and the difference between the inversion value and measured value remained below 0.01 in the range of 0.1~0.3 reflectance, and the reason for the large difference at a low reflectance was analyzed.

In fact, this model is more inclined to be applied in complex environments or situations with large adjacency effects. However, due to the lack of data, it is difficult to prove this now. The follow-up work will focus on this part.

Author Contributions: S.B. was responsible for the data analysis and writing the manuscript. H.C. led the experiment and participated in data processing. L.Z. contributed to the main research ideas. W.H., X.S. and X.W. (Xianhua Wang) reviewed the manuscript and provided valuable suggestions. Y.L., Z.F., Y.C., X.W. (Xinrong Wang) and X.Z. provided image data and valuable suggestions. All authors have read and agreed to the published version of the manuscript.

Funding: This research received no external funding.

Data Availability Statement: Not applicable.

Acknowledgments: The Sentinel image data were downloaded from the ESA website (https://scihub.copernicus.eu/dhus/#/home, accessed on 8 September 2021).

Conflicts of Interest: The authors declare no conflict of interest.

References

1. Yoon, H.; Kacker, R. *Guidelines for Radiometric Calibration of Electro-Optical Instruments for Remote Sensing: Handbook (NIST HB)*; National Institute of Standards and Technology: Gaithersburg, MD, USA, 2015.
2. Tan, K.; Wang, X.; Niu, C.; Wang, F.; Du, P.; Sun, D.X.; Yuan, J.; Zhang, J. Vicarious Calibration for the AHSI Instrument of Gaofen-5 With Reference to the CRCS Dunhuang Test Site. *IEEE Trans. Geosci. Remote Sens.* **2020**, *59*, 3409–3419. [CrossRef]
3. Chander, G.; Markham, B.L.; Barsi, J.A. Revised Landsat-5 Thematic Mapper Radiometric Calibration. *IEEE Geosci. Remote Sens. Lett.* **2007**, *4*, 490–494. [CrossRef]
4. Ma, L.; Zhao, Y.; Woolliams, E.R.; Dai, C.; Wang, N.; Liu, Y.; Li, L.; Wang, X.; Gao, C.; Li, C.; et al. Uncertainty Analysis for RadCalNet Instrumented Test Sites Using the Baotou Sites BTCN and BSCN as Examples. *Remote Sens.* **2020**, *12*, 1696. [CrossRef]
5. Bouvet, M.; Thome, K.; Berthelot, B.; Bialek, A.; Czapla-Myers, J.; Fox, N.P.; Goryl, P.; Henry, P.; Ma, L.; Marcq, S.; et al. RadCalNet: A Radiometric Calibration Network for Earth Observing Imagers Operating in the Visible to Shortwave Infrared Spectral Range. *Remote Sens.* **2019**, *11*, 2401. [CrossRef]
6. Pinto, C.T.; Xin, J.; Leigh, L. Evaluation Analysis of Landsat Level-1 and Level2 Data Products Using In Situ Measurements. *Remote Sens.* **2020**, *12*, 2597. [CrossRef]
7. Yeom, J.-M.; Hwang, J.; Jung, J.-H.; Lee, K.-H.; Lee, C.-S. Initial Radiometric Characteristics of KOMPSAT-3A Multispectral Imagery Using the 6S Radiative Transfer Model, Well-Known Radiometric Tarps, and MFRSR Measurements. *Remote Sens.* **2017**, *9*, 130. [CrossRef]
8. Slater, P.N. Radiometric considerations in remote sensing. *Proc. IEEE* **1985**, *73*, 997–1011. [CrossRef]
9. Sun, B.; Schäfer, M.; Ehrlich, A.; Jäkel, E.; Wendisch, M. Influence of atmospheric adjacency effect on top-of-atmosphere radiances and its correction in the retrieval of Lambertian surface reflectivity based on three-dimensional radiative transfer. *Remote Sens. Environ.* **2021**, *263*, 112543. [CrossRef]
10. Ma, L.; Wang, N.; Liu, Y.; Zhao, Y.; Han, Q.; Wang, X.; Woolliams, E.R.; Bouvet, M.; Gao, C.; Li, C.; et al. An In-Flight Radiometric Calibration Method Considering Adjacency Effects for High-Resolution Optical Sensors Over Artificial Targets. *IEEE Trans. Geosci. Remote Sens.* **2021**, *60*, 1–13. [CrossRef]
11. Vermote, E.F.; Tanré, D.; Deuze, J.L.; Herman, M.; Morcette, J.J. Second simulation of a satellite signal in the solar spectrum-vector (6SV). In *6S User Guide Version*; IEEE: Piscataway, NJ, USA, 2006; Volume 3, pp. 1–55.
12. Slater, P.N.; Biggar, S.F.; Thome, K.J.; Gellman, D.I.; Spyak, P.R. Vicarious radiometric calibrations of EOS sensors. *J. Atmos. Ocean. Technol.* **1996**, *13*, 349–359. [CrossRef]
13. Wang, T.; Du, L.; Yi, W.; Hong, J.; Zhang, L.; Zheng, J.; Li, C.; Ma, X.; Zhang, D.; Fang, W.; et al. An adaptive atmospheric correction algorithm for the effective adjacency effect correction of submeter-scale spatial resolution optical satellite images: Application to a WorldView-3 panchromatic image—ScienceDirect. *Remote Sens. Environ.* **2021**, *259*, 112412. [CrossRef]
14. Richter, R. A fast atmospheric correction algorithm applied to Landsat TM images. *Int. J. Remote Sens.* **1990**, *11*, 159–166. [CrossRef]
15. Kaufman, Y.J. Solution of the equation of radiative transfer for remote sensing over nonuniform surface reflectivity. *J. Geophys. Res.* **1982**, *87*, 4137–4147. [CrossRef]
16. Yan, L.; Hongyao, C.; Zhou, F.; Longfei, L.; Yuanwei, C.; Yongli, H.; Hongqiang, W.; Song, W. On-orbit radiometric calibration and accuracy analysis in dual camera of PRSS-1 Satellite. *Spacecr. Recovery Remote Sens.* **2019**, *40*, 77–88.
17. Gascon, F.; Bouzinac, C.; Thépaut, O.; Jung, M.; Francesconi, B.; Louis, J.; Lonjou, V.; Lafrance, B.; Massera, S.; Gaudel-Vacaresse, A.; et al. Copernicus Sentinel-2 Calibration and Products Validation Status. *Remote Sens.* **2016**, *9*, 584. [CrossRef]
18. Lonjou, V.; Lachérade, S.; Fougnie, B.; Gamet, P.; Marcq, S.; Raynaud, J.L.; Tremas, T. Sentinel-2/MSI absolute calibration: First results. In Proceedings of the SPIE Remote Sensing, Toulouse, France, 21–24 September 2015; International Society for Optics and Photonics: Bellingham, WA, USA, 2015.
19. Angal, A.; Xiong, X.; Shrestha, A. Cross-Calibration of MODIS Reflective Solar Bands with Sentinel 2A/2B MSI Instruments. *IEEE Trans. Geosci. Remote Sens.* **2020**, *58*, 5000–5007. [CrossRef]

remote sensing

Article

Vicarious CAL/VAL Approach for Orbital Hyperspectral Sensors Using Multiple Sites

Daniela Heller Pearlshtien [1,*], Stefano Pignatti [2] and Eyal Ben-Dor [1]

1 Remote Sensing Laboratory, Geography Department, Porter School of the Environment and Earth Sciences, Faculty of Exact Sciences, Tel Aviv University, Tel Aviv 699780, Israel
2 National Research Council (CNR), Institute of Methodologies for Environmental Analysis (IMAA), 85050 Potenza, Italy
* Correspondence: Dh@mail.tau.ac.il; Tel.: +972-545229739

Abstract: The hyperspectral (HSR) sensors Earth Surface Mineral Dust Source Investigation (EMIT) of the National Aeronautics and Space Administration (NASA) and Environmental Mapping and Analysis Program (EnMAP) of the German Aerospace Center (DLR) were recently launched. These state-of-the-art sensors have joined the already operational HSR sensors DESIS (DLR), PRISMA (Italian Space Agency), and HISUI (developed by the Japanese Ministry of Economy, Trade, and Industry METI and Japan Aerospace Exploration Agency JAXA). The launching of more HSR sensors is being planned for the near future (e.g., SBG of NASA, and CHIME of the European Space Agency), and the challenge of monitoring and maintaining their calibration accuracy is becoming more relevant. We proposed two test sites: Amiaz Plain (AP) and Makhtesh Ramon (MR) for spectral, radiometric, and geometric calibration/validation (CAL/VAL). The sites are situated in the arid environment of southern Israel and are in the same overpass coverage. Both test sites have already demonstrated favorable results in assessing an HSR sensor's performance and were chosen to participate in the EMIT and EnMAP validation stage. We first evaluated the feasibility of using AP and MR as CAL/VAL test sites with extensive datasets and sensors, such as the multispectral sensor Landsat (Landsat5 TM and Landsat8 OLI), the airborne HSR sensor AisaFENIX 1K, and the spaceborne HSR sensors DESIS and PRISMA. Field measurements were taken over time. The suggested methodology integrates reflectance and radiometric CAL/VAL test sites into one operational protocol. The method can highlight degradation in the spectral domain early on, help maintain quantitative applications, adjust the sensor's radiometric calibration during its mission lifetime, and minimize uncertainties of calibration parameters. A PRISMA sensor case study demonstrates the complete operational protocol, i.e., performance evaluation, quality assessment, and cross-calibration between HSR sensors. These CAL/VAL sites are ready to serve as operational sites for other HSR sensors.

Keywords: vicarious calibration; validation; cross-calibration

1. Introduction

Sensors, in general, and hyperspectral (HSR) sensors in particular, require periodic radiometric and spectral calibration. Both the calibration and validation of orbital sensors are critical processes in keeping their data reliable for thematic mapping during their operational lifetime, which is especially important in the challenging space environment [1]. Moreover, both the spectral calibration and the sensor's optics can degrade over time, affecting the sensor's spectral profile [2]. Thus, monitoring the spectral performance and radiometric stability periodically throughout the sensor's lifetime is important to ensure accurate products [3].

With the recent launch of HSR sensors such as the Earth Surface Mineral Dust Source Investigation (EMIT) of the National Aeronautics and Space Administration (NASA) [4] and the Environmental Mapping and Analysis Program (EnMAP) of the German Aerospace

Center (DLR) [5] into space, and the increasing number of HSR sensors already in orbit, calibration of the following advanced sensors is vital for the high spectral resolution that they provide: DLR Earth Sensing Imaging Spectrometer (DESIS); PRecursore IperSpettrale della Missione Applicativa (PRISMA) of the Italian Space Agency (ASI); the spaceborne hyperspectral earth imaging system (HISUI) developed by the Japanese Ministry of Economy, Trade and Industry and installed on the International Space Station (ISS) with the collaboration of the Japan Aerospace Exploration Agency (JAXA) module; India's Hyper Spectral Imager (HySIS); and China's Advanced Hyperspectral Imager (AHSI).

Technological advances have been made in the last decade, and the number of spectral bands and spectral ranges has increased to cover most of the sun's radiation. Yet, separating the signals from noise to retrieve reliable physical data remains a complex procedure and requires the maintenance of an accurate sensor and post-calibration process, which is an ongoing challenge [6]. Aside from onboard calibration routines (such as partial-aperture calibrators, standard lamps, and solar diffuser panels [7]), vicarious calibration (VC) practices are important for examining and rectifying the sensor's radiometric and spectral response, and checking its geometric performance [8]. VC comprises techniques that use natural or artificial sites for the post-launch calibration of sensors. It is a significant part of any satellite's lifetime, and space agencies and groups worldwide (e.g., the IEEE P4001, the GEO, CAL/VAL Working Groups of Surface Biology and Geology (SBG) [9], and the RadCalNet [3,10]) are investing much effort in the calibration/validation (CAL/VAL) mission.

There are a few methods to apply VC using lunar views and bright stars, e.g., [11–14]. However, such targets are relatively low radiation sources [15], and they require high-risk spacecraft-platform maneuvers [7]. Other VC methods use non-land earth targets, including atmospheric Rayleigh scattering and gas absorption [16–19], and ocean sunglint [20]. However, the most common way to carry out VC is to use natural ground targets on earth, such as a uniform bright playa [21], and snow and ice fields [22–24]. The VC is applied over a well-known ground site that is stable in space and time. The measurement is conducted during or close to the sensor's overpass time, estimating the top of the atmosphere (TOA) radiance at-sensor.

Aside from examining, and, in some cases, also rectifying a given sensor's radiometric performance, another important practice at VC sites is the cross-calibration of two sensors. Cross-calibration uses data from a well-calibrated sensor to harmonize the information of a less-well-calibrated sensor [25–27]. Cross-calibration is performed on a stable test site (spatially, temporally, and spectrally) that is captured by the sensors at near-simultaneous times [21]. The test sites are referred to as pseudo-invariant calibration test sites (PICS). The main constraint of the cross-calibration approach is to find pairs of scenes for the calibration between two sensors. This is not easy with multispectral sensors, and it is extremely difficult with HSR sensors. Each HSR sensor has its own revisit cycle, and both need to capture the test site without cloud coverage at approximately the same time (usually, up to 3 days apart [21], but for very stable targets, up to 6 days are considered [28]).

There are specific criteria to establish an area as a reliable CAL/VAL site. Radiometric test sites must be uniform in terms of the spectral signal with nearly Lambertian reflectance of the bright surface, have temporal stability over a large area, and be arid regions with a low probability of clouds, scarce vegetation coverage, and a high spectral reflectance (>0.3 for all bands) [24,29]. Spectral and thematic test sites need to have large pure targets with significant spectral signatures across the optical range, which are temporally stable and monitored with frequent ground measurements. There are several radiometric calibration sites worldwide that are recognized by the Committee on Earth Observation Satellites (CEOS)/Working Group on Calibration & Validation (WGCV), and the Infrared and Visible Optical Sensors subgroup (IVOS); six sites are located in the Sahara from Cosnefroy's pseudo-invariant calibration sites' list [30,31]. The four "RadlCalNet" radiometric calibration test sites equipped with automated ground instrumentation that provides continuous measurements of both surface reflectance and local atmospheric conditions for the derivation of TOA reflectance are located in the USA, France, China, and Namibia [3].

Twenty-six years ago, Cosnefroy et al. [31] identified 20 desert sites in North Africa and Saudi Arabia as the PICS for cross-calibration. In 2019, Bacour et al. [30] showed that they are still relevant. However, the six Saharan sites (Algeria 3, Algeria 5, Libya 1, Libya 4, Mauritania 1, and Mauritania 2) have one crucial disadvantage: they are logistically difficult to reach, hindering ground-truth characterization [30].

Satellite images undergo a geometric correction process to connect the image to the ground coordinates. There are various empirical and physically based approaches, and each satellite platform implements its preferred method [32]. Geolocation accuracy affects the quality of the final thematic product; as such, inspection and reporting on the spatial accuracy are strongly needed.

One way to rectify distorted images is to measure ground control points (GCPs) in the field from well-defined targets or high-resolution satellite images [33]. A few geometric calibrations fields are dispersed worldwide, such as the Tsukuba test field in Japan, the Finnish Geodetic Institute test field in Sjokulla, La Crau in France, and more. As with the satellite platforms, each space organization uses its own preferred method. Zheng et al. [34] suggested standardizing the geometric validation process and creating a unified test site for all agencies. To that end, they evaluated the geometric accuracy of the SPOT-6, Pleiades, ALOS, ZY-3, and TH sensors on the Xianning test field in China.

CAL/VAL test sites satisfying all of the calibration and inspection requirements are highly needed, especially for the high-quality HSR sensors. We, therefore, proposed two test sites that fulfill the CAL/VAL criteria to enable spectral, radiometric, geometric, and thematic evaluations. The Amiaz Plain (AP) is a homogeneous bright-surface playa (5×5 km) with high reflectance values (>0.4). Because vegetation is scarce and it is stable in space and time, it is proposed for sensors' radiometric evaluation and cross-calibration (31°04′20.98″N, 35°22′14.83″E). Makhtesh Ramon (MR) is a unique site that is known for its geological formations, and its abundance of many different minerals with distinct spectral features across the 400–2500 nm region (30°36′35.23″N, 34°51′15.22″E). This area is perfect for spectral and thematic validation. The test sites are close to each other (50 km) and very easy to access using well-built roads for routine field measurements. They are situated in the Israeli desert, with a dry climate. Since they are both nature reserves, the landscape is undisturbed and preserved by park rangers. Furthermore, NASA's AErosol RObotic NETwork (AERONET) [35] station is located in"SEDE-BOKER", 20 km from MR and about 45 km from AP.

In this study, we evaluated the characteristics of both test sites using recent HSR data from airborne [36,37] and orbital sensors, accompanied by comprehensive ground-truth and field studies, toward the development of a protocol to monitor the quality assurance (QA) of the HSR data and their related processing chain from space.

2. Materials and Methods

2.1. Test Sites AP and MR

Two areas were selected for the VC sites: AP and MR, located in the south of Israel (Figure 1a). AP is part of the Judean Desert Nature Reserve, and is located within a down-faulted block of the western segment of the Dead Sea fault system that bounds it from the west, and Mount Sodom bounds it from the east [38,39]. AP consists of a homogeneous silty carbonate plain with an elevation of about ~260 m below mean sea level. MR is an anticline formation with an eroded central valley occupying about 200 km^2. The surface valley elevation is about ~500 m above mean sea level, with steep walls that bound it. It is a natural geological park consisting of different geological formations with vast mineralogical exposures, such as dolomite, calcite, kaolinite, hematite, gypsum, montmorillonite, and more. Vegetation is scarce, and it has been widely documented in many geological and ecological studies.

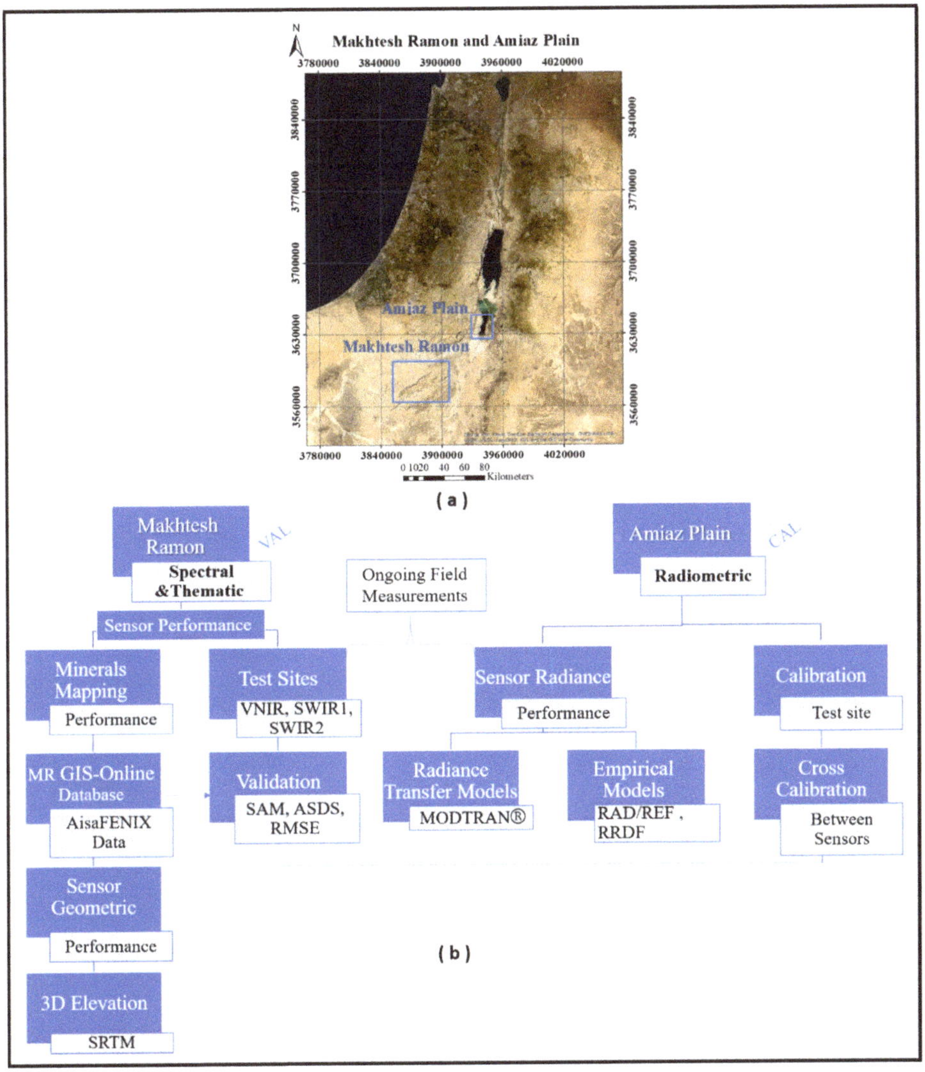

Figure 1. Amiaz Plain (AP) and Makhtesh Ramon (MR) calibration/validation (CAL/VAL) test sites. (**a**) AP and MR locations in Israel. (**b**) Work protocol flowchart.

2.2. Stability of the Test Sites

The most critical CAL/VAL site parameters are spatial and spectral stability over time. To evaluate these parameters for AP and MR, we gathered time-series analyses of Landsat images (Landsat-5 TM (TM5) and Landsat-8 OLI (OLI8)) over 20 and 25 years for AP and MR, respectively. We checked that no rain events had taken place at least 4 weeks before the overpass (this criterion was easily fulfilled due to the test sites' arid climate). For a fair spectral comparison between the 2 different sensors (TM5 and OLI8), only 6 optical bands were used in the analysis (TM5—bands 1–5, 7 and OLI8—bands 2–7). All images were acquired during the summer season under dry conditions and clear skies. The spectral angle mapper (SAM; Equation (2)) [40], an algorithm that determines the spectral similarity between two spectra by calculating the angle between the spectra and treating them as vectors in a space with dimensionality equal to the number of bands; root mean square error (RMSE; Equation (3)); and the average sum of deviations squared (ASDS; Equation (4)) [41] were calculated between the t(0) and t(n) images, where n is the number of years after t(0). Lower values indicate higher spectral similarity.

$$Rr_n = \frac{Rt_n}{Rtr_n}, \quad (1)$$

$$SAM = cos^{-1}\left(\frac{\sum_{i=1}^{n} Rt_{ni}Rtr_{ni}}{\sqrt{\sum_{i=1}^{n} Rt_{ni}{}^2}\sqrt{\sum_{i=1}^{n} Rtr_{ni}{}^2}}\right), \quad (2)$$

$$RMSE = \sqrt{\frac{\sum_{i=1}^{n}(Rtr_{ni} - Rt_{ni})^2}{n}}, \quad (3)$$

$$ASDS = \frac{\sum_{i=1}^{n}(Rr_{ni} - 1)^2}{n}, \quad (4)$$

where Rr_n represents the ratio between the examined and reference spectra, Rt_n is the examined reflectance spectrum, Rtr_n is the reference reflectance spectrum, and n is the number of wavelengths used.

For the spatial examination, we used the "Change Detection Difference Map" method in the ENVI 5.6 program to produce a change-classification image. The difference was computed by subtracting the initial-state image from the final-state image, and the red band (0.66 μm) was selected for the analysis. For AP, t(0) = the year 2000 and t(n) = 2021, and for MR, t(0) = 1996, and t(n) = 2021 (see Section 3.1).

2.3. CAL/VAL Protocol Flowchart

The AP test site was selected to evaluate the sensor's radiance performance and conduct the VC/cross-calibration. The sensor's radiance calibration was evaluated by observing the wavelengths of atmospheric water vapor and trace gases at the sensor's TOA radiance signal. The main absorbances for water vapor (H_2O) are: 720, 820, 940, 1150, 1380, and 1870 nm; for oxygen (O_2): 680 and 760 nm; for ozone (O_3): 550 and 650 nm; for carbon dioxide (CO_2): 2005 and 2060 nm; and for methane: 1660 nm and 2300 nm [30,42,43]. In this study, the simulated TOA radiance transfer signal was obtained using the MODTRAN® version 4 code [42,43], surface reflectance, and the precise atmospheric conditions measured at the AERONET station in Sede Boker. The sensor's TOA signal was compared to the TOA radiance transfer model. Empirical indices—at-sensor radiance-to-reflectance ratio (RAD/REF) and radiance-to-reflectance difference factor (RRDF) [44]—can also be operated as QA indicators for radiance quality (see elaboration in our previous work [36]). We also used this test site for cross-calibration between HSR sensors DESIS, PRISMA, and AisaFENIX (see Sections 3.2.3 and 3.2.4).

MR was used for the spectral QA, validation of atmospheric correction, cross-calibration, and geometric accuracy. The validation process was conducted on MR's online database obtained from comprehensive field measurements and precise airborne HSR AisaFENIX flight products. It included thematic mineral maps of kaolinite, bentonite, hematite, goethite, calcite, and gypsum, and auxiliary MR data, such as geology/geomorphology maps of the area, slope, aspect maps, and a 3D elevation model (created from the SRTM 30 m resolution image of NASA–JPL at a resolution of 1 arc [45]). All of the thematic information, including updated field measurements, is uploaded into the cloud to be utilized by interested parties. Examples of using this database to examine the mapping performance of HSR sensors DESIS and PRISMA are demonstrated in our previous work [36,37]. The MR database is available at the following link: https://storymaps.arcgis.com/stories/bb5bf09ec7414454a012bfe9bf4b8545 (accessed on 11 October 2022) [46].

The main requirement for the reflectance validation test site is that the spectral response be stable in space and time. Accordingly, 6 homogeneous test sites with unique spectral features across the sensor's optical spectral range were used to evaluate the sensor's spectral calibration and to validate the sensor's radiometric re-calibration. The test sites encompassed targets with different spectral ranges: in the visible–near infrared (VNIR) spectral range of 400–1000 nm (Test Sites 1 and 2): 1. Brown questa, a sandstone enriched with iron oxides such as hematite and goethite, with main spectral features at 420–600 nm and 900–1000 nm (30°37′14.26″N, 34°50′32.13″E); 2. Laccolite mineral over a gypsum deposit, a gabbro mineral of plutonic rock, and the laccolite is a very dark target with small absorbance between 800 and 900 nm, a low albedo value (<0.20), and a mostly monotonous signal from 1000 to 2500 nm (30°36′12.23″N, 34°53′42.80″E); in the shortwave infrared (SWIR) 1 spectral range of 1450–1800 nm (Test Sites 3 and 4): 3. Gypsum old mine with sulfate mineral deposits, and a distinctive absorbance of gypsum in that range (30°35′42.24″N, 34°52′21.13″E,); 4. Soil fans, also rich in gypsum (30°36′7.45″N, 34°53′37.08″E); for the SWIR2 spectral range of 2000–2500 nm (Test Sites 5 and 6): 5. Kaolinite old mine (clay mineral), main spectral feature at 2200 nm (30°37′19.85″N, 34°51′0.77″E); 6. Calcite (carbonate) layer, main spectral feature at 2340 nm (30°36′19.72″N, 34°51′49.18″E). Figure 2 shows the average spectral signature and photos of each MR validation test site.

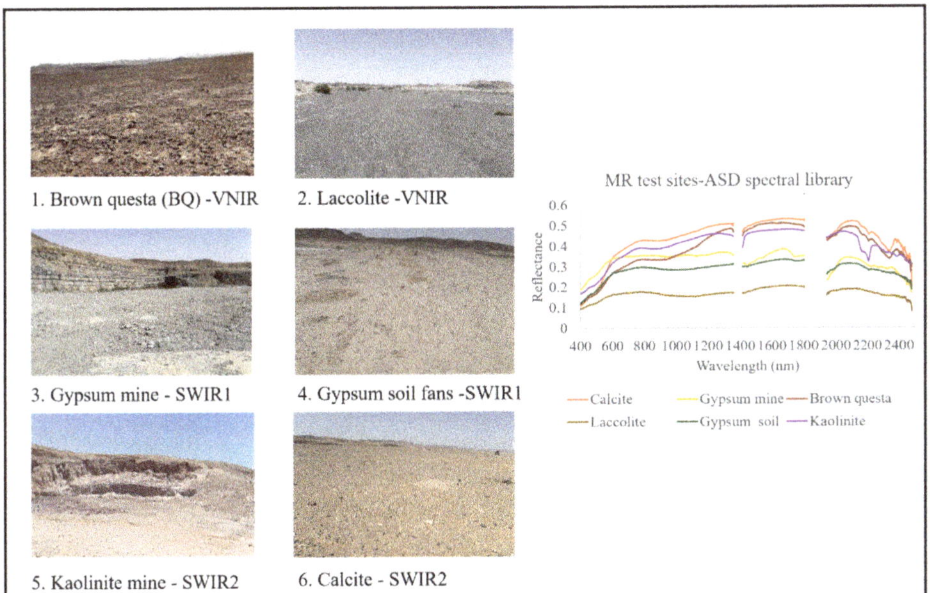

Figure 2. MR test sites. Different mineralogies and spectral signatures characterize the 6 test sites. Spectral signatures are from the spectral library acquired using an Analytical Spectral Devices (ASD) spectroradiometer.

To examine the sensor's spectral accuracy, we compared the orbital sensor's spectra to those of the airborne AisaFENIX and the field ground-truth spectra measured by an Analytical Spectral Devices (ASD) spectroradiometer at each test site. We considered the highly accurate AisaFENIX data to be our benchmark in the validation routine. For that purpose, we resized the AisaFENIX data cube to the sensors' spatial and spectral configuration. The spectral similarity was calculated using SAM (Equation (2)) [40], RMSE (Equation (3)), and ASDS (Equation (4)) [41]. Lower values indicated higher spectral similarity, and the threshold values for good spectral calibration deduced from this study were SAM < 0.1, RMSE < 0.05, and ASDS < 0.1.

In addition, the sensor's geometric accuracy was evaluated by looking at time-series images on well-formed geology targets and roads in MR images (200 points) along with field GPS measurements (see Section 3.2.5). The overall CAL/VAL protocol flowchart for radiometric, spectral, geometric, and thematic assessment is given in Figure 1b.

2.4. Field Measurements

Ongoing field measurements have been carried out at both the MR and AP sites since 2019. We followed the same field protocol at each test site, i.e., 6 MR VAL test sites (Figure 3a) and the CAL test site in AP (Figure 3b), including GPS, spectral measurements, soil sampling, and digital photos. An ASD FieldSpec® model FSP 350–2500 nm spectroradiometer was used for the spectral measurements in the field. The idea of the field protocol was to simulate a pure 1 pixel of a 30 m spatial resolution sensor. At each test site, approximately 32 points were taken along 30 m^2 (CAL and VAL), measuring the reflectance and/or radiance. An X-shaped grid was used (Figure 3c), with 10 measurements for each 30 m line, and then another 12 randomly added measurements around this X shape. The average of all 32 points represented the test site spectral signal (an example for AP is shown in Figure 3d). The GPS coordinates were measured at both ends and the center of each line.

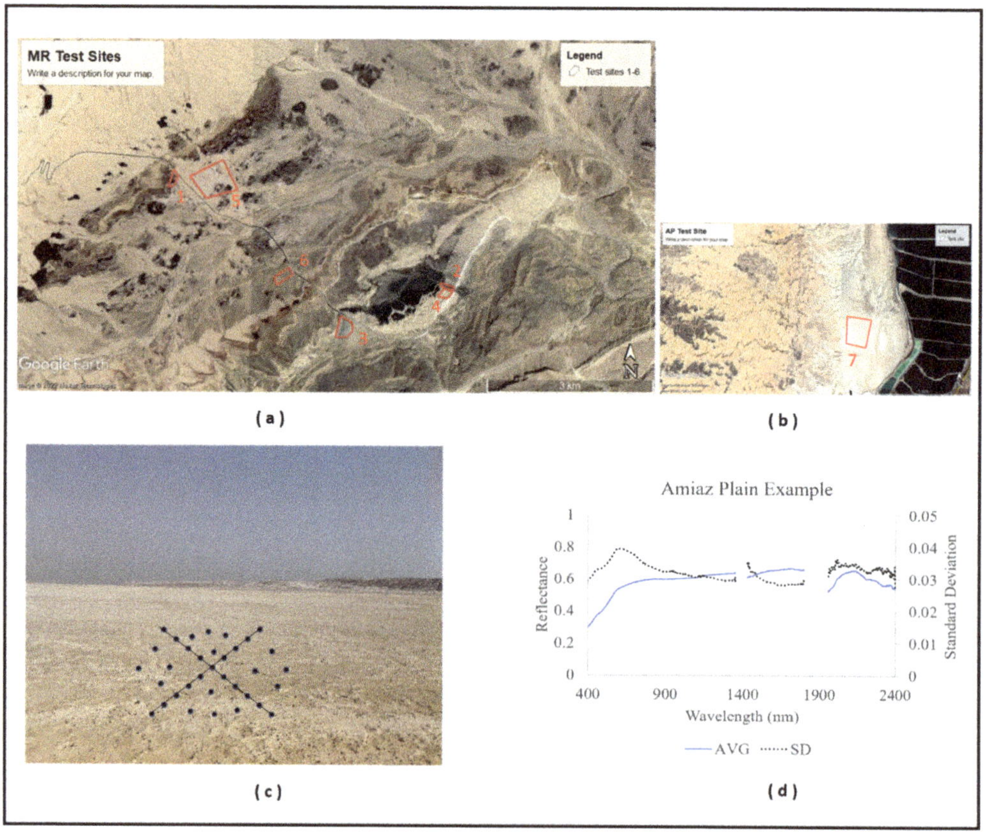

Figure 3. Location of test sites in (**a**) MR, and (**b**) AP. (**c**) Field spectral measurement protocol; example from the AP test site. (**d**) Average spectral signal (AVG, solid blue line) and standard deviation (SD, dashed black line) for AP 30 × 30 cube.

In the case of a satellite with lower spatial resolution (for example, NASA's EMIT with 60 m resolution), we adjusted the grid measurements to simulate a larger pixel: 60 × 60, 90 × 90, etc. It is important to mention that the selected area for the ground-truth measurement covered the most homogeneous flat terrain in the area to minimize any effect of bidirectional reflectance distribution function (BRDF), and the standard deviation of the average spectral reflectance was less than 5%. Table 1 summarizes the general information for each test site (Figure 3(a,b)).

Table 1. Test site information.

Test Site	Number	Altitude (Meters Above Sea Level)	Area (Square Meters)
MR Brown questa	1	521	28,241
MR Laccolite	2	469	17,262
MR Gypsum—old mine	3	498	81,473
MR Gypsum—soil fans	4	469	27,857
MR Kaolinite—old mine	5	502	356,531
MR Calcite	6	503	48,578
AP	7	−258	1,757,943

2.5. Cross-Calibration of HSR Orbital Sensors Using AP Site

Cross-calibration transfers the calibration of a well-calibrated sensor, referred to (in this study) as the motherhood sensor, to a less-well-calibrated sensor. Radiometric calibration is performed using the TOA radiance or TOA reflectance signal. The advantage of using the reflectance values lies in reducing image-to-image variability; it removes the effect of different solar zenith angles and solar irradiance due to the time difference between data acquisitions. Each sensor has a different relative spectral response (RSR), which needs to be normalized by a factor so that the reflectance can be compared. The simulated TOA reflectance is calculated for each sensor by integrating the hyperspectral profile of the surface with the sensor's RSR at each wavelength, weighted by the RSR of the respective sensor. The integral in the numerator calculates the amount of in-band reflectance acquired in the respective RSR and is divided by the integral of the RSR of the sensor, so that there is no gain/loss due to the filter response function. The spectral band adjustment factor (SBAF) is computed from the ratio of the 2 respective simulated sensors' reflectances (Equation (5)). The SBAF is applied to the sensor's original TOA reflectance for calibration (Equation (6)) [21,26,28]. The HSR sensors used in the study were well-calibrated for radiance, as determined by a prior examination conducted by both the space agencies and the authors [36,37]. Moreover, the cross-calibration result (SBAF method) of PRISMA with DESIS using the TOA reflectance did not result in any significant changes for PRISMA (3.2% standard deviation between the original to calibrated PRISMA signal). This led us to our next step, in which the SBAF method was applied on the surface reflectance (obtained from ASI and DLR products) in a cross-calibration exercise for the atmospheric models using AP as a PICS. Both PRISMA and DESIS level 2 atmospheric processors convert TOA spectral radiance into spectral reflectance by using a multidimensional look-up table (LUT) approach with MODTRAN code (6 for PRISMA and 5.4 DESIS). This stage is important for judging the quality of the L2 products over two adjacent sites (AP and MR) spectrally and thematically, to estimate how these two sensors perform together. For the purpose of this exercise, the following parameters were applied to Equation (5):

$$SBAF = \frac{p_{(M)}}{p_{(A)}} = \frac{\frac{\int \rho\lambda h \; RSR(M) d\lambda}{\int RSR(M) d\lambda}}{\frac{\int \rho\lambda h \; RSR(A) d\lambda}{\int RSR(A) d\lambda}}, \tag{5}$$

where RSR (M) and RSR (A) are the relative spectral responses for the reference (motherhood) sensor and the sensor to be calibrated; p (A) is the simulated reflectance for sensor A to be calibrated; p (M) is the simulated reflectance for the well-calibrated sensor M; and $\rho\lambda$h is the accurate hyperspectral reflectance profile of the surface.

$$p'_{(A)} = p_{(A)} \times SBAF, \quad (6)$$

where P' (A) is the new reflectance for sensor A using the SBAF to match the reflectance of sensor M.

We examined the AP site for cross-calibration, calibrating the PRISMA sensor with DESIS and the AisaFENIX (see Sections 3.2.3 and 3.2.4). The "reflectance" used in this study consisted of the average reflectance values of the AP scene obtained directly from the sensor's atmospherically corrected products.

2.6. Sensors' Image Data

A list of the sensor images used in this study is summarized in Table 2. It includes sensor parameters, the number of images, and the purpose of using the selected image.

Table 2. Sensors and images used in the study.

Sensor	Information	Number of Images	Purpose
Landsat 5 Thematic Mapper 5 (NASA/USGS)	7 bands, spectral range visible (VIS)–SWIR 0.45–2.35 μm, 30 m GSD (ground sample distance); thermal infrared (TIR) 10.40–12.50 μm, 120 m GSD	9	Evaluation of spectral and spatial stability of MR and AP for years 1996–2011
Landsat 8 Operational Land Imager (NASA/USGS)	11 bands, spectral range VIS–SWIR 0.43–2.29 μm, 30 m GSD; TIR 10.60–12.51 μm, 100 m GSD; panchromatic 0.5–0.68 μm	6	Evaluation of spectral and spatial stability of MR and AP for years 2015–2021
PRISMA PRecursore IperSpettrale della Missione Applicativa Italian Space Agency (ASI)	234 bands, range of 400–2500 nm, 30 m GSD; full-width half maximum (FWHM) $\leq$12 nm, swath 30 km, sensor altitude 615 km L1 TOA radiometric image, L2B ground radiometric image, and L2D atmosphere- corrected data cube. Acquired from PRISMA's website	15	Validation of CAL/VAL research protocols for AP and MR. Image dates: 2019–2022
DESIS DLR Earth Sensing Imaging Spectrometer German Aerospace Center (DLR)	235 bands, spectral range 400–1000 nm, 30 m GSD; FWHM ~3.5 nm, swath 30 km, sensor altitude 400 km. L1C radiometric georectified image and L2A atmosphere- corrected data cube. Acquired from DESIS EOweb GeoPortal.	6	Validation of CAL/VAL research protocols for AP and MR. Image dates: 2020–2021
AisaFenix 1K (HSR sensor AisaFENIX—Specim, Spectral Imaging Ltd.)	An airborne campaign using the AisaFENIX 1K over MR and AP was carried out on April 5, 2017, covering the entire MR (200 km^2) area and AP (5 km^2). 420 bands, spectral range 375–2500 nm, 1.5 m GSD; FWHM: VIS 3.4 (nm), NIR–SWIR, 6.2 (nm), swath 1.8 km	25 lines on MR 1 line on AP	Establishing high-accuracy reference data for AP and MR, mapping MR main minerals, and summarizing in an online database. Benchmark data for the CAL/VAL protocol [36,46]
ASD FieldSpec model FSP 350–2500 nm (Model 3 and Model 4)	Spectral range of 350–2500 nm with 2151 bands, with 3 nm and 8 nm resolution for the VNIR and SWIR regions	Ongoing	In-situ field measurements of radiance and reflectance for validation. Years 2019–2022
The Shuttle Radar Topography Mission (SRTM)	NASA–JPL at a resolution of 1 arc-s (approximately 30 m) [45]	1	Creating 3D elevation models of MR. Slope and aspect maps

3. Results

3.1. AP and MR Stability

Spatial and spectral change detection offers a straightforward approach to measuring changes between a pair of images that represent an initial and final state. The test sites' spectral stability and spatial stability were examined using USGS Landsat Collection level 2 surface reflectance for Landsat TM5 and OLI8. Seven time-series images from 2000 to 2021 for AP and eight from 1996 to 2021 for MR were selected as representations of that time, all in the summer season. For AP, t(0) = 2000 and t(n) = 2021; for MR, t(0) = 1996, and t(n) = 2021.

Figure 4 shows the spatial change detected from 2000 to 2021 (green pixels); the red box indicates the AP calibration test site area. The Dead Sea area had low anthropogenic activity and is situated in an extreme desert environment; the only significant changes detected were structural and in the water levels of the evaporated ponds of the Dead Sea salt factory. Most of the terrestrial area in general, and the AP area in particular, did not change, remaining stable over 21 years.

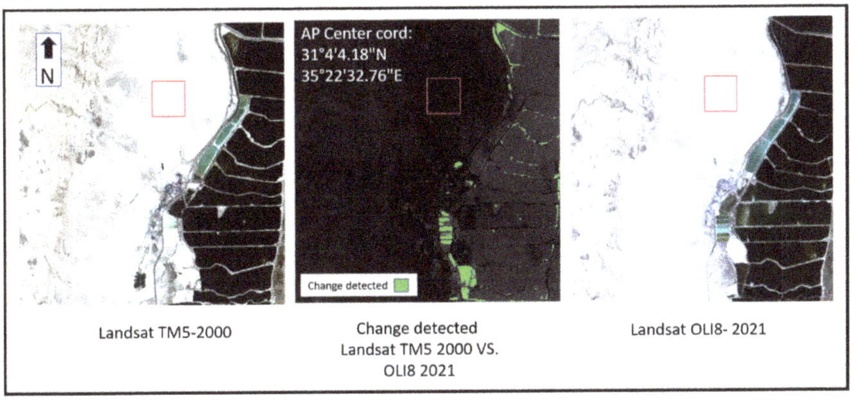

Figure 4. Spatial change detection result for AP, 2000 vs. 2021. Test site is marked in a red box; detected changed pixels are colored in green (center image).

Figure 5a shows a comparison of the mean Landsat reflectance spectra for the years 2000–2021. An average of 200 pixels in each year were used for the comparison. The SAM, ASDS, and RMSE were calculated against the year 2000 (t0) to examine possible spectral reflectance changes. The results are summarized in Figure 5b. The average values of SAM, ASDS, and RMSE were very low (Figure 5a), and these low values were maintained across the years: SAM 0.01–0.04, ASDS 0.0004–0.005, and RMSE 0.008–0.04. Although the spectral resolution was relatively low (six bands), it could be concluded that the spectral reflectance across the 21 years was very stable.

The change detection applied to AP was also applied to MR. In Figure 6, the result of the spatial changes detected in MR for the years 1996–2021 (red pixels) is provided. A significant change was detected over the town of Mizpe Ramon, situated on MR's northern cliff (marked in a dashed green circle). This was mainly due to the massive urban buildup activity in the town in those years. Within the MR crater itself, there were a few changed areas (marked in a turquoise dashed circle). Those changes were part of the Natural Reserve Authority's restoration activities (mostly between 1998 and 2000) to make MR accessible to the public as a touristic National Reserve park. Two camping areas were built, equipment was evacuated from the old mines, and new roads were constructed. The six spectral VAL test sites (marked by yellow Xs) and most of the crater remained stable over 20 years and are expected to remain so in the future.

To examine possible changes in the spectral reflectance of the six test sites at MR, a comparison of the mean Landsat reflectance for the years 1996–2021 was conducted (Figure 7). The SAM, ASDS, and RMSE values were calculated against the year 1996 (t0) at each test site, and the results are shown in Figure 8.

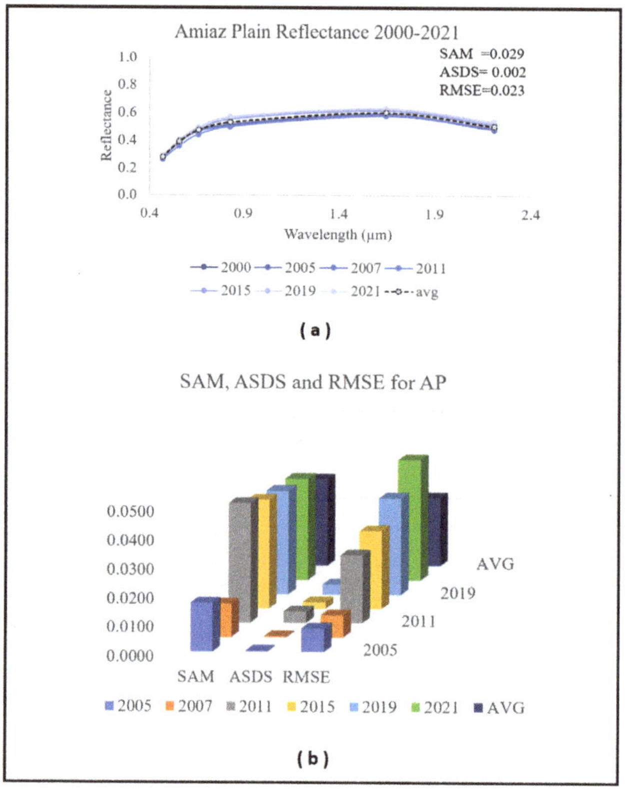

Figure 5. Examination of spectral change in AP. (**a**) Mean Landsat reflectance signal for each year (2000–2021). (**b**) SAM, ASDS, and RMSE for each year against the year 2000 (t0).

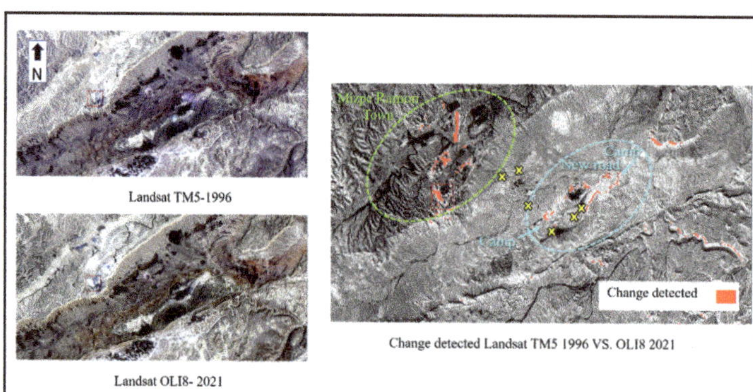

Figure 6. Spatial change detection result for MR, 1996 vs. 2021. Spectral test sites are marked by yellow Xs, detected changed pixels are colored in red.

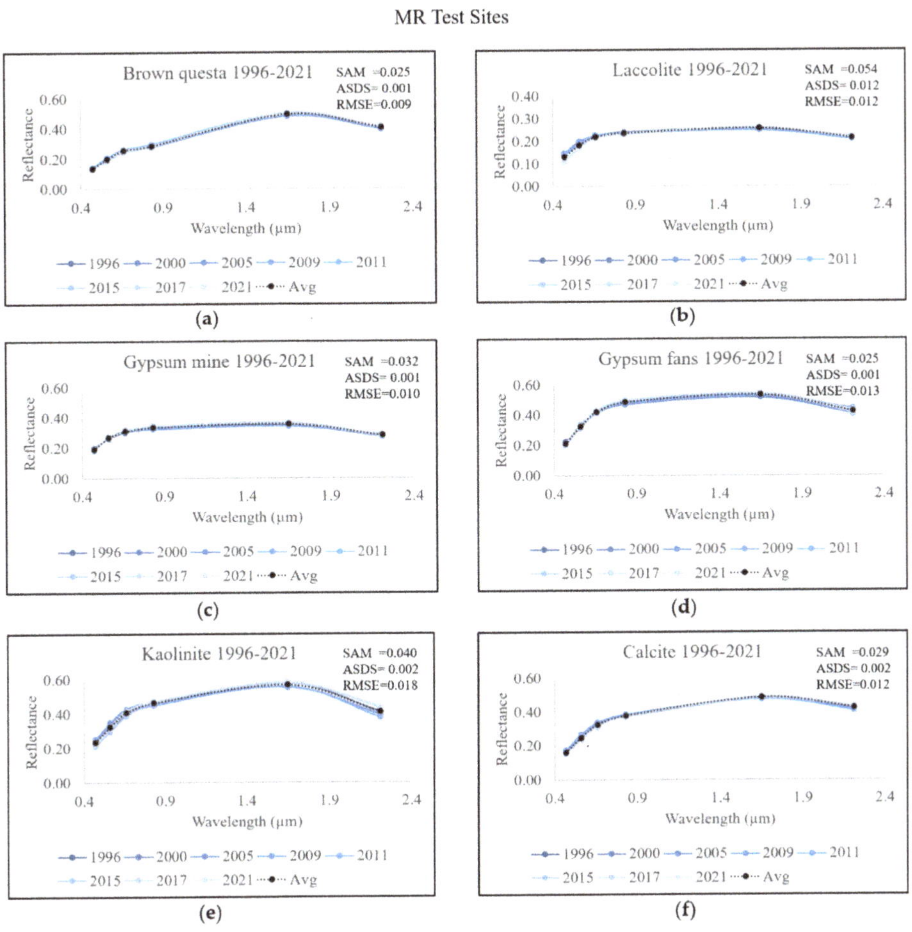

Figure 7. Examination of spectral changes at MR VAL test sites: LANDSAT reflectance signal for each year (1996–2021): (**a**) brown questa; (**b**) laccolite; (**c**) gypsum mine; (**d**) gypsum fans; (**e**) kaolinite; (**f**) calcite. The mean value is marked by black dashed lines.

A very good match was obtained between all spectra in all MR VAL test sites, presenting very low average values of SAM, ASDS, and RMSE: 0.025–0.05, 0.001–0.01, and 0.009–0.018, respectively. The lowest values were obtained for the brown questa and gypsum fans, and the highest values for the kaolinite and laccolite sites. The changes in the laccolite (darkest test site) were mainly in the blue band. However, there was a small variation in the reflectance spectra within the test sites over the years. This may be a result of using different sensors in the calculations (TM5 and OLI8 (from 2015)), where a dark test site is more susceptible to the change (e.g.light dust). Nevertheless, the values are below the significance threshold for good data quality (i.e., SAM < 0.1, ASDS < 0.1, and RMSE < 0.05), indicating high spectral similarity and stability of the reflectance signal over the years.

3.2. AP and MR CAL/VAL Protocol: A Case Study Using the PRISMA HSR Sensor

The working protocol from Figure 1b is described in the following section with the PRISMA sensor as a case study. We used the 15 collected PRISMA images of AP and MR

(2019–2022) in the case study analysis. A field campaign accompanied the November 2021 overpass images; therefore, some of the examples focus on those images.

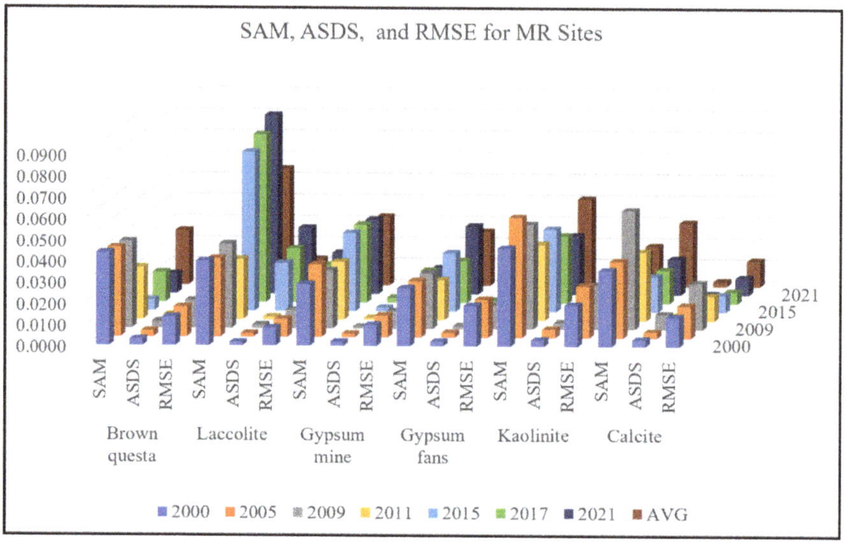

Figure 8. Spectral change examination. SAM, ASDS, and RMSE for each year against the year 1996 (t0).

3.2.1. Evaluate Radiance Performance

On 11 and 24 November 2021, images of AP and MR were acquired by PRISMA and transferred to us by ASI. Field spectral (radiance and reflectance) measurements were taken on 16 November, 2021.

We compared the L1 TOA PRISMA product to a radiance transfer model to evaluate the radiance performance. The TOA radiance model for AP was generated with the MODTRAN4 radiance transfer code and the ASD field reflectance spectrum. Atmospheric and solar radiation was generated to yield simulated TOA radiance values for the PRISMA spectral configuration overpass information (11 November, 2021, azimuth angle 160.8°, zenith angle 50.8°, day of the year 315, desert aerosol, sensor height 615 km); sky conditions, optical depth, and water vapor content were taken from the "SEDE BOKER" AERONET station (water vapor 1.27, angstrom exponent 0.997, ozone 0.2, and coarse /fine AOD in total 0.169). A comparison of the simulated radiance signal (mW m^{-2} sr^{-1} nm^{-1}) and the real PRISMA L1 product signal is shown in Figure 9.

Figure 9 presents the TOA radiance signal of PRISMA (L1) and the simulated TOA radiance using the MODTRAN4 code. In both spectra, the absorbance positions of several gases well-matched their position in the MODTRAN radiance: O_2 at 680 and 760 nm; O_3 at 651 nm; H_2O at 823, 940, and 1131 nm; and CO_2 at 2010 and 2061 nm. There was a slight albedo offset between the PRISMA (L1) and MODTRAN radiance spectra of about 20 mW m^{-2} sr^{-1} nm^{-1} in the visible (VIS) spectral range (400–700 nm). This may be caused by the overfitting of the model when we tried to compensate for AP's topographic elevation of 260 m below mean sea level (the MODTRAN code cannot calculate height below zeroin the model). Regardless, the overall shape and signals were mostly similar.

3.2.2. Evaluate Spectral Performance—PRISMA L2 Product

The PRISMA L2D (atmospherically corrected surface reflectance product) was evaluated at MR in six VAL test sites (Table 1) to determine PRISMA's spectral calibration performance. A comparison of the sensor signal against the ground truth is an early warning step in identifying possible spectral calibration problems. "Problematic" bands were flagged for further examination.

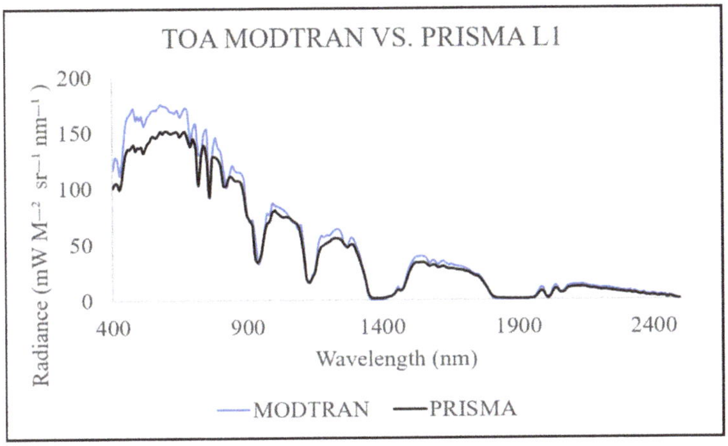

Figure 9. Radiance comparisons. TOA radiance from PRISMA L1 vs. MODTRAN simulation.

It is important to mention that before generating the field-measurement protocol (Section 2.4), a homogeneous examination of the test sites was conducted. Different points (50–100) were randomly measured with a field ASD spectrometer at each test site. The average standard deviation (SD) was calculated for all of the spectral bands (excluding water vapor and the end of the sensor bands, i.e., 2450–2500 nm) and for each spectral range: VNIR, SWIR1, and SWIR2. Table 3 summarizes the results and shows less than 5% deviation when looking at the average result for all bands (350–2450 nm). Moreover, deviation results for the specific spectral ranges showed the lowest SD values for the specific mineral connected to that site's evaluation, i.e., brown questa and laccolite VNIR SD < 0.02, gypsum mine and fans SWIR1 SD < 0.05, and kaolinite and calcite SWIR2 SD < 0.03. For the general evaluation, a comparison between the PRISMA reflectance and the average ASD ground truth was applied (Figure 10). Since PRISMA's L2 atmospheric processor algorithm has changed many times since 2019, we used only five images (years 2021–2022) in the spectral evaluation. The mean reflectance for each MR VAL test site was calculated from the five images of PRISMA; in each image, 10 pixels from the six test sites were obtained from the location at which the field measurements were taken.

Table 3. Homogenous evaluation of each test site for ASD field measurments (SD for each point set).

Test Site	Number of ASD Points	SD VNIR	SD SWIR1	SD SWIR2	SD All Bands
Brown questa	62	0.0161	0.0250	0.0218	0.0180
Laccolite	58	0.0117	0.0173	0.0161	0.0129
Gypsum—mine	61	0.0352	0.0540	0.0627	0.0374
Gypsum—soil fans	50	0.0207	0.0242	0.0318	0.0217
Kaolinite	70	0.0226	0.0363	0.0339	0.0254
Calcite	109	0.0205	0.0222	0.0223	0.0224

The flagged areas of concern (marked in black boxes, Figure 10) from the initial examination of PRISMA were in the VNIR range wavelengths, 920–970 nm, and 1100–1163 nm. This was probably caused by residual atmospheric absorbance (water vapor), which was not entirely removed in the atmospheric correction of the L2D product [47]. The exchange of PRISMA's VNIR to SWIR sensor, which was in this range, may also have contributed. The SWIR2 range above 2320 nm showed a high deviation from the ASD signal; there was a sharp decrease in the spectra, and the calcite absorbance at 2340 nm shifted. This indicates that the performance of the atmospheric correction algorithm applied by the standard L2D processor for the SWIR2 bands is questionable.

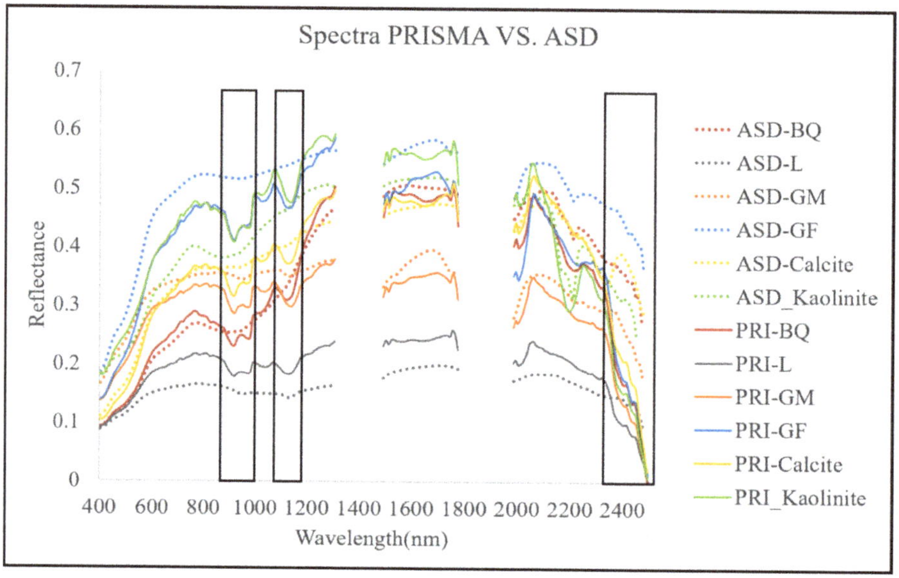

Figure 10. Evaluation of PRISMA L2D spectra (PRI) vs. ASD at the six test sites (PRISMA—solid line, ASD ground truth—dashed line). BQ, brown questa; L, laccolite; GM, gypsum—mine; GF, gypsum—fans.

For a deeper evaluation of the spectral calibration, the MR test sites were divided into specific spectral ranges: VNIR, SWIR1, and SWIR2, according to the major mineral at each VAL site (VNIR—brown questa (iron oxide) and laccolite; SWIR1—gypsum; SWIR2—kaolinite, and calcite). The spectral similarity between PRISMA and AisaFENIX was determined by calculating the SAM, RMSE, and ASDS separately for each range. For that purpose, the AisaFENIX data cube was resized to the 30 m spatial resolution and re-sampled to PRISMA's spectral configuration. It should be noted that the use of AisaFENIX as a benchmark to examine the HSR performance from MR was already established and demonstrated by Heller-Pearlshtien et al. [36].

Table 4 summarizes the results for SAM, ASDS, and RMSE between the PRISMA and AisaFENIX spectra, for each VAL test site-specific spectral range (VNIR, SWIR1, and SWIR2), where PRISMA was the examined reflectance spectrum, and AisaFENIX the reference reflectance spectrum. Lower SAM, ASDS, and RMSE values indicate the similarity between the PRISMA and AisaFENIX spectral signatures. The highest values obtained were for the SWIR2 range (i.e., Test Sites 5 and 6, kaolinite and calcite), which were higher than the threshold for good spectral calibration (SAM and ASDS < 0.1, RMSE < 0.05). They were also considerably higher than the VNIR and SWIR1 test site results (Table 4). We also included the SAM, ASDS, and RMSE results for the entire spectral range (VNIR-SWIR (VSWIR) 400–2400 nm) in Table 4. The spectra differed from VNIR to SWIR, and were much larger at certain test sites. Hence, the potentially problematic spectral bands may be masked by the well-calibrated ones. For example, the result for the calcite test site does not reflect the deterioration after 2300 nm. Therefore, separation into specific spectral ranges in the evaluation is critical. This result and the initial spectral inspection strengthen the suspicion that the PRISMA's SWIR2 spectral bands are less well-calibrated and should be used with caution [47].

Table 4. SAM, ASDS, and RMSE values for each test site.

Test Site	Spectral Range	Wavelength (nm)	SAM	ASDS	RMSE
Brown questa	VNIR	402–998	0.089	0.050	0.032
Laccolite	VNIR	402–998	0.060	0.020	0.020
Gypsum—mine	SWIR1	1480–1794	0.067	0.071	0.027
Gypsum—soil fans	SWIR1	1480–1794	0.066	0.073	0.042
Kaolinite	SWIR2	2001–2400	0.144	0.300	0.103
Calcite	SWIR2	2001–2400	0.155	0.280	0.098
Brown questa	VSWIR	400–2400	0.128	0.220	0.048
Laccolite	VSWIR	400–2400	0.103	0.207	0.026
Gypsum—mine	VSWIR	400–2400	0.089	0.109	0.038
Gypsum—soil fans	VSWIR	400–2400	0.092	0.123	0.075
Kaolinite	VSWIR	400–2400	0. 131	0.180	0.083
Calcite	VSWIR	400–2400	0.101	0.123	0.042

It is important to mention that an entire article was dedicated to examining PRISMA's mineral mapping capability (i.e., goethite, hematite, gypsum, kaolinite, and calcite) and validated against the airborne AisaFENIX database [36]. It was concluded that the sensor provides a very accurate spectral-based mapping of minerals over the MR site (excluding a poor result for the calcite mineral). Therefore, this step in the CAL/VAL protocol was excluded from the current paper.

3.2.3. Cross-Calibration and Validation between PRISMA and DESIS Level-2 Products

Cross-calibration between spaceborne sensors can be achieved by using PICS. These sites are chosen for their high surface homogeneity and spectral, spatial, and temporal stability [30]. AP was found to be a very stable area over 20 years (Section 3.1); therefore, it can be used as a vicarious calibration test site and for cross-calibration as a PICS. In this section, we conducted a cross-calibration between PRISMA and DESIS on the atmospheric-correction models obtained from their products (i.e., PRISMA L2D, DESIS L2A) using their shared spectral region, i.e., the VNIR range. The images were captured over AP 3 days apart (PRISMA 3 June, 2021, 8:26 UTC, and DESIS 30 May, 2021, 8:53 UTC). The DESIS sensor had 235 bands in the VNIR spectral range (400–1000 nm) and 30 m ground sampling distance (GSD), and was found in previous work to perform well (above 450 nm) [37]. The PRISMA sensor had 63 bands (450–998 nm) and 30 m GSD. The simulated reflectance for each sensor was obtained by integrating the AisaFENIX hyperspecral profile of the AP surface with the sensor's RSR. Beforehand, the DESIS and AisaFENIX spectral signals were converted to PRISMA's VNIR 63-band configuration. The SBAF was obtained by dividing the sensor's simulated reflectance (Equation (5)), which was then operated on PRISMA's original L2D signal, to create a new reflectance (Equation (6)). The calculation was created by implementing the following data into Equations (5) and (6):

p (sensor A)—simulated reflectance for the sensor to be calibrated (PRISMA);
p (sensor M)—simulated reflectance for the well-calibrated (motherhood) DESIS sensor;
$\rho\lambda h$—AisaFENIX's accurate hyperspectral reflectance profile of the surface.

The result of the PRISMA cross-calibration in AP (VNIR bands) is shown in Figure 11: the original (before cross-calibration) PRISMA L2D spectral signal (black line) against the newly calibrated signal (blue line), and the ground-truth field ASD reflectance (dashed red line).

We calculated the RMSE of PRISMA's VNIR signal to validate our result. The RMSE decreased from 8% to 3% after the cross-calibration (Figure 11a), thereby improving the reflectance signal accuracy for PRISMA's VNIR bands.

To further validate the cross-calibration result, we applied the AP's SBAF to PRISMA's image of MR (taken at the same time as the AP image). MR's brown questa test site had the most pronounced spectral signature in the VNIR range due to the rich iron-oxide composition. Figure 11b shows the result of applying AP's SBAF on PRISMA's brown questa signal after calibration (blue line) against the original signal (black line), and the ground-truth ASD spectrum (dashed red line). Though the brown questa spectra from the original PRISMA signal, the calibrated signal, and the ASD spectra were similar, it can be seen that the noticeable artifacts in the original L2D signal (before calibration, black line) at 900–1000 nm were less pronounced after the cross-calibration (blue line).

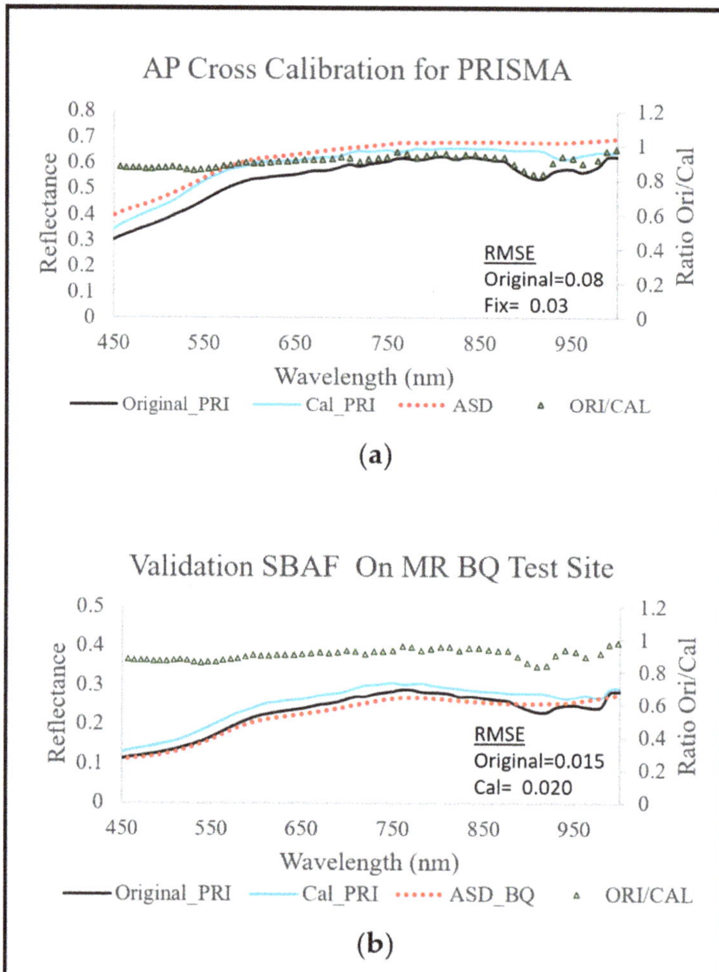

Figure 11. (**a**) AP cross-calibration results for PRISMA (PRI); original (before calibration; black line), after SBAF cross-calibration (Cal, blue line) against ASD ground truth (red dotted line), ratio of original (ORI) to calibrated (green triangle). (**b**) Validation of AP SBAF on MR VNIR test site using brown questa (BQ).

Uncertainty Analysis

Different sources of uncertainty influence the cross-calibration accuracy, including the inherent variability of the sensors, and the process and techniques involved in the measurement [21]. In this study, the SBAF was calculated with only one pair of images (not a dataset), showing the potential of the cross-calibration method using PRISMA and DESIS over the AP CAL site. Consequently, the calculation of uncertainty was limited. We considered the following parameters as the main sources of uncertainty: the site's spatial variability (non-uniformity), the sensors' temporal drift (when capturing the site area), and the calibration model error. The site's spatial variability uncertainty (σ^2 spatial) manifested both the spatial and temporal drift components and was obtained by calculating the mean of the standard deviation of each sensor's region of interest (AP test site area) reflectance (i.e., PRISMA, DESIS, and AisaFENIX, Equation (7)). The calibration model uncertainty (σ^2 calibration) was calculated by the RMSE of the model to predict the surface reflectance, which is the predicted reflectance against the measured ASD ground-truth reflectance at each wavelength (Equation (8)). The obtained total uncertainty for the cross-calibration model (Equation (9)), combined from the above, was 0.047.

$$Uncertainty\sigma\ spatial = \sqrt{(\sigma^2 spatial(A) + \sigma^2 spatial(M) + \sigma^2 spatial(\text{ph}))}\,, \tag{7}$$

$$Uncertainty\sigma\ calibration = \sqrt{\frac{\sum_{i=1}^{n}(ASD_{ni-}\ Predict_{ni})^2}{n}}\,, \tag{8}$$

$$Uncertainty\ \sigma\ total = \sqrt{\sigma^2 spatial + \sigma^2 calibration}, \tag{9}$$

$$Uncertainty\ \sigma\ total = \sqrt{(spatial(0.015^2 + 0.023^2 + 0.024^2) + calibration(0.03)^2)} = 0.047$$

where n is 63, the number of wavelengths used for the cross-calibration; A is PRISMA; M is DESIS; and ph is AisaFENIX.

3.2.4. Cross-Calibration against a Fixed High-Quality Sensor Image

The main constraint of classical cross-calibration is finding pairs of scenes that have been captured within a short time interval (up to 6 days in a stable PICS). This is easier with multispectral sensors, but enormously difficult with HSR sensors. Each HSR sensor has its own revisit cycle, and both need to capture the test site with no cloud cover. It may take a long time to accumulate dozens of datasets.

A potential solution to this is to perform an empirical cross-calibration process with a fixed high-quality motherhood sensor, where the HSR images are not captured simultaneously, and to use reflectance to reduce the effect of the different time acquisitions (such as solar irradiance, solar zenith angles, and variation in the earth–sun distance). Accurate ground-truth measurements of the test site, to be integrated into the calculation of the sensors' simulated reflectance (motherhood and the sensor to be calibrated), can help normalize the signals for the time difference. Needless to say, the areas should not undergo surface changes in each overpass.

Accordingly, we performed a cross-calibration of PRISMA's sensor with the well-calibrated AisaFENIX HSR airborne data as the motherhood (calibrated) sensor, using four PRISMA images of AP. The field ASD measurements provided an accurate spectral profile of the test site.

We re-sampled the AisaFENIX and ASD spectral signals to PRISMA's 220-band configuration (400–2500 nm, excluding water vapor bands), and re-sized AisaFENIX to a 30 m spatial resolution. The SBAF (Equation (5)) was calculated for four different PRISMA AP images dated April, June, November, and December 2021, implementing the following:

p (sensor A)—simulated reflectance for PRISMA (to be calibrated);
p (sensor M)—simulated reflectance for the well-calibrated (motherhood) sensor AisaFENIX;
$\rho\lambda$h—ASD field hyperspectral profile of the surface (resampled to PRISMA bands).

The simulated reflectance for each sensor (PRISMA and AisaFENIX) was obtained by integrating the ASD profile with their RSRs. The SBAF was obtained by dividing the sensor's simulated reflectance.

The SBAFs for each band, resulting from the cross-calibration of PRISMA with AisaFENIX, are shown in Figure 12. The averaged SBAF is indicated by a black line, the gain factors ranged between 0.97 and 1.78 up to wavelength 2350 nm, and the values increased above this wavelength. The average standard deviation for all bands' SBAFs was 13%. The high variation implies that the dataset is not large enough to create a generic SBAF for PRISMA's images; more images are needed. Still, the SBAF for each month can be used to calibrate PRISMA's images near their timeline.

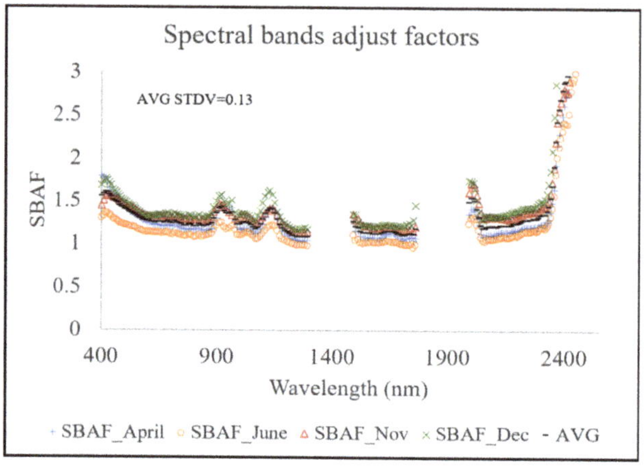

Figure 12. The spectral band adjustment factors (SBAFs) for PRISMA in different acquisition months (April—blue; June—orange; November—red; December—green; average SBAF—black).

To validate the concept of this method, we applied the November AP SBAF on the original PRISMA AP image. Figure 13 shows the result for the November image: the original PRISMA L2D (before calibration) signal (black line) against the calibrated SBAF (blue line), and the ground-truth ASD (dashed red line). The artifacts in the VNIR bands (900–1000 nm) were removed, and the absorbance at 2340 nm (carbonates) was better obtained in the PRISMA signal after the calibration. The RMSE of the calibrated November signal against the ASD reflectance was 0.05 vs. the original (L2D product before calibration) 0.15. The total σ uncertainty for the model was calculated from Equations (7)–(9), where n is 220, the number of wavelengths used for the cross-calibration; A is PRISMA (sensor to be calibrated); M is AisaFENIX (well-calibrated sensor); and ph is ASD, and was 0.06.

$$
\begin{aligned}
\textit{Uncertainty total } \sigma &= \sqrt{\sigma^2 Spatial + \sigma^2 calibration} \\
&= \sqrt{Spatial[(PRISMA(0.015)^2 + ASD(0.020)^2 + AISA(0.023)^2)] + calibration(0.05)^2} = 0.060
\end{aligned}
$$

To further validate the SBAF model, the AP's November SBAF was applied to the MR's November PRISMA image (acquired at the same time). The SAM, ASDS, and RMSE of the calibrated PRISMA reflectance (blue line, Figure 14) against the field ASD reflectance (dashed black line, Figure 14) were calculated. The results for the six MR VAL test sites are presented in Figure 14.

After applying the SBAF, the corrected reflectance well-matched the field ASD spectra in most test sites. The SAM results in all test sites had values <0.1 (the threshold for good spectral calibration). For the ASDS and RMSE, all test site results were below the threshold (ASDS < 0.1, RMSE < 0.05) except for Test Sites 2 (laccolite) and 6 (calcite) (Figure 14b,f), and the ASDS value for Test Site 1 (Figure 14a) was slightly higher from the thresholds (value of 0.107). This is because even though the signals for Test Sites 1, 2, and 6 were similar to the ASD, the RMSE and ASDS were more susceptible to albedo differences than the SAM.

Overall, the calibrated results for the model were good; the artifacts in the PRISMA's original VNIR signal (around 900–1000 nm) decreased, especially in Test Site 1 (brown questa; Figure 14a, marked by box). Another significant impact of the SBAF correction can be seen in the calcite (Test Site 6; Figure 14f, marked by box), where the significant deterioration across the end of SWIR2 of PRISMA was well-corrected and the calcite main absorbance feature at 2340 nm was back to its theoretical position.

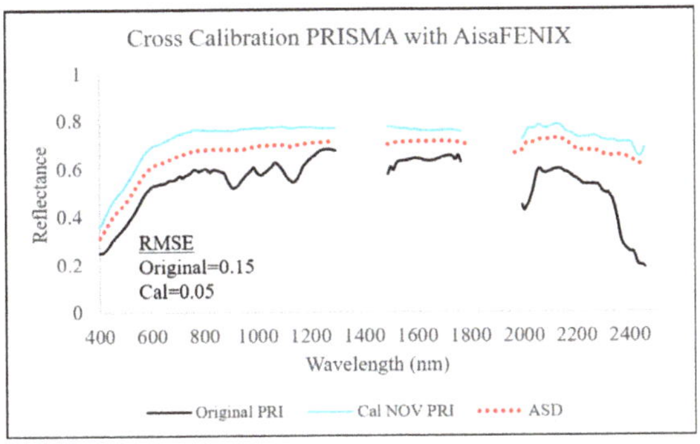

Figure 13. The AP cross-calibration result for PRISMA (PRI) with well-calibrated AisaFENIX; original reflectance (before calibration, black line), after SBAF calibration (blue line) against ASD ground truth (red dotted line).

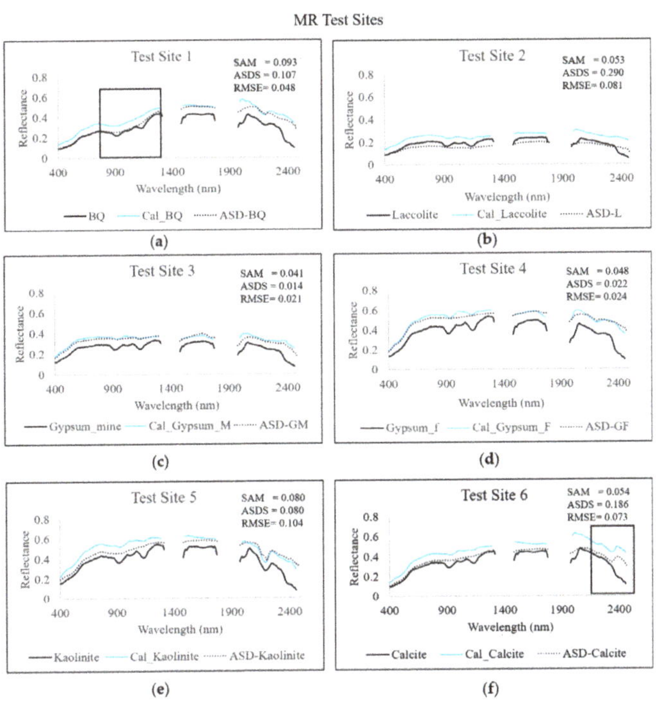

Figure 14. The MR test sites of PRISMA original (before calibration) reflectance (black line) against corrected reflectance (after applying the SBAF, blue line), and the field ASD spectra (dashed black line); (**a**) brown questa (BQ), (**b**) laccolite, (**c**) gypsum mine, (**d**) gypsum fans, (**e**) kaolinite, (**f**) calcite.

3.2.5. Evaluate Sensor's Thematic Stability

The geometric rectification of an image is an essential stage in geolocating surface targets and generating an accurate thematic map. The performance of the sensor's geometric correction is crucial when time-series images are used, either from the same sensor or when different sensors are integrated. MR's unique geological formations are stable in time, visually well-defined, and can be exploited to evaluate the geometric accuracy of an image. Accordingly, we examined the geometric precision performance of PRISMA's sensor over 4 years of operation (2019–2022). In general, PRISMA declares a geolocation accuracy of 200 m [32], whereas ASI is intended to provide a better-rectified product of up to 15 m using well-defined GCPs.

Four PRISMA MR images (one per year) were used. In each image, 200 randomly spread GCPs were selected across the image, and their location coordinates were compared to the previous year's image. In addition, the initial and final years (2019, 2022) were examined for any drifts in the rectification process over 4 years of operation. The results are summarized in Table 5. Figures 15 and 16 show examples of the geometric drift in the PRISMA image. Figure 15 shows the thematic change between the years 2019 and 2022 at MR. There was a shift of about 8 pixels (238 m) in the longitudinal direction and 3 pixels (95 m) in the latitudinal one. Several places are marked with red arrows to emphasize the drifts in 2019 (solid line) vs. 2022 (dashed line), such as the main road around the crater, basalt deposit, and dirt road. Figure 16 zooms in on three of our VAL test sites: kaolinite (top panels), gypsum soil fans, and laccolite (bottom panels). The red cross indicates the same coordinate position.

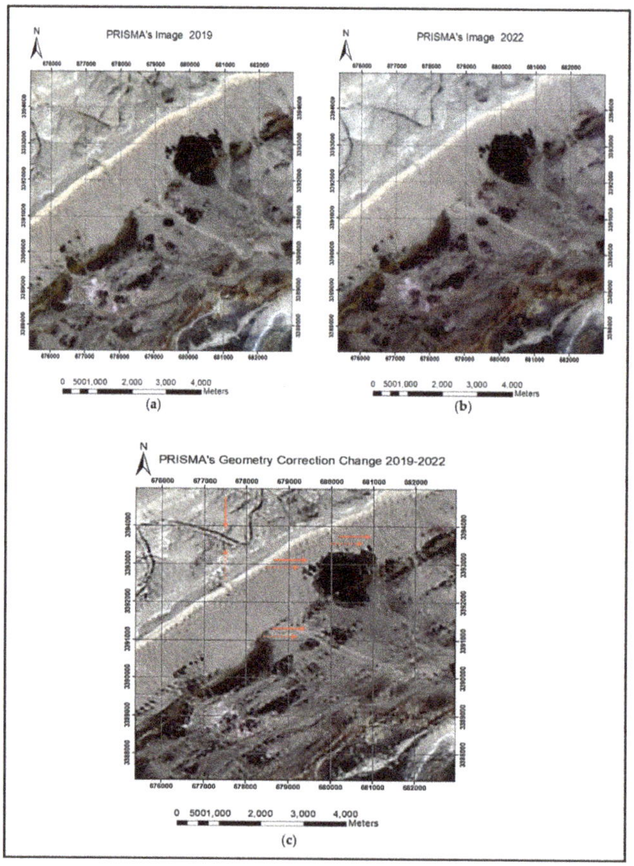

Figure 15. Comparison of PRISMA's 2019 and 2022 geometric corrections: (**a**) 2019 MR image; (**b**) 2022 MR image; (c) 2022 image overlaid on 2019 image. Several drifts in the geolocation are marked with red arrows (2019—solid; 2022—dashed).

It can be seen that the highest drift occured between 2019 and 2022, and the least change in position was noticed between images 2019 and 2020. Still, the maximum drift stayed around PRISMA's expected 200 m accuracy.

Table 5. Shifts in longitudinal (X) and latitudinal (Y) location for 200 GCPs at MR, between years 2019 and 2022. Average error and standard deviation were calculated.

Years Compared	X Error (m)	Y Error (m)	SD X	SD Y
2019–2020	16.8	19.7	0.69	0.37
2020–2021	243.1	66.9	0.75	1.13
2021–2022	16.9	18.3	0.43	0.23
2019–2022	238.6	95.1	0.51	1.18

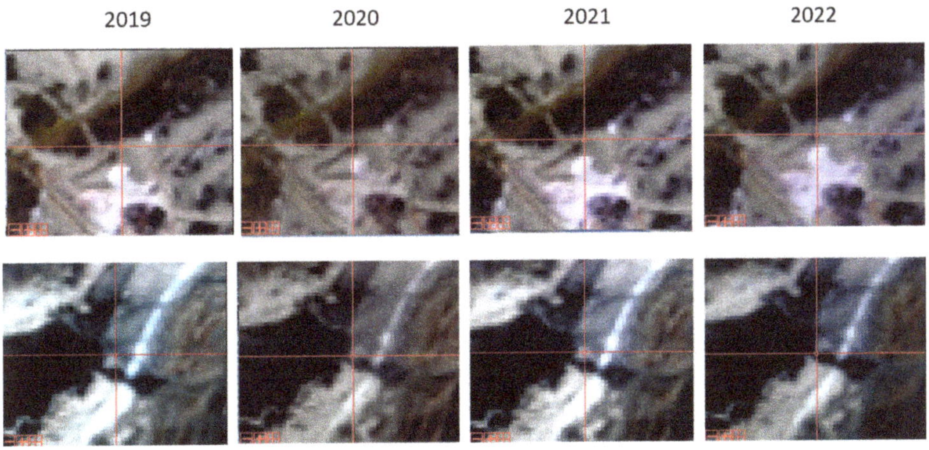

Figure 16. Comparison of the same coordinates (red cross) for the years 2019–2022 at three MR test sites: kaolinite (top panels), gypsum soil fans, and laccolite (bottom panels). There is a slight shift in the locations.

4. Discussion

The primary requirement for CAL/VAL test sites is that they be stable spectrally and spatially in space and time. We demonstrated that AP and MR fulfill this requirement using Landsat time-series images over 20 and 25 years, respectively. No spatial changes were detected across the years 2000–2021 at the AP CAL site (Figure 4). Furthermore, the spectral signals were stable in time as judged by the low values for SAM, ASDS, and RMSE (0.029, 0.002, and 0.023, respectively; Figure 5). At MR (VAL site), during the years 1996–2021, small spatial changes were observed due to restoration work toward establishing MR as a touristic nature park (Figure 6). However, this activity did not affect the geology or landscape of the six selected VAL test sites, as evidenced by the low values of SAM, ASDS, and RMSE (Figures 7 and 8). It should be remembered that both AP and MR are located in desert areas where the sky is clear most of the year, vegetation is scarce, and precipitation is very limited (25 mm annually). Since both areas have been declared nature reserves, vigorous enforcement by rangers to keep the park's landscape undisturbed means that only designated paths are available to tourists. Researchers require special permits to work in the park. This action will maintain the stability of the parks in the coming years. Other advantages are that both CAL/VAL sites can be captured in the same acquisition overpass, access to the sites is simple, and the distance between AP and MR is about a 1 h drive, enabling ground-truth measurements during or close to the overpasses. In addition, NASA's AERONET station, which collects daily atmospheric aerosol data, is located at

Sede Boker, close to both MR (20 km) and AP (45 km). These make AP and MR ideal areas for CAL/VAL sites: spectrally, spatially, and thematically.

We demonstrated the use of AP as a cross-calibration PICS between HSR sensors. Figure 11 shows the reflectance cross-calibration results between DESIS and the PRISMA VNIR bands, with an uncertainty of 0.047. After applying the SBAF calibration, the RMSE of the ground-truth ASD measurements against the original PRISMA L2D reflectance signal decreased from 8% to 3%.

It is important to note that in practice, TOA radiance or TOA reflectance information is used for cross-calibration between sensors. Our "cross-calibration" (SBAF method) was conducted using surface reflectance-calibrated products to demonstrate the advantages of using two adjacent sites: AP for reflectance cross-calibration examination, and MR for spectral cross-validation, which is actually a QA procedure of space agencies' products and is vital to ensuring that the data from multiple sensors can be used to provide a consistent set of measurements. Still, in the operational process, dozens of pairs of images are used to create a generic SBAF between two sensors, and not just one pair, as done in this study. Gathering a dataset of HSR image pairs can be challenging and time-consuming. The images must be taken almost simultaneously with no clouds, while synchronizing the HSR sensors' revisit cycle.

Therefore, a new potential approach for cross-calibration against a fixed high-quality sensor when the images are not taken at the same time was presented. We used the well-calibrated AisaFENIX AP image as the benchmark and calibrated PRISMA's AP images at different times. The SBAF was applied and validated on six MR (VAL) test sites, showing an improvement in the PRISMA's calibrated signal: the artifacts in the original L2D PRISMA signal (900–1000 nm) were quite markedly diminished at the brown questa site (Figure 14a). Moreover, at the calcite site, the mineral spectral signature returned to its accurate position at 2340 nm (Figure 14f). PRISMA's longer wavelength (SWIR2) demonstrated a significant variation from the field ASD measurement, suggesting that the PRISMA L2D performance in this spectral region should be regarded with caution. This is in accordance with the results of [47,48]. We showed only a primary result of this empirical fixed high-quality sensor cross-calibration method; further study and examination with different sensors and timelines are needed.

The geometric correction accuracy of HSR images is a fundamental parameter affecting the quality of the products. Scientists worldwide put effort into finding new validation methods for sensor geometric correction and increased precision. Because the HSR images are used for quantitative applications, their geometric accuracy is critical. For time-series applications, precision is crucial. MR's stability with time enables the selecting of GCPs to check the geometric accuracy and rectify it. PRISMA's images from 2019–2022 were examined, and the drifts between the geometric locations were calculated (Table 4). The highest shift was approximately 8 pixels (238 m) in the longitudinal direction and 3 pixels (95 m) in the latitudinal one (2019 to 2022), which is within the expected PRISMA accuracy of about 200 m. In the near future, ASI plans to improve the geometric precision to half a pixel by implementing GCPs in the geometric correction [32].

With this new era of launching sophisticated HSR sensors (e.g., EMIT, EnMAP, and DESIS), the demand for more accurate and challenging applications has increased. The HSR sensors need to provide high-end products that are strongly related to the quality of the retrieved reflectance and the radiance calibration of the sensors.

We performed a case study on ASI's PRISMA sensor, demonstrating the unique combination of radiance (AP CAL site), reflectance, and thematics (MR VAL site) for monitoring, calibrating, and then validating sensor performance. Using these ideal CAL/VAL sites may improve the day-to-day utilization of HSR sensors while identifying problems first-hand, such as the deterioration of specific spectral wavelengths and radiance calibration, and rectifying them accordingly.

5. Conclusions

This research demonstrated that AP and MR are excellent sites for radiometric, spectral, and geometric/thematic validation and calibration. Both sites are very stable spectrally and spatially and can be used as CAL/VAL sites. AP is ideal for the vicarious radiometric calibration or pseudo-invariant calibration for cross-calibration between sensors. MR's unique geological features, encompassing various minerals, are ideal for spectral assessment across the 400–2500 nm optical range. Six VAL test sites were found for the validation process, and the MR web database can be used to evaluate and validate the HSR sensor's mapping performance. The advantage of using MR for determining the sensors' geometric accuracy and stability was also established.

Using these dual test sites simultaneously can improve the assessment of HSR sensor radiance, and spectral and thematic performance. The suggested generic CAL/VAL protocol can help maintain a precise, accurate signal during the sensor's mission lifetime, thus assisting in the challenge of maintaining the sensor's high-end products.

Author Contributions: Conceptualization, D.H.P. and E.B.-D.; methodology, D.H.P. and E.B.-D.; software, D.H.P.; validation, D.H.P.; formal analysis, D.H.P. and E.B.-D.; investigation, D.H.P., S.P. and E.B.-D.; resources, S.P. and E.B.-D.; data curation, S.P. and D.H.P.; writing—original draft preparation, D.H.P. and E.B.-D.; writing—review and editing, D.H.P., E.B.-D. and S.P.; visualization, D.H.P.; supervision, E.B.-D.; funding acquisition, E.B.-D. and S.P. All authors have read and agreed to the published version of the manuscript.

Funding: This research was partially funded by the Israel Ministry of Science and Space project, grant number 3-17510, and by the Italian Space Agency ASI, PRISCAV project (grant agreement n. 2019-5-HH.0).

Data Availability Statement: The MR database developed for this study is available at the following link: https://storymaps.arcgis.com/stories/bb5bf09ec7414454a012bfe9bf4b8545 (accessed on 11 October 2022) [46].

Conflicts of Interest: The authors declare no conflict of interest.

References

1. Mckenzie, I.; Karafolas, N. Fiber optic sensing in space structures: The experience of the European Space Agency (Invited Paper). In Proceedings of the 17th International Conference on Optical Fibre Sensors, Bruges, Belgium, 23 May 2005; Volume 5855, p. 262. [CrossRef]
2. Chander, G.; Helder, D.L.; Aaron, D.; Mishra, N.; Shrestha, A.K. Assessment of spectral, misregistration, and spatial uncertainties inherent in the cross-calibration study. *IEEE Trans. Geosci. Remote Sens.* **2013**, *51*, 1282–1296. [CrossRef]
3. Bouvet, M.; Thome, K.; Berthelot, B.; Bialek, A.; Czapla-Myers, J.; Fox, N.P.; Goryl, P.; Henry, P.; Ma, L.; Marcq, S.; et al. RadCalNet: A radiometric calibration network for earth observing imagers operating in the visible to shortwave infrared spectral range. *Remote Sens.* **2019**, *11*, 2401. [CrossRef]
4. EMIT Homepage. Available online: https://earth.jpl.nasa.gov/emit/instrument/overview/ (accessed on 10 August 2022).
5. EnMAP Homepage. Available online: https://www.enmap.org/ (accessed on 10 August 2022).
6. Dubovik, O.; Schuster, G.L.; Xu, F.; Hu, Y.; Bösch, H.; Landgraf, J.; Li, Z. Grand Challenges in Satellite Remote Sensing. *Front. Remote Sens.* **2021**, *2*, 619818. [CrossRef]
7. Thenkabail, P. *Remote Sensing Handbook Volume 1: Remotely Sensed Data Characterized, Classification and Accuracies*; Taylor & Francis Group, LLC: Abingdon, UK, 2016.
8. Xu, H.; Zhang, L.; Huang, W.; Li, X.; Si, X.; Xu, W.; Song, Q. On-Board Absolute Radiometric Calibration and Validation Based on Solar Diffuser of HY-1C SCS. *Guangxue Xuebao/Acta Opt. Sin.* **2020**, *40*, 30015–30034. [CrossRef]
9. CAL/VAL Working Groups—Surface Biology and Geology. Available online: https://sbg.jpl.nasa.gov/groups (accessed on 10 August 2022).
10. RadCalNet Portal. Available online: https://www.radcalnet.org/#!/ (accessed on 22 July 2022).
11. Kieffer, H.H.; Wildey, R.L. Absolute calibration of Landsat instruments using the moon. *Photogramm. Eng. Remote Sens.* **1985**, *51*, 1391–1393.
12. Barnes, R.A.; Eplee, R.E.; Patt, F.S.; Kieffer, H.H.; Stone, T.C.; Meister, G.; Butler, J.J.; McClain, C.R. Comparison of SeaWiFS measurements of the Moon with the U.S. Geological Survey lunar model. *Appl. Opt.* **2004**, *43*, 5838–5854. [CrossRef] [PubMed]
13. Kouyama, T.; Kato, S.; Kikuchi, M.; Sakuma, F.; Miura, A.; Tachikawa, T.; Tsuchida, S.; Obata, K.; Nakamura, R. Lunar calibration for ASTER VNIR and TIR with observations of the Moon in 2003 and 2017. *Remote Sens.* **2019**, *11*, 2712. [CrossRef]

14. Kieffer, H.H.; Stone, T.C.; Barnes, R.A.; Bender, S.C.; Eplee, R.E., Jr.; Mendenhall, J.A.; Ong, L. On-orbit radiometric calibration over time and between spacecraft using the Moon. In Proceedings of the Sensors, Systems, and Next-Generation Satellites VI, Crete, Greece, 8 April 2003; Volume 4881, p. 287.
15. Chang, I.-L.; Dean, C.; Li, Z.; Weinreb, M.; Wu, X.; Swamy, P.A.V.B. Refined algorithms for star-based monitoring of GOES Imager visible-channel responsivities. In Proceedings of the Earth Observing Systems XVII, San Diego, CA, USA, 15 October 2012; Volume 8510, p. 85100R.
16. Vermote, E.; Santer, R.; Deschamps, P.Y.; Herman, M. In-flight calibration of large field of view sensors at short wavelengths using Rayleigh scattering. *Int. J. Remote Sens.* **1992**, *13*, 3409–3429. [CrossRef]
17. Kaufman, Y.J.; Holben, B.N. Calibration of the AVHRR visible and near-IR bands by atmospheric scattering, ocean glint and desert reflection. *Int. J. Remote Sens.* **1993**, *14*, 21–52. [CrossRef]
18. Robert, S.F.; Kaufman, Y.J. Calibration of satellite sensors after launch. *Appl. Opt.* **1986**, *25*, 1177–1185.
19. Meygret, A.; Briottet, X.; Henry, P.J.; Hagolle, O. Calibration of SPOT4 HRVIR and Vegetation cameras over Rayleigh scattering. In Proceedings of the Earth Observing Systems V, San Diego, CA, USA, 15 November 2000; Volume 4135, p. 302.
20. Luderer, G.; Coakley, J.A.; Tahnk, W.R. Using sun glint to check the relative calibration of reflected spectral radiances. *J. Atmos. Ocean. Technol.* **2005**, *22*, 1480–1493. [CrossRef]
21. Khakurel, P.; Leigh, L.; Kaewmanee, M.; Pinto, C.T. Extended pseudo invariant site-based trend-to-trend cross-calibration of optical satellite sensors. *Remote Sens.* **2021**, *13*, 1545. [CrossRef]
22. Loeb, N.G. In-flight calibration of NOAA AVHRR visible and near-IR bands over Greenland and Antarctica. *Int. J. Remote Sens.* **1997**, *18*, 477–490. [CrossRef]
23. Tahnk, W.R.; Coakley, J.A. Updated calibration coefficients for NOAA-14 AVHRR channels 1 and 2. *Int. J. Remote Sens.* **2001**, *22*, 3053–3057. [CrossRef]
24. Ling, W.; Xiuqing, H.; Yupeng, L.; Zhizhao, L.; Min, M. Selection and Characterization of Glaciers on the Tibetan Plateau as Potential Pseudoinvariant Calibration Sites. *IEEE J. Sel. Top. Appl. Earth Obs. Remote Sens.* **2019**, *12*, 424–436. [CrossRef]
25. Kwiatkowska, E.J.; Franz, B.A.; Meister, G.; McClain, C.R.; Xiong, X. Cross calibration of ocean-color bands from Moderate Resolution Imaging Spectroradiometer on Terra platform. *Appl. Opt.* **2008**, *47*, 6796–6810. [CrossRef] [PubMed]
26. Chander, G.; Mishra, N.; Helder, D.L.; Aaron, D.B.; Angal, A.; Choi, T.; Xiong, X.; Doelling, D.R. Applications of spectral band adjustment factors (SBAF) for cross-calibration. *IEEE Trans. Geosci. Remote Sens.* **2013**, *51*, 1267–1281. [CrossRef]
27. Obata, K.; Tsuchida, S.; Yamamoto, H.; Thome, K. Cross-calibration between ASTER and MODIS visible to near-infrared bands for improvement of aster radiometric calibration. *Sensors* **2017**, *17*, 1793. [CrossRef]
28. Barsi, J.A.; Alhammoud, B.; Czapla-Myers, J.; Gascon, F.; Haque, M.O.; Kaewmanee, M.; Leigh, L.; Markham, B.L. Sentinel-2A MSI and Landsat-8 OLI radiometric cross comparison over desert sites. *Eur. J. Remote Sens.* **2018**, *51*, 822–837. [CrossRef]
29. Scott, K.P.; Thome, K.J.; Brownlee, M.R. Evaluation of Railroad Valley playa for use in vicarious calibration. In Proceedings of the Multispectral Imaging for Terrestrial Applications, Denver, CO, USA, 4 November 1996; Volume 2818, p. 158.
30. Bacour, C.; Briottet, X.; Bréon, F.M.; Viallefont-Robinet, F.; Bouvet, M. Revisiting Pseudo Invariant Calibration Sites (PICS) over sand deserts for vicarious calibration of optical imagers at 20 km and 100 km scales. *Remote Sens.* **2019**, *11*, 1166. [CrossRef]
31. Cosnefroy, H.; Leroy, M.; Briottet, X. Selection and characterization of Saharan and Arabian desert sites for the calibration of optical satellite sensors. *Remote Sens. Environ.* **1996**, *58*, 101–114. [CrossRef]
32. Baiocchi, V.; Giannone, F.; Monti, F. How to Orient and Orthorectify PRISMA Images and Related Issues. *Remote Sens.* **2022**, *14*, 1991. [CrossRef]
33. Berthelot, B.; Santer, R. *Calibration Test Sites Selection and Characterisation Site Equipment and Auxiliary Data*; Vega Technique SAS: Erstein, France, 2008.
34. Zheng, X.; Huang, Q.; Wang, J.; Wang, T.; Zhang, G. Geometric accuracy evaluation of high-resolution satellite images based on Xianning test field. *Sensors* **2018**, *18*, 2121. [CrossRef]
35. Aerosol Robotic Network (AERONET) Homepage. Available online: https://aeronet.gsfc.nasa.gov/ (accessed on 17 July 2022).
36. Heller-Pearlshtien, D.; Pignatti, S.; Greisman-ran, U.; Ben-Dor, E. PRISMA sensor evaluation: A case study of mineral mapping performance over Makhtesh Ramon, Israel. *Int. J. Remote Sens.* **2021**, *42*, 5882–5914. [CrossRef]
37. Heller-Pearlshtien, D.; Ben-Dor, E. CalVal Evaluation of DESIS products in Amiaz Plain and Makhtesh Ramon Test sites. In Proceedings of the 1st DESIS User Workshop—Imaging Spectrometer Space Mission, Calibration and Validation, Applications, Methods, Virtual, 28 September–1 October 2021; pp. 13–21.
38. Zak, R.; Freund, R. Strain Measurements in Eastern Marginal Shear Zone of Mount Sedom Salt Diapir, Israel. *Am. Assoc. Pet. Geol. Bull.* **1980**, *64*, 568–581. [CrossRef]
39. Weinberger, R.; Bar-Matthews, M.; Levi, T.; Begin, Z.B. Late-Pleistocene rise of the Sedom diapir on the backdrop of water-level fluctuations of Lake Lisan, Dead Sea basin. *Quat. Int.* **2007**, *175*, 53–61. [CrossRef]
40. Kruse, F.A.; Lefkoff, A.B.; Boardman, J.W.; Heidebrecht, K.B.; Shapiro, A.T.; Barloon, P.J.; Goetz, A.F.H. The spectral image processing system (SIPS)-interactive visualization and analysis of imaging spectrometer data. *Remote Sens. Environ.* **1993**, *44*, 145–163. [CrossRef]
41. Ben-Dor, E.; Kindel, B.; Goetz, A.F.H. Quality assessment of several methods to recover surface reflectance using synthetic imaging spectroscopy data. *Remote Sens. Environ.* **2004**, *90*, 389–404. [CrossRef]

42. Kneizys, F.; Abreu, L.; Anderson, G.; Chetwynd, J.; Shettle, E.; Berk, A.; Bernstein, L.; Robertson, D.; Acharya, P.; Rothman, L.; et al. *The MODTRAN 2/3 Report and LOWTRAN-7 Model*; Contract F19628-91-C-0132; Ontar Corporation: North Andover, MA, USA, 1996.
43. MODTRAN®. Available online: http://modtran.spectral.com/modtran_index (accessed on 15 September 2022).
44. Brook, A.; Ben-Dor, E. Supervised vicarious calibration (SVC) of hyperspectral remote-sensing data. *Remote Sens. Environ.* **2011**, *115*, 1543–1555. [CrossRef]
45. Farr, T.G.; Rosen, P.A.; Caro, E.; Crippen, R.; Duren, R.; Hensley, S.; Kobrick, M.; Paller, M.; Rodriguez, E.; Roth, L.; et al. The Shuttle Radar Topography Mission. *Rev. Geophys.* **2007**, *45*, RG2004. [CrossRef]
46. Makhtesh Ramon Cal/Val Site. Available online: https://storymaps.arcgis.com/stories/bb5bf09ec7414454a012bfe9bf4b8545 (accessed on 23 July 2022).
47. Pignatti, S.; Amodeo, A.; Carfora, M.F.; Casa, R.; Mona, L.; Palombo, A.; Pascucci, S.; Rosoldi, M.; Santini, F.; Laneve, G. PRISMA L1 and L2 Performances within the PRISCAV Project: The Pignola Test Site in Southern Italy. *Remote Sens.* **2022**, *14*, 1985. [CrossRef]
48. Cogliati, S.; Sarti, F.; Chiarantini, L.; Cosi, M.; Lorusso, R.; Lopinto, E.; Miglietta, F.; Genesio, L.; Guanter, L.; Damm, A.; et al. The PRISMA imaging spectroscopy mission: Overview and first performance analysis. *Remote Sens. Environ.* **2021**, *262*, 112499. [CrossRef]

remote sensing

MDPI

Article

Vicarious Radiometric Calibration of the Multispectral Imager Onboard SDGSAT-1 over the Dunhuang Calibration Site, China

Zhenzhen Cui [1,2], Chao Ma [1], Hao Zhang [2,*], Yonghong Hu [3,4], Lin Yan [3,4], Changyong Dou [3,4] and Xiao-Ming Li [3,5]

1 School of Surveying and Land Information Engineering, Henan Polytechnic University, Jiaozuo 454003, China; 211804010010@home.hpu.edu.cn (Z.C.); mac@hpu.edu.cn (C.M.)
2 Airborne Remote Sensing Center, Aerospace Information Research Institute, Chinese Academy of Sciences, Beijing 100094, China
3 International Research Center of Big Data for Sustainable Development Goals, Beijing 100094, China; huyonghong@aircas.ac.cn (Y.H.); yanlin@aircas.ac.cn (L.Y.); doucy@aircas.ac.cn (C.D.); lixiaoming@aircas.ac.cn (X.-M.L.)
4 Key Laboratory of Digital Earth Science, Aerospace Information Research Institute, Chinese Academy of Sciences, Beijing 100094, China
5 Hainan Aerospace Information Research Institute, Wenchang 571399, China
* Correspondence: haozhang@aircas.ac.cn

Abstract: The multispectral imager (MII), onboard the Sustainable Development Science Satellite 1 (SDGSAT-1), performs detailed terrestrial change detection and coastal monitoring. SDGSAT-1 was launched at 2:19 UTC on 5 November 2021, as the world's first Earth science satellite to serve the United Nations 2030 Sustainable Development Agenda. A vicarious radiometric calibration experiment was conducted at the Dunhuang calibration site (Gobi Desert, China) on 14 December 2021. In-situ measurements of ground reflectance, aerosol optical depth (AOD), total columnar water vapor, radiosonde data, and diffuse-to-global irradiance (DG) ratio were performed to predict the top-of-atmosphere radiance by the reflectance-, irradiance-, and improved irradiance-based methods using the moderate resolution atmospheric transmission model. The MII calibration coefficients were calculated by dividing the top-of-atmosphere radiance by the average digital number value of the image. The radiometric calibration coefficients calculated by the three calibration methods were reliable (average relative differences: 2.20% (reflectance-based vs. irradiance-based method) and 1.43% (reflectance-based vs. improved irradiance-based method)). The total calibration uncertainties of the reflectance-, irradiance-, and improved irradiance-based methods were 2.77–5.23%, 3.62–5.79%, and 3.50–5.23%, respectively. The extra DG ratio measurements in the latter two methods did not improve the calibration accuracy for AODs ≤ 0.1. The calibrated MII images were verified using Landsat-8 Operational Land Imager (OLI) and Sentinel-2A MultiSpectral Instrument (MSI) images. The retrieved ground reflectances of the MII over different surface types were cross-compared with those of OLI and MSI using the FAST Line-of-sight Atmospheric Analysis of Hypercubes software. The MII retrievals differed by <0.0075 (7.13%) from OLI retrievals and <0.0084 (7.47%) from MSI retrievals for calibration coefficients from the reflectance-based method; <0.0089 (7.57%) from OLI retrievals and <0.0111 (8.65%) from MSI retrievals for the irradiance-based method; and <0.0082 (7.33%) from OLI retrievals and <0.0101 (8.59%) from MSI retrievals for the improved irradiance-based method. Thus, our findings support the application of SDGSAT-1 data.

Keywords: vicarious calibration; reflectance-based method; irradiance-based method; Dunhuang site; SDGSAT-1

1. Introduction

The Sustainable Development Science Satellite 1 (SDGSAT-1) is the world's first Earth science satellite specifically dedicated to collecting data related to the Sustainable Development Goals (SDGs) by 2030 [1]. It is also the first Earth science satellite of the Chinese

Academy of Sciences (CAS). The satellite was developed by the Earth Big Data Science and Engineering Project of the CAS and is the first satellite planned by the International Research Center of Big Data for Sustainable Development Goals. It carries three main payloads: the multispectral imager (MII), the glimmer imager (GLI), and the thermal infrared spectrometer (TIS), which are aimed at meeting the needs for monitoring, evaluation, and research of the global SDGs. Through the joint and continuous observation by the three imagers, SDGSAT-1 provides ample data support for the study of indicators representing human–nature interactions, the fine characterization of "traces of human activities", and the implementation of the global SDGs. SDGSAT-1 is expected to lead in reducing the global imbalance in sustainable development and the digital divide between regions. Most importantly, all SDGSAT-1 data are shared globally and freely, and they can be downloaded from the SDGSAT-1 Data Open System (http://124.16.184.48:6008, accessed on 9 June 2022).

Radiometric calibration is an essential assurance for the quantitative application of remote sensing data. Although the performance specifications of the satellite sensors were precisely evaluated in the laboratory before launch, the spectral and radiometric performance of the sensors are affected by post-launch changes, including space environment effects and the radiometric degradation of calibration equipment and the imaging sensor. To guarantee the MII's radiometric precision, vicarious radiometric calibration in the post-launch stage is required [2]. Vicarious radiometric calibration is most widely performed using in-situ experiments [3]. The reflectance-, irradiance-, and radiance-based methods, which were first proposed and applied in radiometric calibration tests conducted by the Remote Sensing Group of the University of Arizona, are three representative in-situ vicarious radiometric calibration methods [4]. According to data published by the United States, France, and other countries, the accuracy of absolute radiometric calibration using the current vicarious radiometric calibration methods in the visible and near-infrared bands is 3–5% [5]. The reflectance-based method depends on ground-based measurements, including ground reflectance measurements and atmospheric parameters. In addition, the top-of-atmosphere (TOA) radiance is obtained by entering these parameters into a radiative transfer model. The irradiance-based method is similar to the reflectance-based method; the main difference is the addition of the diffuse-to-global irradiance (DG) ratio measurement to reduce errors introduced by aerosol model assumption. The radiance-based method requires a strictly calibrated radiometer onboard the airplane and an uncalibrated sensor to obtain the radiance of the target under nearly simultaneous and consistent observation conditions. The measured radiance is corrected according to the atmospheric influence between the radiometer and the sensor.

Since the 1980s, numerous satellite radiometric calibration sites have been constructed in many different nations, consisting of White Sands Missile Range [3], Railroad Valley Playa [6], Lunar Lake Playa [7], Rogers Dry Lake [8], and Ivanpah Playa [9] in the United States; La Crau [10] in France; Newell Country [11] in Canada; Tinga Tingana [12], Uardry [13], and Lake Frome [14] in Australia; and Dunhuang [15], Qinghai Lake [16], and Baotou [17] in China. These calibration sites have provided solid support for the radiometric calibration of numerous optical sensors, including the thermal mapper [18,19], the enhanced thermal mapper plus [20,21], and the operational land imager (OLI) [22–24] onboard Landsat; the high-resolution visible onboard SPOT [25]; the moderate resolution imaging spectroradiometer onboard Terra and Aqua [26,27]; the multispectral sensor onboard the ZY satellite [28]; and the hyperspectral imager onboard the HJ-1A satellite [29]. Calibration results have shown that the radiance-based method has the maximum radiometric calibration accuracy among the three methods and that the higher the airplane's altitude, the higher the accuracy of this method. Nevertheless, the high cost of manpower and material resources, the strictly simultaneous measurements between the airplane and the satellite, the high accuracy of the radiometer, and the low success rate greatly limit the application of the radiance-based method. The irradiance-based method has lower accuracy than the radiance-based method but is superior to the reflectance-based method, particularly under poor atmospheric conditions with large aerosol optical depth (AOD). To solve the problem of low calibration accuracy under unstable atmospheric conditions, novel methods have been put forward in the past years, including the supervised vicarious [30] and the improved irradiance-based [31] methods, which have been effectively used for the vicarious radiometric calibration of an unmanned aerial vehicle hyperspectral sensor [31], airborne hyperspectral sensors [32], and the SPARK-01/02 satellites [33].

To further validate these methods, an experiment was performed at the Dunhuang calibration site in China on 14 December 2021, to calibrate the MII. This was the first in-situ calibration experiment of the MII and is expected to provide reliable radiometric calibration coefficients. Calibration uncertainty analysis is discussed in detail in this article. Two reference sensors, i.e., the OLI onboard Landsat-8 and the MSI onboard Sentinel-2A, were selected to cross-validate the results of the vicarious radiometric calibration.

The remainder of this article is divided into the following sections: the data, measurements, and vicarious radiometric calibration methods employed in this investigation are thoroughly described in Section 2. The MII calibration results are listed in Section 3. Finally, in Section 5, we summarize the key findings of our research.

2. Materials and Methods

2.1. Materials

2.1.1. Overview of SDGSAT-1

SDGSAT-1, developed jointly by the Aerospace Information Research Institute, Changchun Institute of Optics, Fine Mechanics and Physics, the Shanghai Institute of Technical Physics, and the Innovation Academy for Microsatellites (CAS, China), was successfully launched by the CZ-6 rocket at the Taiyuan Satellite Launch Center at 2:19 UTC on 5 November 2021. SDGSAT-1 carries three sensors (MII, GLI, and TIS) to observe the Earth's surface continuously during the day and night. The MII is anticipated to deliver multispectral data over inland water bodies and terrestrial surfaces. The GLI and TIS were designed to acquire global night data and thermal infrared data, respectively. Table 1 lists the specific parameters of SDGSAT-1. For detailed information on SDGSAT-1, please visit http://www.sdgsat.ac.cn/ (accessed on 29 July 2022).

Table 1. Major technical parameters of SDGSAT-1.

SDGSAT-1	Parameters
Payloads	MII, GLI, TIS
Spectral bands	MII: 7 bands ranging from 374 to 911 nm GLI: 4 bands ranging from 424 to 910 nm TIS: 3 bands ranging from 8 to 12.5 μm
Swath	300 km
Orbit	Type: sun-synchronous orbit Altitude: 505 km Inclination: 97.5°
Spatial resolution	MII: 10 m @ 505 km GLI: 10 m @ 505 km (PAN), 40 m @ 505 km (B, G, R) TIS: 30 m @ 505 km
Revisit period	11 days

2.1.2. SDGSAT-1 MII

The MII instrument, designed by the Changchun Institute of Optics, Fine Mechanics and Physics and the Aerospace Information Research Institute CAS, comprises two cameras, A and B, with eight detectors per camera. Seven spectral bands are set in the visible and near-infrared bands: B1 (coastal/aerosol 1), B2 (coastal/aerosol 2), B3 (blue), B4 (green), B5 (red), B6 (red edge), and B7 (near infrared). The MII instrument was designed to have a large swath width of 300 km at a nominal altitude of 505 km. The spatial resolution is approximately 10 m in all seven bands, providing high-resolution data support for SDGs

2, 6, 11, 13, 14, and 15 [34]. Figure 1 depicts the relative spectral response function of the SDGSAT-1 MII, while Table 2 displays the major technical parameters.

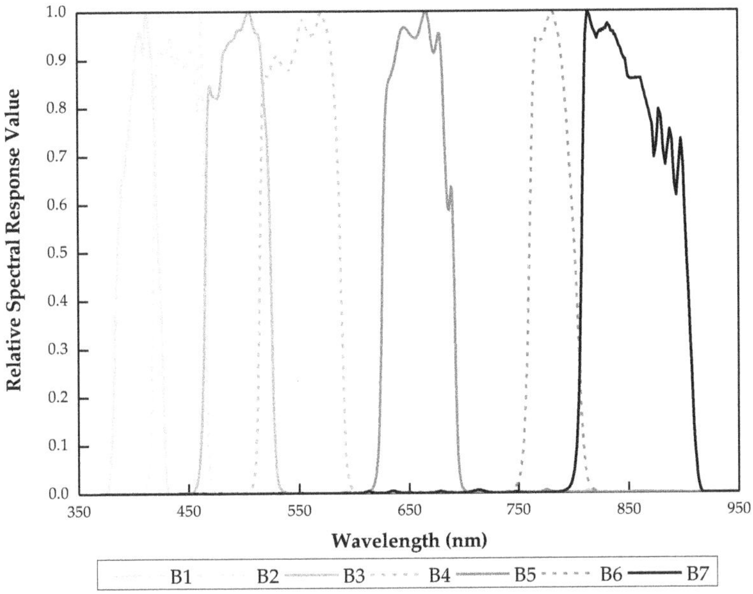

Figure 1. Relative spectral response function of the SDGSAT-1 MII.

Table 2. Major technical parameters of the SDGSAT-1 MII.

SDGSAT-1 MII	Description	Technical Parameters
Spectral range	B1	374–427 nm
	B2	410–467 nm
	B3	457–529 nm
	B4	510–597 nm
	B5	618–696 nm
	B6	744–813 nm
	B7	798–911 nm
Spatial resolution	B1–B7	10 m @ 505 km
Signal-to-noise ratio	B1	≥130
	B2–B7	≥150
Swath width	B1–B7	300 km
Modulation transfer function	Static	≥0.23
	Dynamic	≥0.10
Dynamic range	B1–B7	≥60 dB
Digitalizing bits	B1–B7	12 bits

2.1.3. Calibration Site and Image Data Acquisition

The Dunhuang calibration site, established at the end of the 20th century, is one of the China Radiometric Calibration Sites for the vicarious radiometric calibration of satellite sensors, as well as an internationally recognized radiometric calibration site with flat terrain, uniform surface, stable and measurable ground objects, and good directional characteristics [2]. The Dunhuang calibration site is located in northwest China on the regeneration alluvial fan of the Danghe River more than 20 km northwest of Dunhuang City. The calibration site has a total area of approximately 30 km × 30 km; its geographical coordinates are 40.04–40.28°N and 94.17–94.5°E; and its altitude is 1160 m. As shown in Figure 2, the region utilized for vicarious radiometric calibration tests of high- and medium-spatial resolution sensors is roughly 500 m × 500 m and is located in the middle of the alluvial fan. The entire alluvial fan is fairly flat, and the surface is mainly evenly distributed gravel with diameters of 0.2–8 cm [35]. The ground surface of the Dunhuang site is relatively flat, with low vegetation coverage and high spectral stability. The atmosphere is dry and clean, with a low AOD (in the absence of sandstorms, the local AOD is 0.1–0.2) and low total columnar water vapor (CWV). The sunshine duration is long. Therefore, the Dunhuang site is conducive for calibration experiments. Vicarious radiometric calibration of multiple domestic satellites, including BJ-1 [36], FY-1C [37], HJ-1A [29], CBERS-02 [38], and SPARK-01/02 [33], was carried out at the Dunhuang calibration site. The SDGSAT-1 MII data over the Dunhuang calibration site were acquired at 03:45:17 UTC on 14 December 2021. Table 3 lists the observation geometries on the SDGSAT-1 overpass date.

Figure 2. Sentinel-2A/MSI image of the Dunhuang calibration site acquired on 17 December 2021.

Table 3. Information on the capture of SDGSAT-1 MII image at the Dunhuang calibration site.

Date	Overpass Time (UTC)	Solar Zenith (°)	Solar Azimuth (°)	View Zenith (°)	View Azimuth (°)
14 December 2021	03:45:17	68.5554	152.2536	18.1581	304.6388

2.1.4. Simultaneous Measurement of Ground Reflectance

The internationally popular FieldSpec-4 ASD spectroradiometer (ASD Inc., Longmont, CO, USA), with a wavelength range of 350–2500 nm, was used for simultaneous ground reflectance measurements. The calibration site was a 500 m × 500 m square region, with SDGSAT-1 MII coverage of approximately 50 cross-track pixels and 50 along-track pixels. Ground reflectance measurements were performed in the square region (Figure 2) from 10:30 to 12:30 UTC + 8. Ground reflectance was measured along a designed route, as shown in Figure 3b. Adjacent measurement points were ~20 m apart. Five measurements were taken around each point, and five spectra were collected for each measurement. The total measured data exceeded 400 groups, with nearly 2000 spectral data points of the Dunhuang calibration site. After removing abnormal and erroneous measurements, all valid measurements were averaged to represent the final reflectance of the Dunhuang calibration site (Figure 4).

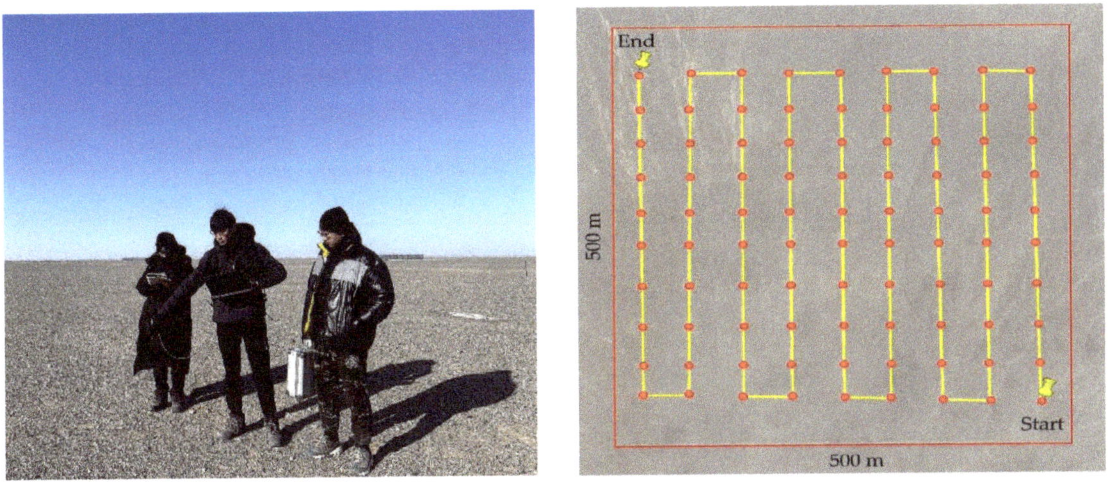

(a) (b)

Figure 3. (**a**) Photograph of the in-situ ground reflectance measurement at Dunhuang calibration site. (**b**) Scheme of the ground surface measurement route.

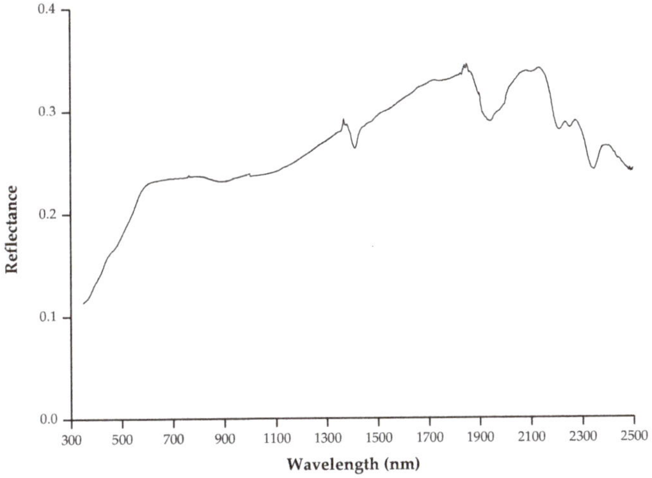

Figure 4. Simultaneous ground reflectance measurements performed over the Dunhuang calibration site on 14 December 2021, using FieldSpec-4 ASD spectroradiometer.

2.1.5. Simultaneous Measurement of Atmospheric Parameters

Simultaneous atmospheric parameters (AOD, CWV, the DG ratio, and the atmospheric vertical profile) were acquired on 14 December 2021. Total AOD and CWV were measured using an automated CIMEL CE318 sun photometer (Cimel Electronique, Paris, France) and retrieved using an improved Beer–Lambert–Bouguer method [39] and a four-parameter method [40]. The AOD values at the 440 nm and 670 nm channels were employed to derive the AOD at the 550 nm channel via logarithmic interpolation, using Equation (1). Figure 5 shows the 550 nm AOD and CWV on the SDGSAT-1 overpass date.

$$\tau_\alpha(\lambda) = \beta \cdot \lambda^{-\alpha} \tag{1}$$

$$\alpha = -\frac{\ln(\mathrm{AOD}(\lambda_m))/\mathrm{AOD}(\lambda_n))}{\ln(\lambda_m/\lambda_n)} \tag{2}$$

$$\beta = \frac{\mathrm{AOD}(\lambda_m)}{\lambda_m{}^{-\alpha}} = \frac{\mathrm{AOD}(\lambda_n)}{\lambda_n{}^{-\alpha}} \tag{3}$$

where $\tau_\alpha(\lambda)$ is the AOD at wavelength λ (in μm), α is the Ångstrom wavelength exponent, and β is the atmospheric turbidity coefficient.

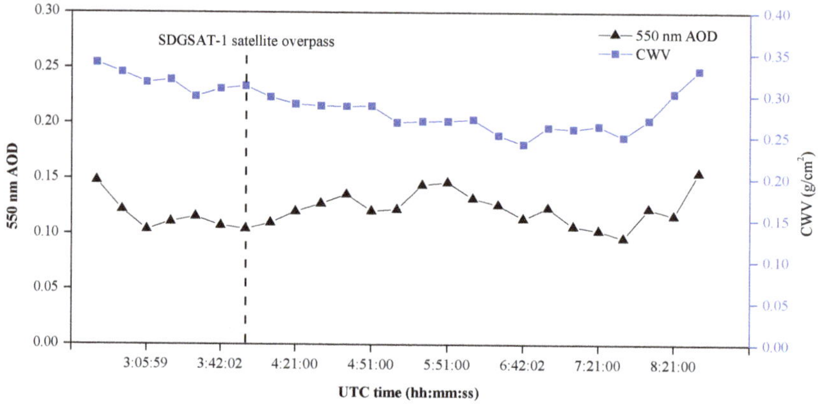

Figure 5. The 550 nm AOD and CWV retrieved from the CE318 sun photometer on 14 December 2021.

Data retrieved from the CE318 sun photometer within 5 min before and after the overpass time were averaged to obtain the 550 nm AOD and CWV at the overpass time. A level-3 ozone data product (https://acdisc.gsfc.nasa.gov/data/Aura_OMI_Level3/OMDOAO3e.003/2021/, accessed on 21 December 2021) retrieved from the ozone monitoring instrument onboard the Aura satellite was used as the columnar ozone content at the overpass time. Table 4 presents the simultaneous measurements of atmospheric parameters at the SDGSAT-1 overpass time.

Table 4. Simultaneous measurements of atmospheric parameters.

Atmospheric Parameters	Simultaneous Measurements at Overpass Time
550 nm AOD	0.1045
CWV	0.3114 g/cm^2
Columnar ozone content	301.6 DU

An automated spectral radiometer was used to measure the DG ratio within the Dunhuang calibration site. The DG ratio was recorded at 6 min intervals during the daytime. Figure 6a,b show the DG ratios at 550 nm throughout the day of the SDGSAT-1 overpass and the entire spectrum at the SDGSAT-1 overpass time, respectively. Abnormal data were eliminated to derive a smoothed DG ratio curve, which indicated a stable atmospheric condition most of the time, with some exceptions (e.g., at around 2:17 and 7:48 UTC), on 14 December 2021.

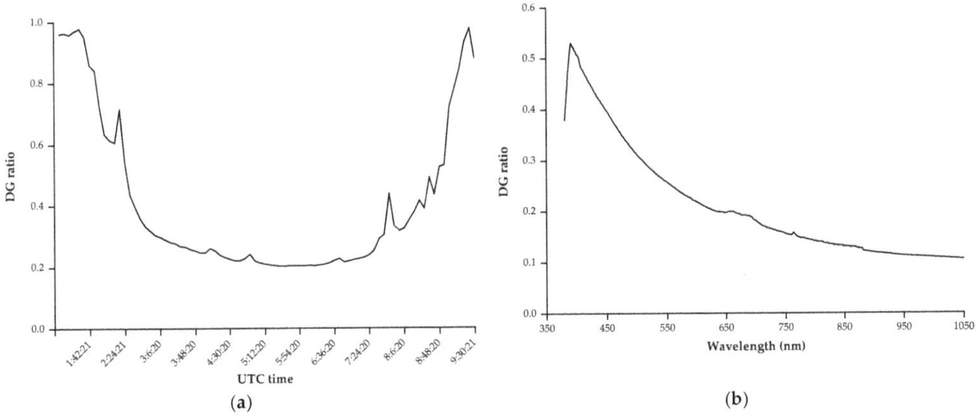

Figure 6. Measured DG ratios (**a**) at 550 nm on the date of the SDGSAT-1 overpass and (**b**) the entire spectrum at the time of the SDGSAT-1 overpass.

A radiosonde balloon was used to obtain the atmospheric vertical profiles of pressure, temperature, and humidity on the SDGSAT-1 overpass date. The balloon was released from the Dunhuang National Reference Climate Station at 07:15 a.m. (Beijing time) on 14 December 2021. Figure 7 shows the measured atmospheric vertical profiles of pressure, temperature, and relative humidity, which varied with altitude, on 14 December 2021.

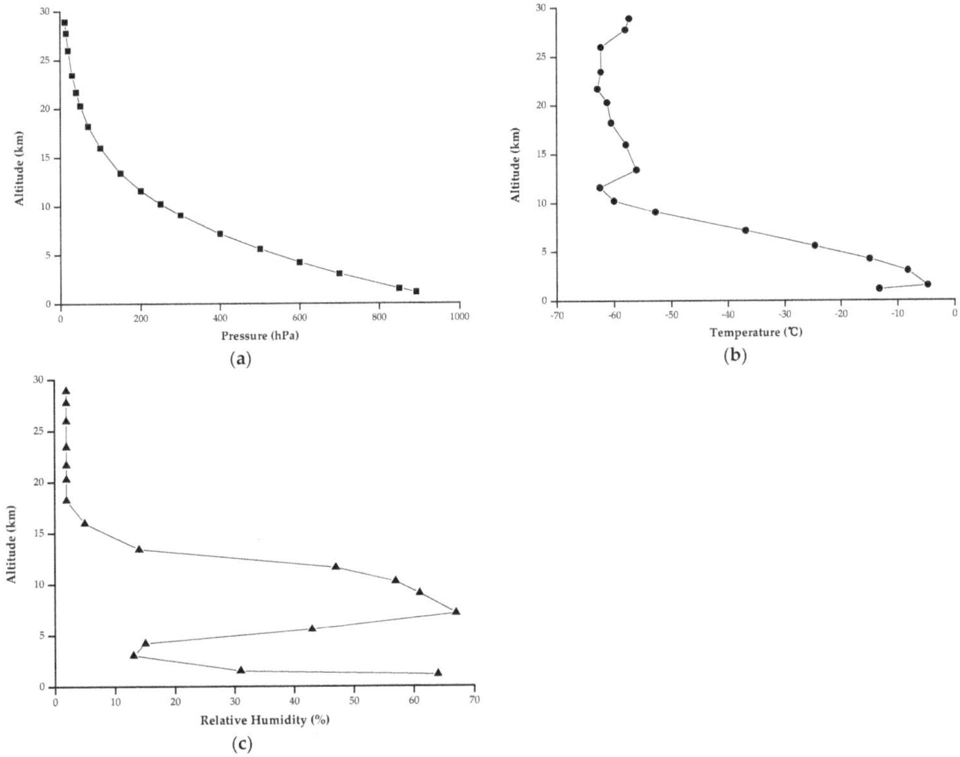

Figure 7. Atmospheric vertical profiles of (**a**) pressure, (**b**) temperature, and (**c**) relative humidity obtained by the radiosonde balloon released on 14 December 2021.

2.2. Methods

2.2.1. Radiometric Calibration Method

The widely used reflectance- [4], irradiance- [18], and improved irradiance-based [31,33] methods were adopted to predict the TOA radiance. All three methods rely on the accurate simultaneous measurements of the ground reflectance and atmospheric parameters. The reflectance-based method for radiometric calibration is generally used in clear and cloudless weather conditions. It requires fewer surface observations and less labour input than the other methods and is easy to operate. However, the reflectance-based method requires high visibility and a stable atmosphere to diminish the calibration uncertainty caused by aerosol model assumptions. The total calibration uncertainty significantly increases in the case of a high AOD [33]. The irradiance-based method is an enhanced reflectance-based method that requires simultaneous measurements of the ground reflectance and the DG ratio. The latter parameter is utilized to replace the aerosol model assumption in the reflectance-based method. The aerosol model assumption in the irradiance-based method only affects path radiance and the hemisphere albedo in Equation (7), while large uncertainties in the upward and downward scattering transmittance are minimized by using the measured DG ratio. The reflectance-based method has a total uncertainty of 4.9% [41], and the simultaneous ground reflectance in in-situ measurements and the aerosol model assumption in the radiative transfer calculation are its main sources of uncertainty. The irradiance-based method generally outperforms the reflectance-based method with an uncertainty of 3.5% [41], and the DG ratio in in-situ measurements is its main source of uncertainty. However, in the case of sunny weather and a low AOD, the calibration accuracy of the reflectance-based method is comparable to that of the irradiance-based method [31]. The vicarious radiometric calibration equations for the reflectance- and irradiance-based methods used for the satellite sensors are Equations (4) and (7), respectively. The TOA radiance can be converted using Equation (8) [33]. Equations (9) and (10) were used to determine the spectral radiometric calibration coefficients on the assumption that the sensor response was linear.

$$\rho^*(\theta_s,\theta_v,\varphi_{v-s}) = \rho_a(\theta_s,\theta_v,\varphi_{v-s}) + \frac{\rho_t}{1-\rho_t \times s} \times T(\theta_s) \times T(\theta_v) \tag{4}$$

$$T(\theta_s) = T_{dir}(\theta_s) + T_{dif}(\theta_s) = (1-\rho_t \times s) \times \frac{e^{-\tau/\mu_s}}{1-\alpha_s} \tag{5}$$

$$T(\theta_v) = T_{dir}(\theta_v) + T_{dif}(\theta_v) = (1-\rho_t \times s) \times \frac{e^{-\tau/\mu_v}}{1-\alpha_v} \tag{6}$$

$$\rho^*(\theta_s,\theta_v,\varphi_{v-s}) = \rho_a(\theta_s,\theta_v,\varphi_{v-s}) + \frac{e^{-\tau/\mu_s}}{1-\alpha_s} \times \rho_t \times (1-\rho_t \times s) \times \frac{e^{-\tau/\mu_v}}{1-\alpha_v} \tag{7}$$

$$L = \rho^* \times \cos(\theta_s) \times E_0 / \left(d^2 \times \pi\right) \tag{8}$$

$$L_i(\lambda) = \frac{\int L(\lambda) \cdot RSR_i(\lambda) d\lambda}{\int RSR_i(\lambda) d\lambda} \tag{9}$$

$$L_i = DN_i \cdot Gain_i + Bias_i \tag{10}$$

where θ_s and θ_v are the sun zenith angle and the view zenith angle, respectively; φ_{v-s} is known as the relative azimuth angle between the view azimuth angle and the sun azimuth angle. ρ_t is the measured ground reflectance, and ρ_a is the atmospheric intrinsic reflectance. s is the atmospheric hemisphere reflectance. $T(\theta_s)$ and $T(\theta_v)$ are the total transmittances of both the downward direction (solar path) and the upward direction (view path), respectively; $T_{dir}(\theta_s)$ and $T_{dif}(\theta_s)$ are the direct and diffuse transmittances in the downward direction, and $T_{dir}(\theta_v)$ and $T_{dif}(\theta_v)$ are the direct and diffuse transmittances in the upward direction. ρ^* and L are the TOA reflectance and TOA radiance of the surface target, respectively. μ_s and μ_v refer to the values of $\cos(\theta_s)$ and $\cos(\theta_v)$, respectively; α_s and α_v represent the DG ratios of the solar direction and viewing direction, respectively.

d is the Sun–Earth distance in astronomical units and E_0 is the TOA solar irradiance. L_i represents the TOA radiance for the i-th band. RSR_i is the relative spectral response for the i-th band. DN_i is the digital number (DN) derived from the L1A image for the i-th band, and $Gain_i$ and $Bias_i$ are the calibration coefficients for the i-th band.

Even though the measured DG ratio used in the irradiance-based method would greatly reduce the uncertainty error caused by the aerosol model assumption, it is difficult to directly measure the DG ratio in the observation direction (0° of view zenith angle in most cases) in in-situ experiments. Therefore, it has to be extrapolated by fitting the values at different solar zenith angles. There is a linear relationship if the atmospheric conditions are stable, as shown in Equation (11) [4].

$$ln(1-\alpha_s) = ln(1-\rho_t s) - (1-b)\tau m \tag{11}$$

where m is the relative optical air mass (i.e., the inverse of the cosine of solar zenith, $1/\mu_s$) and $-(1-b)\tau$ and $ln(1-\rho_t s)$ are the slope and intercept of the linear fitting equation, respectively.

Ideal stable atmospheric conditions are not guaranteed in in-situ experiments. Therefore, when the atmospheric condition is unstable, but DG ratio data are used for satellite vicarious radiometric calibration, only α_s is employed to substitute for the scattering effect in the reflectance-based method, as shown in Equation (12) [31], which is the improved irradiance-based method. This method has been used for vicarious calibration of the SPARK-01/-02 satellites [33] and an unmanned aerial vehicle hyperspectral sensor [31].

$$\rho^*(\theta_s, \theta_v, \varphi_{v-s}) = \rho_a(\theta_s, \theta_v, \varphi_{v-s}) + \frac{\rho_t \times e^{-\tau/\mu_s}}{1-\alpha_s} \times T(\theta_v) \tag{12}$$

The relationship between the ratio of $ln(1-\alpha_s)$ and the relative optical air mass (m) at 550 nm measured on 14 December 2021 is represented by a scatter plot (Figure 8), and it revealed a nearly linear relationship of the measurements, with an R^2 value of 0.9965, indicating a relatively stable atmospheric condition on this date. Figure 9 shows goodness-of-fit (R^2) statistics for the DG ratio measurements in the entire visible to near infrared spectral range. The DG ratios in both the solar and viewing directions at the SDGSAT-1 overpass time are shown in Figure 10.

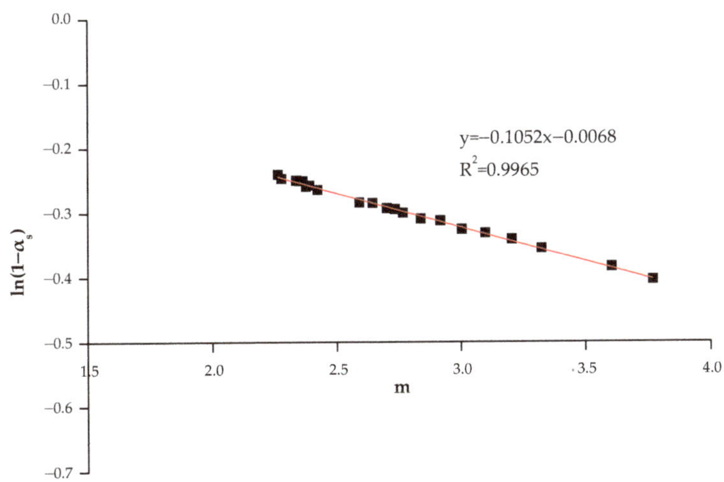

Figure 8. Scatter plot of $ln(1-\alpha_s)$ versus m at 550 nm.

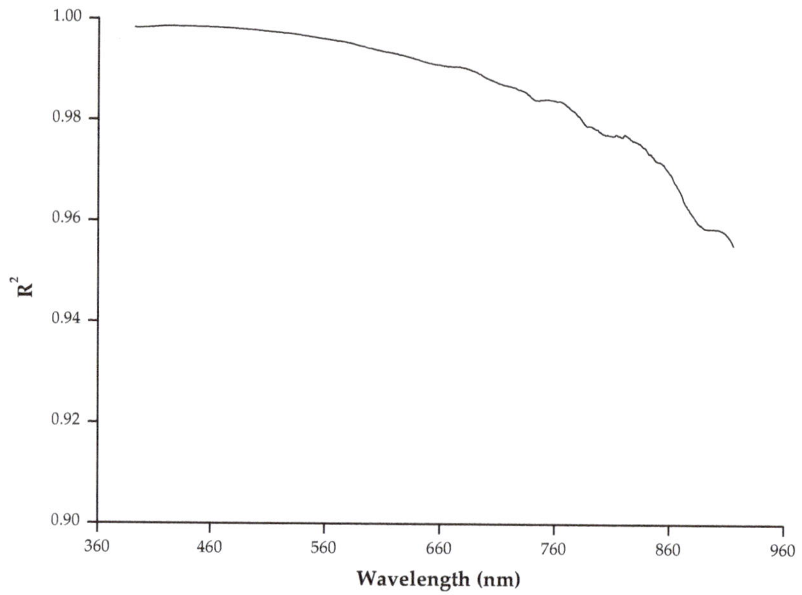

Figure 9. Goodness-of-fit statistics for DG ratio measurements according to Equation (11).

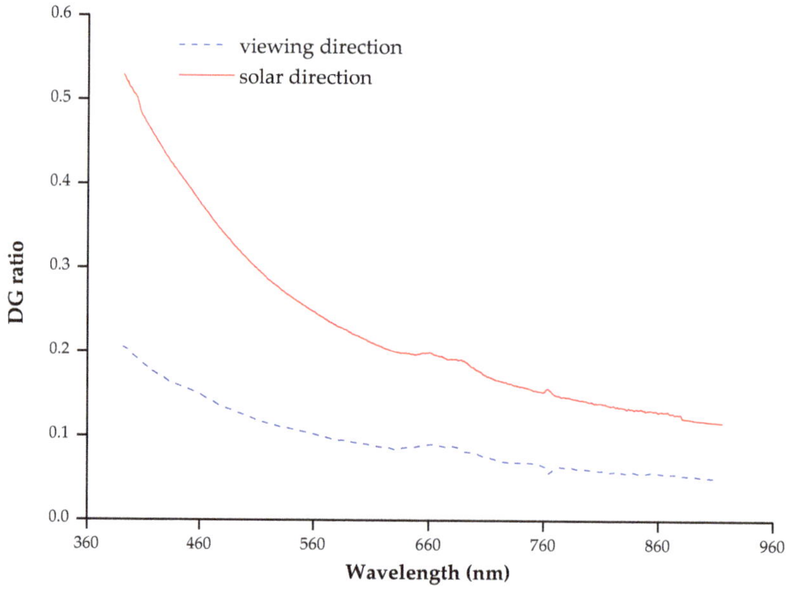

Figure 10. DG ratios in solar and viewing directions during the SDGSAT-1 overpass.

2.2.2. Radiative Transfer Calculations

The moderate resolution atmospheric transmission (MODTRAN) model is currently one of the most frequently used radiative transfer models for radiometric calibration. The spectral range of MODTRAN is 0.3–100 μm, which allows the calculation of atmospheric radiative transfer from the visible, near-infrared, mid-infrared, and far-infrared bands [42]. The measured ground reflectance, spectral response function of the SDGSAT-1 MII sensor,

550 nm AOD, CWV, ozone content, image geometric parameters, and other parameters were input into MODTRAN v.5.2.1 software. The Ångstrom index was used to constrain the band scattering characteristics of the default aerosol type in the original radiative transfer model, and then the band-equivalent TOA spectral radiance of the MII sensor was obtained. Assuming that each MII channel had a linear response and that the bias was zero because of the low dark current of the instrument (Figure 11), the TOA radiance and the average DN value of the image of each channel were substituted into Equation (13) to obtain the calibration coefficient $Gain_i$ of each channel.

$$Gain_i = \frac{L_i}{\overline{DN_i}} \quad (13)$$

where L_i is the band-equivalent TOA spectral radiance ($\mathrm{W \cdot m^{-2} \cdot sr^{-1} \cdot \mu m^{-1}}$), $\overline{DN_i}$ is the average DN of the selected area of the satellite image in band i, and $Gain_i$ is the radiometric calibration coefficient $Gain$ in band i.

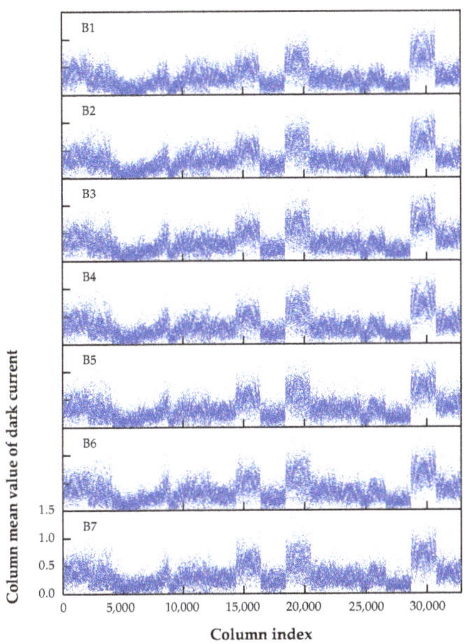

Figure 11. The column mean value of dark current acquired by imaging the ocean at night on 14 December 2021.

2.2.3. Calibration Uncertainty Estimation

The relative difference is used to represent the vicarious radiometric calibration uncertainty, and the uncertainty of each factor on the calibration coefficient ε_i is calculated according to Equation (14). $L_{Original_i}$ is the reference spectral TOA radiance, and L_{New_i} is the spectral TOA radiance after changing the input conditions. The final total uncertainty is expressed as the square root of the sum of the squares of each error uncertainty [31], as described by Equation (15).

$$\varepsilon_i = \left| \frac{L_{New_i} - L_{Original_i}}{L_{Original_i}} \right| \times 100\% \quad (14)$$

$$\varepsilon_{Total} = \sqrt{\varepsilon_1{}^2 + \varepsilon_2{}^2 + \varepsilon_3{}^2 + \ldots + \varepsilon_n{}^2} \quad (15)$$

2.2.4. Cross Validation

Given that the relative spectral response of each band differs among sensors, the ground reflectance of the "reference sensor" has to be transferred to the "target sensor" using a spectral band adjustment factor (SBAF). The divergences in the relative spectral response between various satellite sensors were computed by the SBAF using Equation (16):

$$SBAF = \frac{\hat{\rho}_{Reference}}{\hat{\rho}_{Target}} = \frac{\frac{\int \rho_{in-situ}(\lambda) RSR_{Reference}(\lambda) d\lambda}{\int RSR_{Reference}(\lambda) d\lambda}}{\frac{\int \rho_{in-situ}(\lambda) RSR_{Target}(\lambda) d\lambda}{\int RSR_{Target}(\lambda) d\lambda}} \tag{16}$$

where $\rho_{in-situ}(\lambda)$ is the per-wavelength in-situ ground reflectance at the Dunhuang calibration site, $RSR_{Reference}$ and RSR_{Target} are the per-wavelength relative spectral response curves from the "reference sensor" and the "target sensor," and $\hat{\rho}_{Reference}$ and $\hat{\rho}_{Target}$ are the in-situ ground reflectance values based on the integration of the "reference sensor" and "target sensor" values. The ground reflectance of the "reference sensor" was transferred to the "target sensor" using the SBAF according to Equation (17):

$$\rho'_{Target} = \rho_{Target} \times SBAF \tag{17}$$

3. Results

The SDGSAT-1 TOA radiance simulated by MODTRAN 5.2.1 according to the reflectance-, irradiance-, and improved irradiance-based methods for the SDGSAT-1 MII is shown in Figure 12. Slight differences in the TOA spectral radiance were observed among the three methods.

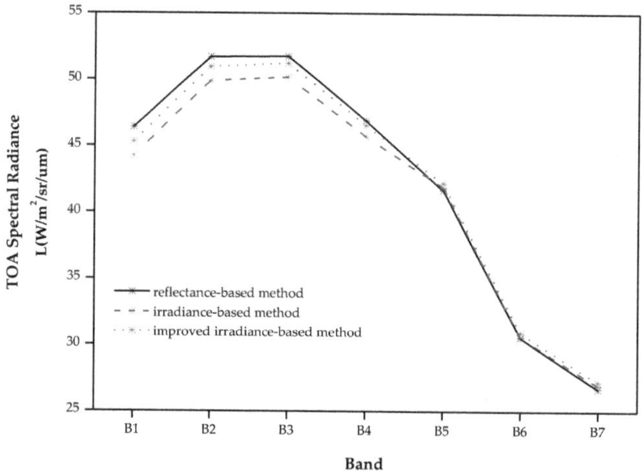

Figure 12. MODTRAN-simulated TOA spectral radiances for the SDGSAT-1 MII calculated by the reflectance-, irradiance-, and improved irradiance-based methods.

Table 5 lists the relative differences in TOA spectral radiances simulated by MODTRAN between the reflectance-based method and the irradiance- and improved irradiance-based methods. The relative differences between the reflectance- and irradiance-based methods varied from 0.12% to 4.61% in different bands, while differing from 0.73% to 2.31% between reflectance- and improved irradiance-based methods. In addition, the relative differences for B1 to B3 were larger than that of B4 to B7, indicating a higher radiometric calibration accuracy for B4 to B7 compared with that of B1 to B3. As expected, the strong aerosol scattering effects in the shorter spectral range cause much larger calibration uncertainty in the first three bands than in the other bands. In general, the reflectance-based method has a similar accuracy with the irradiance- and improved irradiance-based methods when the AOD is small. However, the maximum difference between the reflectance- and irradiance-based methods can reach up to 4.61%, which was attributed either to the measurement uncertainties of the DG ratio in the solar direction or the extrapolation uncertainties in the viewing direction. Such kind of difference was seldom mentioned in the previous literature and will be discussed in later sections. As a comparison, the difference between reflectance- and improved irradiance-based methods was lower because only the DG ratio in the solar direction was used in the latter method. The average relative difference between the reflectance- and improved irradiance-based methods (1.43%) was lower than that between the reflectance- and irradiance-based methods (2.20%) because only the DG ratio in the solar direction was used in the improved irradiance-based method. In addition, the relative differences for B1 to B3 were larger than those for B4 to B7, indicating a higher radiometric calibration accuracy for the latter bands.

Table 5. Relative differences in TOA spectral radiances between the reflectance-based method and the irradiance- and improved irradiance-based methods.

Band	Relative Difference between Reflectance- and Irradiance-Based Methods	Relative Difference between Reflectance- and Improved Irradiance-Based Methods
B1	4.61%	2.31%
B2	3.44%	1.41%
B3	3.00%	0.99%
B4	2.54%	0.73%
B5	0.63%	1.35%
B6	0.12%	1.18%
B7	1.09%	2.04%
Average relative difference	2.20%	1.43%

The calibration coefficients of the MII were calculated by dividing the TOA radiance by the average DN of the 40 × 40 pixels. Table 6 lists the vicarious radiometric calibration coefficients derived by the three different methods.

Table 6. Vicarious radiometric calibration coefficients for the SDGSAT-1 MII (unit: $W \cdot m^{-2} \cdot sr^{-1} \cdot \mu m^{-1} \cdot DN^{-1}$).

Band	Reflectance-Based Method	Irradiance-Based Method	Improved Irradiance-Based Method
B1	0.051616908	0.049237722	0.050422238
B2	0.036291910	0.035042783	0.035781749
B3	0.023327113	0.022627800	0.023095942
B4	0.015849453	0.015446672	0.015733488
B5	0.016096157	0.016197140	0.016314004
B6	0.019731394	0.019754163	0.019963674
B7	0.013811256	0.013961794	0.014092721

4. Discussion

4.1. Uncertainty Analysis

The uncertainty errors associated with in-situ simultaneous measurements, data processing, and calibration method selection can be directly or indirectly attributed to the total uncertainty in the satellite sensor calibration calculation results [35,43,44]. In this section, we mainly focus on the uncertainties caused by aerosol model assumption, atmospheric model assumption, AOD retrieval, water vapor retrieval, ground reflectance measurements, viewing geometry error, radiative transfer model-inherent errors, and DG ratio measurement. The calibration uncertainty contributed by each parameter was estimated from the difference between the TOA radiance predicted when errors were added to each parameter and the original measurement (Equation (14)).

4.1.1. Uncertainty Analysis of Aerosol Model Assumptions

A previous study revealed that the natural aerosol type at the Dunhuang calibration site is similar to the rural and desert types in MODTRAN [33]. To estimate the uncertainty caused by different aerosol types, desert, urban, and maritime aerosol types were input in sequence while other parameters remained unchanged during the MODTRAN calculation, and the TOA spectral radiance of the output for the three aerosol types was compared with that of the rural type. The Ångstrom exponent coefficients determined by multiband aerosol optical thickness derived from CE318 measurement data can be used as input to constrain the band scattering characteristics of the default aerosol type in the original radiative transfer model to improve the calculation accuracy of aerosol scattering in each band [41]. The Ångstrom index of SDGSAT-1 is 0.7938. The relative differences in the TOA spectral radiance for different aerosol types for the reflectance-, irradiance-, and improved irradiance-based methods are exhibited in Figure 13. The average relative differences are provided in Table 7.

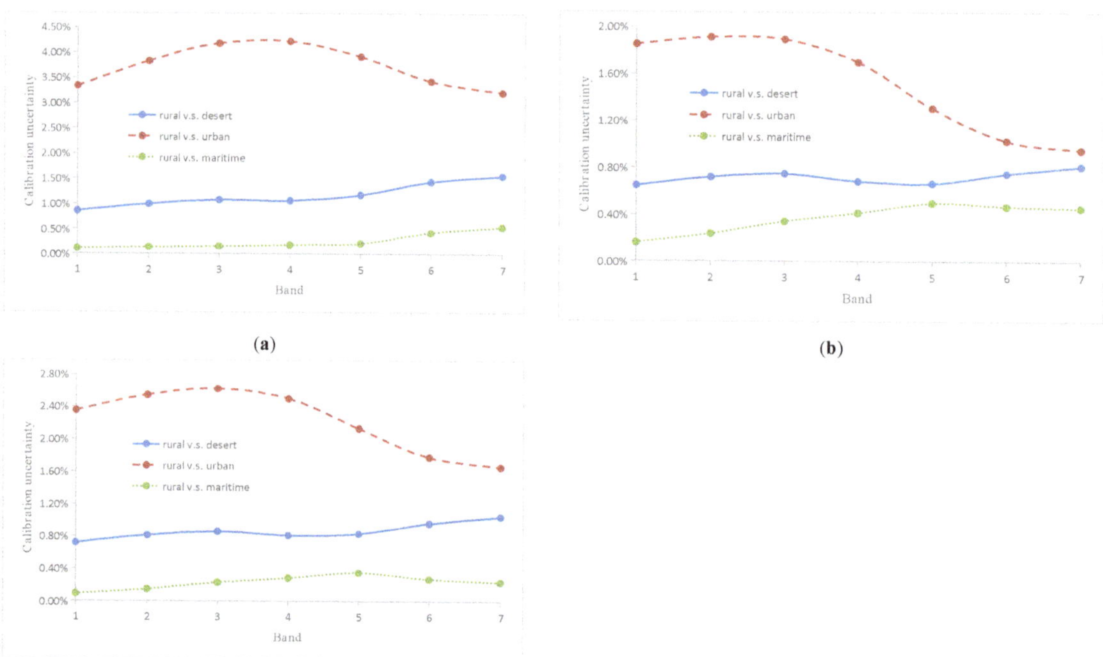

(c)

Figure 13. Calibration uncertainty caused by the assumption of different aerosol types for SDGSAT-1 MII calibration on 24 December 2021, utilizing the (**a**) reflectance-, (**b**) irradiance-, and (**c**) improved irradiance-based methods.

Table 7. Average relative differences in TOA radiance for the SDGSAT-1 MII.

Method	Rural vs. Desert	Rural vs. Urban	Rural vs. Maritime
Reflectance-based method	1.17%	3.74%	0.25%
Irradiance-based method	0.72%	1.52%	0.37%
Improved irradiance-based method	0.86%	2.23%	0.23%

As shown in Figure 13 and Table 7, the urban aerosol type had the largest average relative difference from the rural aerosol type (3.74%), whereas the average relative differences for desert and maritime aerosol model assumptions were relatively minor, particularly for the maritime type (<0.54% in all seven bands). For the reflectance-based method, aerosol model assumption was the largest error source affecting the calibration accuracy, with the largest uncertainty of 3.74%. Both the irradiance- and improved irradiance-based methods adopt the DG ratio to reduce the uncertainty associated with aerosol model assumptions, and their uncertainties were significantly reduced to within 1.52% and 2.23%, respectively.

We further analysed the relative differences between the rural and the other three aerosol types for the reflectance-, irradiance-, and improved irradiance-based methods under different AOD conditions (550 nm AOD = 0.05, 0.1, 0.2, 0.3, 0.4, and 0.5) (Figure 14). The results confirmed that aerosol model assumption is the major factor affecting the calibration accuracy of the reflectance-based method under different AOD conditions. The urban aerosol type had the largest uncertainty when compared with the rural aerosol type under different AOD conditions, leading to the large uncertainty of the reflectance-based method. The irradiance-based method is the first choice for in-situ experiments for the desert and urban aerosol types. In contrast, the improved irradiance-based method is preferred for the rural and maritime aerosol types.

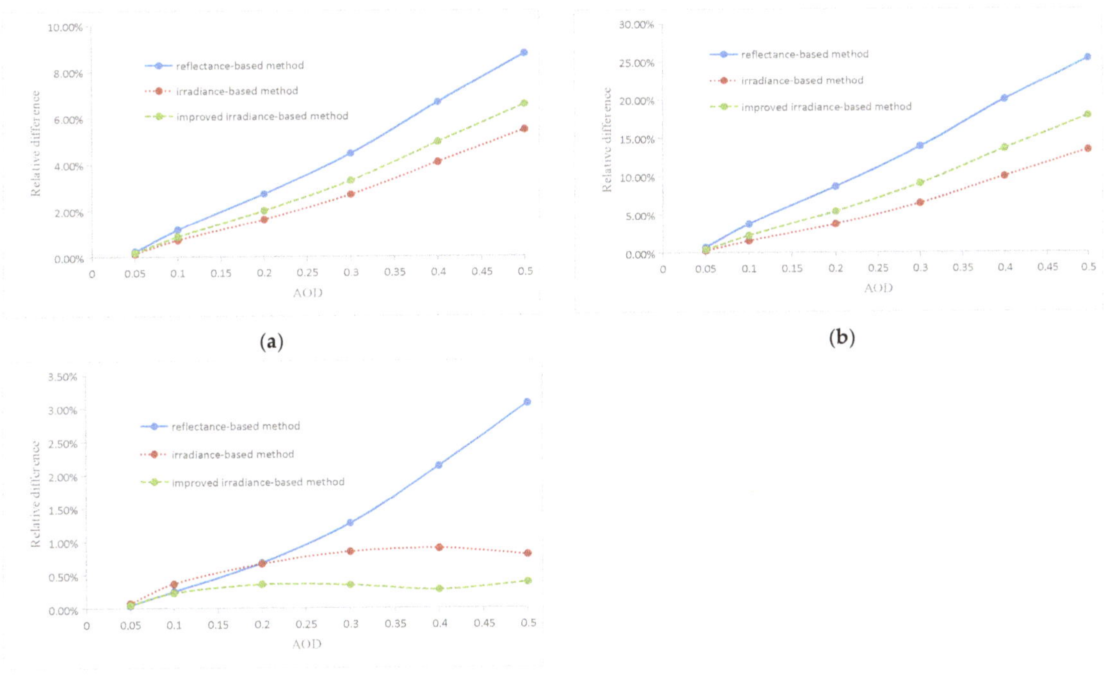

Figure 14. Relative differences between the (**a**) rural and desert aerosol types, (**b**) rural and urban aerosol types, and (**c**) rural and maritime aerosol types for the reflectance-, irradiance-, and improved irradiance-based methods under different AOD conditions (AOD = 0.05, 0.1, 0.2, 0.3, 0.4, and 0.5).

4.1.2. Uncertainty Analysis of Atmospheric Profile Measurements

Temperature, altitude, relative humidity, and air pressure data combined with other atmospheric components are utilized to obtain the TOA radiance in MODTRAN v.5.2.1. In this study, a radiosonde balloon was used to acquire the vertical atmospheric profile parameters as inputs for MODTRAN to simulate the TOA radiance. It should be noted that there was a time difference of approximately 4 h between the radiosonde balloon release time (around 07:15 Beijing time) and the actual overpass time (around 11:45 Beijing time). To explore the uncertainty caused by atmospheric measurements, three MODTRAN-embedded atmospheric models (mid-latitude summer [MLS], mid-latitude winter, [MLW], and 1976 US standard atmosphere [US]) were used to simulate TOA radiances, which were compared with the results obtained by using the atmospheric profile measured by the radiosonde, as shown in Figure 15.

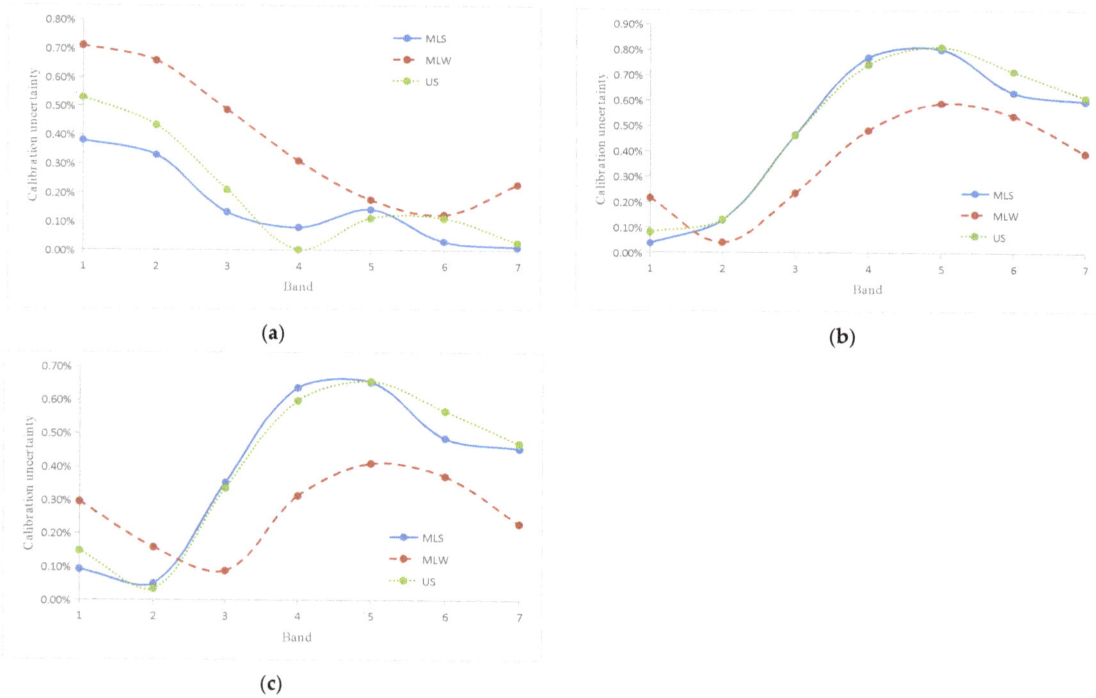

Figure 15. Calibration uncertainty caused by atmospheric profile measurements for SDGSAT-1 MII calibration on 24 December 2021, utilizing the (**a**) reflectance-, (**b**) irradiance-, and (**c**) improved irradiance-based methods.

In comparison to the reflectance-based method, the irradiance- and improved irradiance-based methods usually produced slightly higher uncertainties due to the atmospheric profile; however, the uncertainties in all bands of the two methods were below 0.81%. For the reflectance-based method, the predicted TOA using the MLS model appeared to be the closest to that obtained using the radiosonde measurements. In contrast, for the irradiance- and improved irradiance-based methods, the MLW model appeared to better reflect the radiosonde measurements and actual conditions. Therefore, we adopted the largest relative differences, derived from replacing the radiosonde measurements with the US and MLS models, as the vicarious radiometric calibration uncertainty caused by different atmospheric model assumptions.

4.1.3. Uncertainty Analysis of AOD Retrieval

The total uncertainty for the AOD, which is retrieved from CE318 data, is 0.01–0.02 [44]. In this experiment, the AOD at 550 nm was interpolated from CE318 measurements at 440 and 670 nm. As shown in Figure 5, the AOD at 550 nm at the Dunhuang radiometric calibration site varied from 0.096 to 0.155. To predict the calibration uncertainty owing to variation in the AOD at 550 nm, the original AOD at 550 nm value of 0.1045 was replaced with values of 0.1245 and 0.0845 in radiative transfer simulation. The calibration uncertainty due to variation in the AOD at 550 nm was estimated by comparing the predicted TOA radiance at different levels with the base value set to 0.02. Figure 16 exhibits the calibration uncertainty owing to AOD measurements.

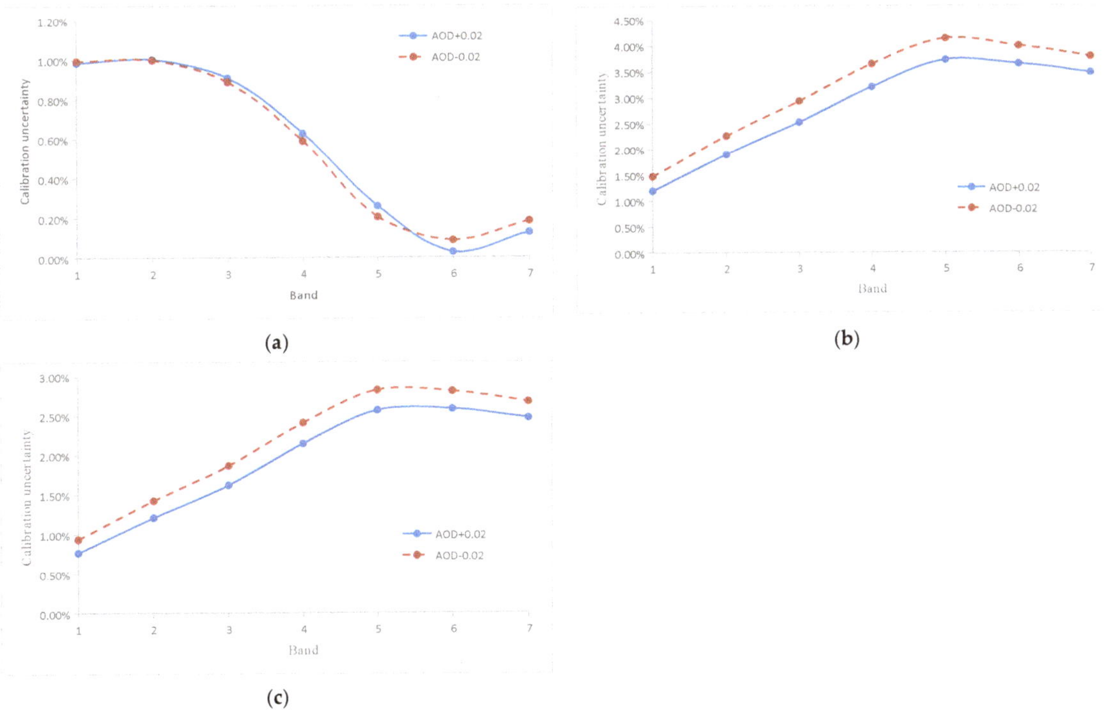

Figure 16. Calibration uncertainty caused by AOD retrieval for SDGSAT-1 MII calibration on 24 December 2021, utilizing the (**a**) reflectance-, (**b**) irradiance-, and (**c**) improved irradiance-based methods.

As shown in Figure 16, although the same difference was considered in the AOD measurements, the relative differences differed for the three methods. For the reflectance-based method, a variation of ±0.02 in the AOD added little uncertainty in the simulated TOA radiance, with a maximum value of 1.00%. In addition, when the AOD changed by ±0.02, the calibration uncertainty decreased with increasing wavelength. However, for the other two methods, a variation of ±0.02 in the AOD introduced significantly larger uncertainties in the simulated TOA radiance than for the reflectance-based method, with maximum values of 4.15% and 2.83%, respectively. Moreover, when the AOD changed by ±0.02, the calibration uncertainty became larger with increasing wavelength, which was the opposite of the uncertainty trend for the reflectance-based method.

A change in the AOD led to opposite trends for direct and diffuse transmittances, revealing an inverse relationship, and, as the same difference was considered in AOD retrieval, the relative difference was symmetrical (Figure 17). For the reflectance-based method (Equation (4)), both the direct transmittance and the diffuse transmittance ($T_{dir}(\theta_s)$, $T_{dif}(\theta_s)$, $T_{dir}(\theta_v)$, and $T_{dif}(\theta_v)$) were affected by AOD errors according to Equations (5) and (6). The direct transmittance increased or decreased with a change in the AOD, while the diffuse transmittance changed in the opposite direction, revealing a "when one falls, another rises" relation. This is why the calibration uncertainty caused by AOD errors was relatively low when using the reflectance-based method, in which the variation in the direct transmittance partially compensates that of the diffuse transmittance in both downward and upward directions. As a comparison, the AOD variation mainly affected the direct transmittance in both downward and upward directions ($e^{-\tau/\mu_s}$ and $e^{-\tau/\mu_v}$, that is $T_{dir}(\theta_s)$ and $T_{dir}(\theta_v)$) in the irradiance-based method. With increasing wavelength, direct transmittances in both downward and upward directions also increased. The improved irradiance-based method is optimal for downward diffuse transmittance, as an additional compensating factor is included in the TOA radiance calculation. Consequently, the calibration uncertainty introduced by AOD errors is small for the reflectance-based method and large for the irradiance-based method compared to the improved irradiance-based method.

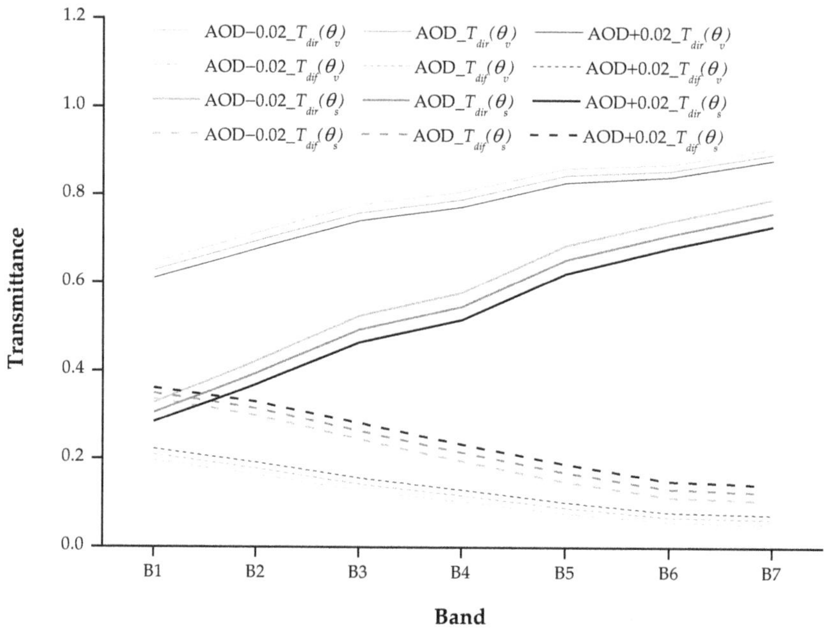

Figure 17. Changes in direct and diffuse transmittances for both downward and upward directions under different 550 nm AOD.

From Figure 17, we can infer that when the AOD is small, the irradiance- and improved irradiance-based methods are inferior to the reflectance-based method. To verify this, various AOD values (0.1, 0.2, 0.3, 0.4, and 0.5) were input into MODTRAN to obtain transmittance values (Figure 18). As expected, the direct transmittance decreased, while the diffuse transmittance increased with increasing AOD. The larger the AOD was, the smaller the direct transmittance and the larger the diffuse transmittance were. In addition, we calculated the corresponding transmittance errors when the AOD changed by ± 0.02 under the different AOD conditions for the three methods, and the results (Table 8) confirmed our hypothesis. For the reflectance-based method, the transmittance error caused by an AOD change was significantly smaller than those for the irradiance- and the improved irradiance-based methods under different AOD conditions. Based on these findings, we concluded that when the AOD is low (≤ 0.1), the calibration accuracy of the reflectance-based method is higher than that of the irradiance- and improved irradiance-based methods.

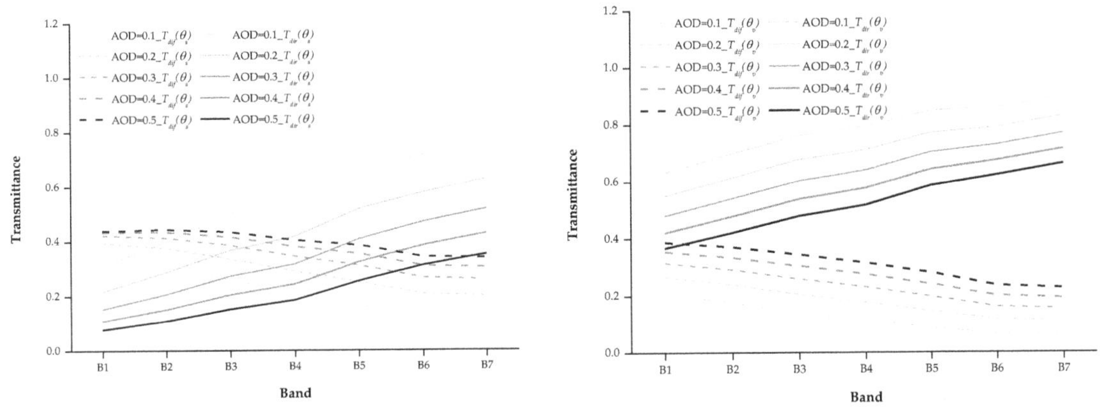

(a) (b)

Figure 18. Changes in direct transmittance and transmittances under different 550 nm AOD conditions, including direct and diffuse transmittances in (**a**) downward and (**b**) upward directions.

Table 8. Transmittance errors when the AOD changes by ±0.02 under different AOD conditions for the reflectance-, irradiance-, and improved irradiance-based methods.

AOD		Reflectance-Based Method ($T(\theta_s) \times T(\theta_v)$)	Irradiance-Based Method ($T_{dir}(\theta_s) \times T_{dir}(\theta_v)$)	Improved Irradiance-Based Method ($T_{dir}(\theta_s) \times T(\theta_v)$)
AOD = 0.1	AOD + 0.02	1.90%	7.40%	5.80%
	AOD − 0.02	1.97%	8.05%	6.20%
AOD = 0.2	AOD + 0.02	1.77%	7.33%	5.75%
	AOD − 0.02	1.83%	7.95%	6.12%
AOD = 0.3	AOD + 0.02	1.62%	6.93%	5.44%
	AOD − 0.02	1.65%	7.39%	5.72%
AOD = 0.4	AOD + 0.02	1.59%	7.15%	5.62%
	AOD − 0.02	1.62%	7.68%	5.94%
AOD = 0.5	AOD + 0.02	1.55%	7.29%	5.74%
	AOD − 0.02	1.58%	7.86%	6.08%

4.1.4. Uncertainty Analysis of Water Vapor Measurement

Like the AOD at 550 nm, the CWV is retrieved from CE318 data, and the error of CWV retrieval is within 10% [33]. Therefore, an uncertainty of ±10% was replaced with the CWV measured by CE318 in MODTRAN 5.2.1. As shown in Figure 19, the effect of a change in CWV on calibration uncertainty was nearly negligible in all bands, with a maximum value of 0.32%. Given that the Dunhuang calibration site is located in an arid area with dry atmospheric conditions, the water vapor measurement errors have little impact on the calibration uncertainty.

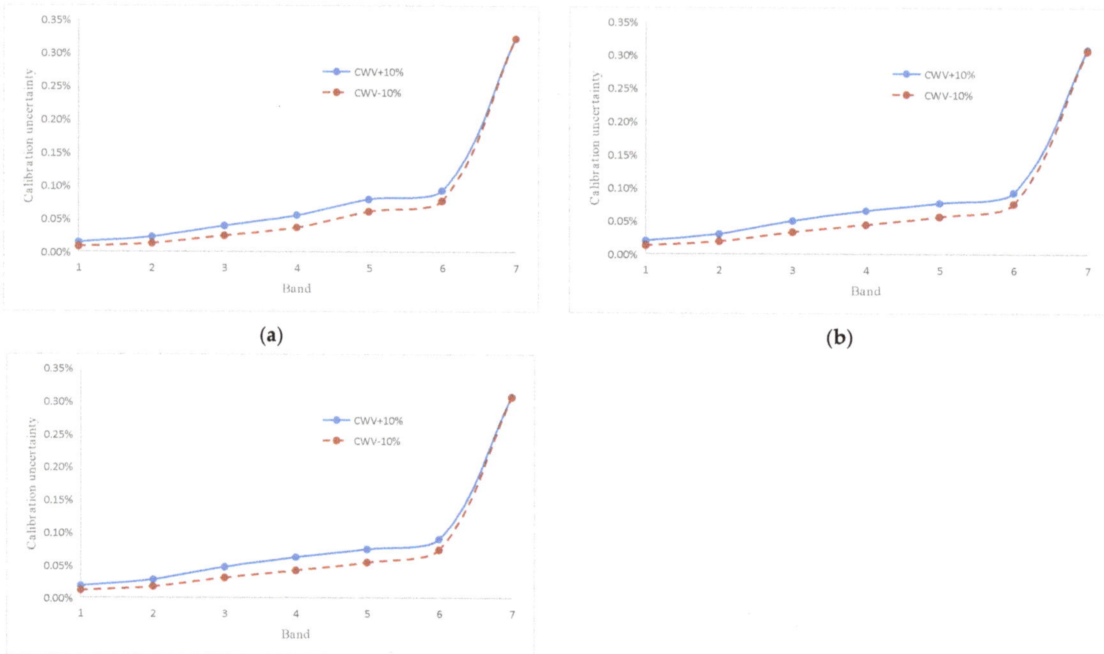

(a) **(b)** **(c)**

Figure 19. Calibration uncertainty caused by CWV retrieval for SDGSAT-1 MII calibration on 24 December 2021, utilizing the (**a**) reflectance-, (**b**) irradiance-, and (**c**) improved irradiance-based methods.

4.1.5. Uncertainty Analysis of Simultaneous Ground Reflectance Measurements

The Dunhuang calibration site is one of China's main radiometric calibration sites for satellites with visible to near-infrared band sensors [2]. The stable ground reflectance of the site guarantees high radiometric calibration accuracy. Both the reflectance- and irradiance-based methods rely on simultaneous ground reflectance measurements. The accuracy of the simultaneous ground reflectance measurement, as MODTRAN simulation input, directly affects the radiative transfer accuracy. Multiple calibration studies have indicated that the error of ground reflectance measurement is approximately 2% [33,43,45,46]. The ground reflectance at the Dunhuang calibration site fluctuated by $\pm1.5\%$ in the last 15 years, showing good stability [47]. In addition, the measurement procedure and instrument specifications meet the requirements of in-situ calibration experiments; therefore, the calibration uncertainty caused by ground reflectance errors was set to 1.5% in this study.

4.1.6. Uncertainty Analysis of Viewing Geometry

The solar zenith angle, solar azimuth angle, view zenith angle, and view azimuth angle are all parts of the satellite viewing geometry. They determine the amount of energy reflected from the ground and the energy received by the sensor entrance pupil. The solar angles can be calculated according to the satellite transit time and the longitude and latitude of the calibration site with relatively high accuracy. The viewing angles can be obtained from information in the image itself. To assess the uncertainty brought by the viewing geometry, we assumed that the calculation errors of the solar angles were 0.1° because of high calculation precision, whereas those of the viewing angles were set to 1° [42]. By sequentially changing the solar and viewing angles, the TOA radiance under the new viewing geometry was obtained and compared with the actual observation results. The relative differences between the simulated results under different angles and the actual observation results, calculated using Equation (14), were regarded as the calibration uncertainty caused by a variation in the viewing geometry (Figure 20).

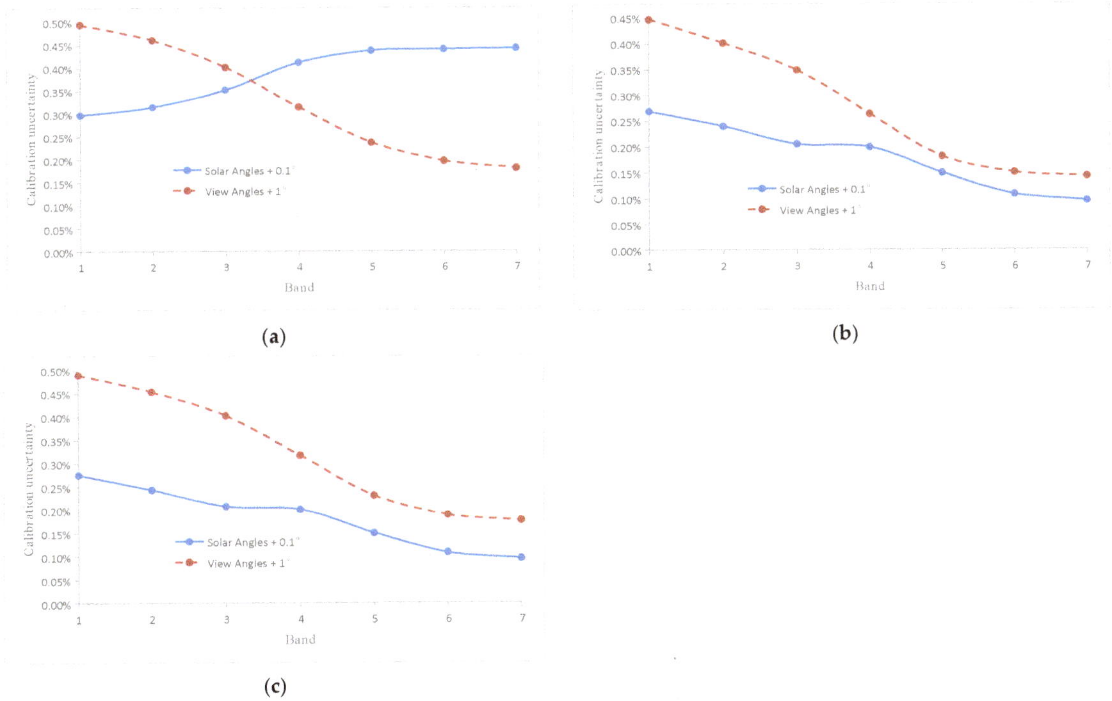

Figure 20. Calibration uncertainty caused by viewing geometry errors for SDGSAT-1 MII calibration on 24 December 2021, utilizing the (**a**) reflectance-, (**b**) irradiance-, and (**c**) improved irradiance-based methods.

Changes in the solar and viewing angles had little impact on the TOA radiance. When the solar angles increased by 0.1°, the variation in the TOA radiance was below 0.44%. When the viewing angles increased by 1°, the variation in the TOA radiance was below 0.50%. Therefore, the calibration uncertainty caused by viewing geometry errors was set to 0.50%.

4.1.7. Total Calibration Uncertainty Estimation

Previous studies have demonstrated that the calibration uncertainties that come from other sources are relatively stable [29,31,33,48,49]. The calibration uncertainties caused by the ozone measurement [49,50], BRDF error [36], image misregistration errors, radiative transfer code accuracy [51,52], and DG ratio measurement [48] were estimated to be 0.6%, 2.0%, 0.2%, 1.0%, and 2.0%, respectively. The total vicarious radiometric calibration uncertainty caused by all of the above factors was calculated according to Equation (12). The total calibration uncertainties for the three methods are presented in Table 9.

According to previous studies, the accuracy of the irradiance- and improved irradiance-based methods is generally better than that of the reflectance-based method in the case of large AOD. However, the data in Table 9 makes it abundantly evident that the total uncertainties of both the irradiance- and improved irradiance-based methods were slightly higher than that of the reflectance-based method. This is because the DG ratio is an indispensable parameter in the former methods. In addition to the same parameters as the reflectance-based method, the irradiance-based method also uses DG ratios in both the solar and viewing directions. However, it is difficult to directly measure the DG ratio in the viewing direction, which has to be extrapolated by fitting the values under different solar zenith angles. During the calibration experiment on 14 December 2021, because of the small variation range of the solar zenith angle within the effective observation time, it was difficult to obtain a relatively accurate DG ratio in the viewing direction after linear extrapolation, which is one of the main reasons why the total uncertainty of the irradiance-based method was higher than that of the reflectance-based method in this study. In the improved irradiance-based method, only the DG ratio measured in the solar direction is included in the calculation to avoid the error introduced by linear extrapolation. Therefore, this method is more accurate than the irradiance-based method. Overall, the irradiance- and improved irradiance-based methods strongly rely on the precision of the direct transmittance measurements and the DG ratios; consequently, improving their measurement accuracy is key to high-precision calibration.

Table 9. Absolute calibration uncertainties of reflectance-, irradiance-, and improved irradiance-based methods.

Source of Uncertainty	Reflectance-Based Method (%)	Irradiance-Based Method (%)	Improved Irradiance-Based Method (%)
Assumption of aerosol type	0.11–4.23	0.16–1.91	0.09–2.63
Assumption of atmospheric model	0.01–0.71	0.04–0.81	0.09–0.66
550 nm AOD retrieval	0.03–1.00	1.20–4.15	0.77–2.83
Water vapor retrieval	0.01–0.32	0.01–0.31	0.01–0.31
Ozone measurement	0.6	0.6	0.6
Ground reflectance measurement	1.5	1.5	1.5
BRDF error	2.0	2.0	2.0
Viewing geometries	0.18–0.50	0.10–0.45	0.10–0.49
Image misregistration errors	0.2	0.2	0.2
Radiative transfer code accuracy	1.0	1.0	1.0
DG ratio measurement	/	2.0	2.0
Total uncertainty (Root sum of squares)	2.77–5.23	3.62–5.79	3.50–5.23

The above results were obtained under a low AOD, and we concluded that the largest calibration uncertainty in the reflectance-based method stemmed from the aerosol model assumption, whereas the retrieval uncertainty of the 550 nm AOD was the primary calibration uncertainty for the other two methods. We analysed and discussed the reasons for this conclusion in Sections 4.1.1 and 4.1.3, respectively. The actual aerosol type in Dunhuang is similar to the rural and desert types, and the uncertainty of the aerosol model assumption was expected to not lead to such a significant relative discrepancy between the urban and rural aerosol types. Therefore, we adopted the relative difference between the rural and desert aerosol types as the uncertainty of the aerosol model assumption. The total uncertainties of the three methods under different AOD conditions are provided in Table 10. When the AOD was 0.1, the reflectance-based method performed better than the other two methods, with a total uncertainty of 3.10%. When the AOD was 0.2, the irradiance-based method performed slightly better than the other two methods, with a total uncertainty of 3.84%, and the accuracy of the reflectance-based method was basically the same as that of the improved irradiance-based method. The irradiance-based method performed best when the AOD was 0.3, 0.4, or 0.5.

Table 10. The total uncertainties of reflectance-, irradiance-, and improved irradiance-based methods under different AOD conditions.

Total Uncertainty	Reflectance-Based Method (%)	Irradiance-Based Method (%)	Improved Irradiance-Based Method (%)
AOD = 0.1	3.10	3.55	3.58
AOD = 0.2	3.96	3.84	4.00
AOD = 0.3	5.28	4.37	4.76
AOD = 0.4	7.24	5.35	6.03
AOD = 0.5	9.21	6.47	7.42

The three methods each have their advantages and applicable conditions. Our research results suggest that when the 550 nm AOD is low ($\leq$0.1), the calibration accuracy of the reflectance-based method may be higher than that of the other two methods, but when the 550 nm AOD becomes higher (>0.1), the irradiance-based method is the first choice. Therefore, it is essential to use multiple separate methods to calibrate a sensor in a vicarious calibration experiment, to mutually verify the results and detect possible errors caused by a single method. Furthermore, if conditions permit, calibration experiments should be repeated as often as possible to reduce the uncertainty of calibration accuracy.

4.2. Validation

Validation, i.e., testing and affirming the calibration results, is an important step in calibration experiments. Only after radiometric calibration and validation research can data products obtained from remote sensing inversion be credible and used in quantitative applications. In this section, we first compare the reflectance retrieved using the FLAASH atmospheric correction software with the in-situ measured reflectance over the Dunhuang calibration site. Then we utilize the well-calibrated Landsat-8 OLI and Sentinel-2A MSI as reference sensors to cross-validate the SDGSAT-1 MII results over different surface types. Finally, we calculate the relative difference in the retrieved ground reflectance between MII and the other two sensors.

4.2.1. Comparison with Measured Reflectance over the Dunhuang Calibration Site

We used the reflectance inversion method [53] to verify the rationality of the radiometric calibration methods in this paper. The calibration coefficients obtained in Section 3 were used in conjunction with FLAASH to obtain the inverse reflectance, which was then compared with the measured reflectance over the Dunhuang calibration site. A comparison of the reflectance inversion results is shown in Table 11. The inversed reflectance was largely consistent with the measured reflectance, and the absolute difference between them was within 0.001, demonstrating the rationality of the calibration methods.

Table 11. Comparison of reflectance inversion results.

Band	Inversion	Measurement	Absolute Difference
B1	0.136456	0.136229	0.00023
B2	0.155277	0.154839	0.00044
B3	0.175996	0.177304	0.00131
B4	0.206255	0.207094	0.00084
B5	0.233118	0.233125	0.00001
B6	0.236585	0.236509	0.00008
B7	0.233967	0.233836	0.00013

4.2.2. Ground Reflectance Validation Cross-Compared with Landsat-8 OLI and Sentinel-2A MSI over Different Surface Types

Because of the lack of spectral measurements of typical features, we utilized remote sensors with high radiometric calibration accuracy as references to cross-validate the sensor to be calibrated. We selected the Landsat-8 OLI and Sentinel-2A MSI as reference sensors to cross-validate the SDGSAT-1 MII. The details of the near-coincident Landsat-8 OLI and Sentinel-2A MSI images are provided in Table 12. The MII, OLI, and MSI overpass times over the Dunhuang calibration site differ. Therefore, we had to consider the influence of atmospheric conditions. The retrieved ground reflectance was obtained using the atmospheric parameters measured at the Dunhuang calibration site and the FLAASH atmospheric model according to the different radiometric calibration coefficients.

Table 12. Near-coincident Landsat-8 OLI and Sentinel-2A MSI data used in this study.

Sensor	Data Acquisition Time (UTC + 8)	Center Coordinate	Solar Zenith	Solar Azimuth	550 nm AOD	CWV (g/cm^2)
SDGSAT-1 MII	14 December 2021, 11:45:17	40.0913°N, 94.3938°E	68.5021°	152.2956°	0.1045	0.3114
Landsat-8 OLI	14 December 2021, 12:20:23	40.3329°N, 95.0782°E	65.8780°	161.3918°	0.1199	0.2899
Sentinel-2A MSI	17 December 2021, 12:32:11	40.1419°N, 94.8185°E	65.0575°	164.5149°	0.1229	0.2880

Figure 21 shows the relative spectral response curves of comparable bands for SDGSAT-1 MII, Landsat-8 OLI, and Sentinel-2A MSI. Table 13 shows the visible and near-infrared band information for the three sensors. To directly compare the spectral ground reflectance retrieved by FLAASH between SDGSAT-1 MII, Landsat-8 OLI, and Sentinel-2A MSI, the spectral band differences must be accounted for according to Equations (16) and (17).

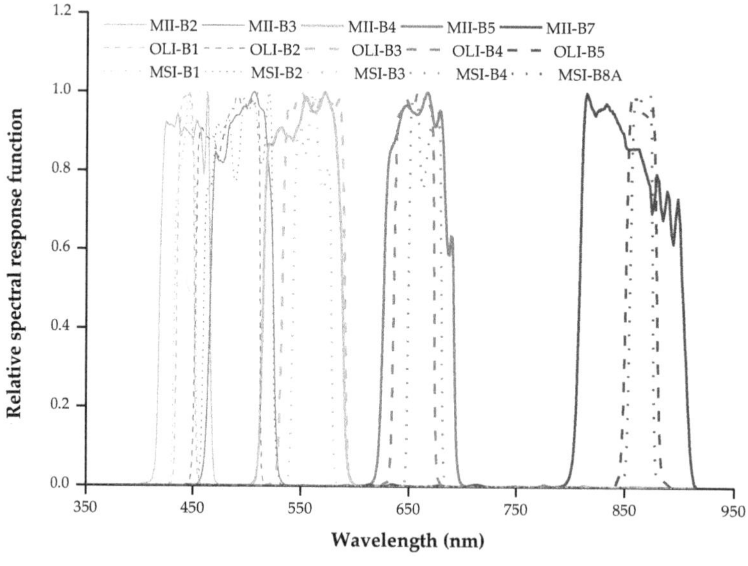

Figure 21. Relative spectral responses of corresponding bands for SDGSAT-1 MII, Landsat-8 OLI, and Sentinel-2A MSI.

Table 13. Spectral ranges of SDGSAT-1 MII, Landsat-8 OLI, and Sentinel-2A MSI.

SDGSAT-1 MII			Landsat-8 OLI			Sentinel-2A MSI		
Band	Center Wavelength (nm)	Spectral Range (nm)	Band	Center Wavelength (nm)	Spectral Range (nm)	Band	Center Wavelength (nm)	Spectral Range (nm)
B2	462	410–467	B1	445	435–451	B1	443	431–454
B3	506	457–529	B2	509	452–512	B2	520	458–527
B4	571	510–597	B3	550	533–590	B3	560	504–602
B5	666	618–696	B4	656	636–673	B4	654	649–680
B7	814	798–911	B5	859	851–879	B8A	871	855–875

Different surface types, including water, bare land, desert, and dry farmland, around Dunhuang City in Gansu Province, China, were selected to evaluate the vicarious calibration coefficients according to the reflectance-, irradiance-, and improved irradiance-based methods. Figure 22 shows the water, bare land, desert, and dry farmland sites selected for measurements on 14 December 2021. The ground reflectance in the regions of interest was calculated based on the different calibration coefficients, measured atmospheric parameters, and FLAASH. Using this approach, the vicarious radiometric calibration coefficients of SDGSAT-1 MII were validated for all selected surface types, as shown in the scatter plots and curve fits in Figure 23.

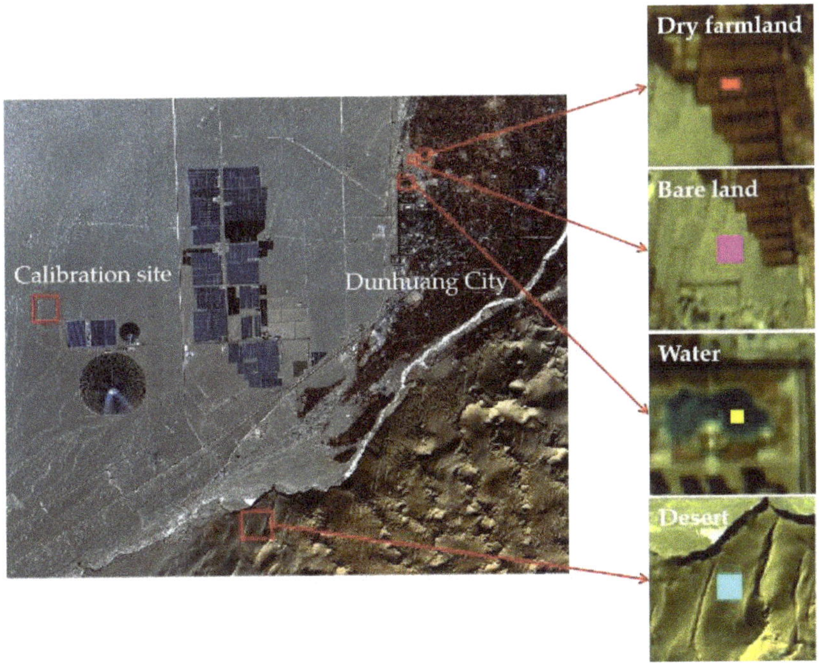

Figure 22. Different surface types selected for validation.

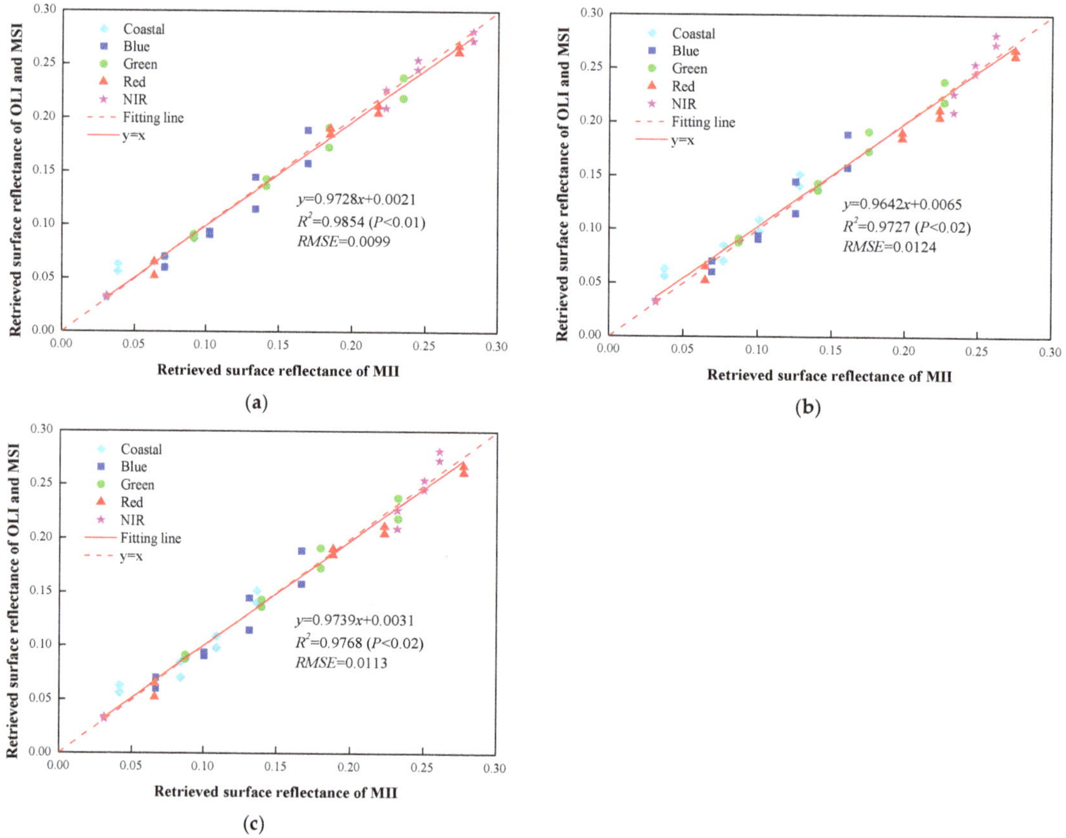

Figure 23. Comparison of retrieved ground reflectances obtained by the indicated sensors according to the (**a**) reflectance-, (**b**) irradiance-, and (**c**) improved irradiance-based methods.

Figure 23 shows the cross-comparison results of the reflectances for the selected surface types obtained using the different satellite sensors and the reflectance-, irradiance-, and improved irradiance-based methods. y = x is the reference line; the closer the data points are to y = x, the closer the retrieved ground reflectance of MII is to that of OLI and MSI and, thus, the higher accuracy of the calibration method. All data points were linearly fitted, and the linear fitting equation expression and goodness-of-fit (R^2) were obtained. The larger the R^2 value, the higher the degree of fit. In addition, the root mean square error (RMSE) of each calibration method was calculated. The smaller the RMSE, the higher the accuracy of the calibration method.

Overall, the retrieved ground reflectance of the MII was in good agreement with that of the OLI and MSI sensors for all three methods ($R^2 > 97\%$, $p < 0.02$, RMSE < 0.013). The results revealed that the calibrated SDGSAT-1 MII image was highly consistent with the Landsat-8 OLI and Sentinel-2A MSI images. Interestingly, the calibration accuracy of the three methods was also reflected by the R^2 values (reflectance-based, 0.9854, $p < 0.01$; improved irradiance-based, 0.9768, $p < 0.02$; and irradiance-based, 0.9727, $p < 0.02$) and RMSE values (0.0124, 0.0113, and 0.0099, respectively). The cross-comparison results agreed with the results exhibited in Section 4.1, with the reflectance-based method having a slightly higher accuracy than the other two methods.

To quantitatively verify the calibration accuracy of the different methods, the absolute and relative differences between MII and OLI/MSI were calculated. Table 14 lists the results. The absolute (relative) differences in most bands were below 0.009 (5%). The average absolute (relative) differences between MII and OLI and between MII and MSI were all within 0.01 (10%). This demonstrated that the calibration accuracy of the MII coincides with that of the OLI and MSI sensors, which is consistent with the above results.

Table 14. Average absolute (relative) differences in retrieved reflectances over different surface types between MII and OLI/MSI utilizing the reflectance-, irradiance-, and improved irradiance-based methods.

Method	Band	Absolute (Relative) Difference between MII and OLI	Absolute (Relative) Difference between MII and MSI
Reflectance-based method	Coastal/aerosol	0.0111 (14.04%)	0.0086 (12.84%)
	Blue	0.0133 (13.76%)	0.0099 (7.12%)
	Green	0.0081 (4.46%)	0.0041 (2.78%)
	Red	0.0033 (1.83%)	0.0103 (8.82%)
	Near infrared	0.0016 (1.57%)	0.0091 (5.81%)
	Average difference	0.0075 (7.13%)	0.0084 (7.47%)
Irradiance-based method	Coastal/aerosol	0.0165 (18.45%)	0.0101 (13.56%)
	Blue	0.0078 (8.86%)	0.0137 (9.10%)
	Green	0.0046 (3.16%)	0.0076 (3.92%)
	Red	0.0082 (4.09%)	0.0129 (10.28%)
	Near infrared	0.0074 (3.31%)	0.0111(6.38%)
	Average difference	0.0089 (7.57%)	0.0111 (8.65%)
Improved irradiance-based method	Coastal/aerosol	0.0097 (15.54%)	0.0110 (15.04%)
	Blue	0.0101 (10.12%)	0.0113 (8.13%)
	Green	0.0068 (4.09%)	0.0052 (2.78%)
	Red	0.0062 (2.95%)	0.0126 (10.54%)
	Near infrared	0.0082 (3.94%)	0.0106 (6.48%)
	Average difference	0.0082 (7.33%)	0.0101 (8.59%)

The average absolute (relative) differences between MII and OLI were in the order reflectance-based method (0.0075 (7.13%)) < improved irradiance-based method (0.0082 (7.33%)) < irradiance-based method (0.0089 (7.57%)); those between MII and MSI were in the order reflectance-based method (0.0084 (7.47%)) < improved irradiance-based method (0.0101 (8.59%)) < irradiance-based method (0.0111 (8.65%)). The absolute (relative) differences in the retrieved ground reflectance between MII and OLI/MSI provided strong evidence that the calibration accuracy of the reflectance-based method is basically the same as that of irradiance- and improved irradiance-based methods, and it is even better when the AOD is low ($\leq$0.1).

Notably, the relative differences were the largest in the coastal/aerosol and blue bands for all three methods (all > 10%). This is partly attributed to the low reflectance of water and the low response of the blue channel; therefore, a small difference will be amplified. Moreover, the relative difference in the red band was approximately 10% due to the low reflectance of water, which introduces a large error, resulting in a large difference over all surface types in the red band.

Overall, our research revealed that the calibration accuracy of the SDGSAT-1 MII is highly consistent with that of the Landsat-8 OLI and Sentinel-2A MSI, and we demonstrated that the calibration coefficients for the SDGSAT-1 MII are reliable and highly accurate. The different calibration methods were found to have different calibration accuracy; when the AOD is low ($\leq$0.1), the calibration accuracy of the reflectance-based method is similar to or even higher than that of the irradiance- and improved irradiance-based methods.

5. Conclusions

We comprehensively described the first in-situ vicarious radiometric calibration experiment of the SDGSAT-1 MII at the Dunhuang calibration site on 14 December 2021. In-situ measurements, including ground reflectance, atmospheric parameters, and radiosonde data, were acquired during the satellite overpass date. Reflectance-, irradiance-, and improved irradiance-based calibration methods were utilized to predict the TOA spectral radiances using MODTRAN v.5.2.1 software. The vicarious radiometric calibration coefficients for the SDGSAT-1 MII were directly determined by dividing the TOA spectral radiances by the averaged DN over the 500 × 500 m calibration site. The calibration uncertainties were analysed in detail in this paper. The key findings are summarized as follows:

(1) The radiometric calibration coefficients obtained by the three vicarious calibration methods were reliable, with average relative differences of 2.20% (between the reflectance- and irradiance-based methods) and 1.43% (between the reflectance- and improved irradiance-based methods). The total calibration uncertainties of the reflectance-, irradiance-, and improved irradiance-based methods were 2.77–5.23%, 3.62–5.79%, and 3.50–5.23%, respectively. The largest calibration uncertainty in the reflectance-based method stemmed from the aerosol model assumption, while the retrieval uncertainty of the AOD at 550 nm was the primary calibration uncertainty in the other two methods.

(2) To verify the calibration accuracy of SDGSAT-1 MII, we utilized well-calibrated sensors, Landsat-8 OLI and Sentinel-2A MSI, as reference sensors for cross-comparison with the SDGSAT-1 MII over different surface types. MII retrievals differed less than 0.0075 (7.13%) from OLI retrievals and less than 0.0084 (7.47%) from MSI retrievals when applying calibration coefficients from the reflectance-based method; less than 0.0089 (7.57%) from OLI retrievals and less than 0.0111 (8.65%) from MSI retrievals for the irradiance-based method; and less than 0.0082 (7.33%) from OLI retrievals and less than 0.0101 (8.59%) from MSI retrievals for the improved irradiance-based method. The cross-comparison results showed that the SDGSAT-1 MII calibration accuracy was consistent with those of Landsat-8 OLI and Sentinel-2A MSI, and they showed that the calibration coefficients for the SDGSAT-1 MII are reliable and highly accurate.

(3) Unexpected calibration errors caused by measurement uncertainties in in-situ calibration experiments are hard to detect. Therefore, using different calibration methods and cross-comparing their results is important. In the case of a low AOD ($\leq$0.1), the irradiance- and improved irradiance-based methods showed no obvious advantage over the reflectance-based method, which is therefore recommended under such conditions due to its superiorities of easy operations, low cost, and high accuracy. If DG ratios are not measured, using only the reflectance-based method for calibration under ideal weather conditions is reliable.

(4) The performances of the three calibration methods varied for different aerosol types under different AOD conditions. For calibration sites with ground and atmospheric conditions similar to the Dunhuang calibration site, we suggest using the rural and desert aerosol types. When the AOD was 0.1, the reflectance-based method was the best choice, with a total uncertainty of 3.10%, whereas when the AOD was 0.2, 0.3, 0.4, or 0.5, the irradiance-based method had a higher accuracy, with total uncertainties of 3.84%, 4.37%, 5.35%, and 6.47%, than the other two methods. The improved irradiance-based method is recommended when the DG ratio cannot be obtained under a wide-angle measurement (for example, in winter, the range of angle variation is small) or when the atmospheric condition is unstable.

The vicarious absolute radiometric calibration coefficients of SDGSAT-1 MII obtained based on this experiment have been provided to users at http://www.sdgsat.ac.cn/, (accessed on 9 June 2022). This research needs to be extended to other radiometric calibration sites, especially, the Radiometric Calibration Network (RadCalNet). Furthermore, extensive evaluation and validation campaigns should be performed to monitor the long-term radiometric performance of SDGSAT-1 MII.

Author Contributions: Conceptualization, H.Z and Z.C.; investigation, Z.C., Y.H. and L.Y.; data curation, Z.C.; formal analysis, H.Z., C.D. and C.M.; validation, Z.C.; writing—original draft preparation, Z.C.; writing—review and editing, H.Z. and X.-M.L.; project administration, H.Z., X.-M.L. and C.M. All authors have read and agreed to the published version of the manuscript.

Funding: This research was funded by the CAS Strategic Priority Research Program (grant number XDA19010402) and the National Natural Science Foundation of China (grant number 41771397, U21A20108).

Data Availability Statement: Not applicable.

Acknowledgments: The authors thank the personnel at the Dunhuang National Climate Station for providing the AOD, water vapor content, and radiosonde data, as well as Ning Wang, Yongzheng Ren, and Hujun Liang for their help with the ground reflectance measurements.

Conflicts of Interest: The authors declare no conflict of interest.

References

1. Guo, H.D.; Liang, D.; Sun, Z.C.; Chen, F.; Wang, X.Y.; Li, J.S.; Zhu, L.; Bian, J.H.; Wei, Y.Q.; Huang, L.; et al. Measuring and evaluating SDG indicators with Big Earth Data. *Sci. Bull.* **2022**, *67*, 1792–1801. [CrossRef] [PubMed]
2. Tan, K.; Wang, X.; Niu, C.; Wang, F.; Du, P.J.; Sun, D.X.; Yuan, J.; Zhang, J. Vicarious Calibration for the AHSI Instrument of Gaofen-5 With Reference to the CRCS Dunhuang Test Site. *IEEE Trans. Geosci. Remote Sens.* **2021**, *59*, 3409–3419. [CrossRef]
3. Slater, P.N.; Biggar, S.F.; Holm, R.G.; Jackson, R.D.; Mao, Y.; Moran, M.S.; Palmer, J.M.; Yuan, B. Reflectance- and radiance-based methods for the in-flight absolute calibration of multispectral sensors. *Remote Sens. Environ.* **1987**, *22*, 11–37. [CrossRef]
4. Biggar, S.F. In-Flight Methods for Satellite Sensor Absolute Radiometric Calibration. Ph.D. Thesis, College of Optical Sciences, University of Arizona, Tucson, AZ, USA, 1990.
5. Xu, W.B.; Shi, J.M.; Zheng, X.B.; Yi, W.N. Design and Application of Global Calibration-Site Network Database. *Acta Opt. Sin.* **2014**, *34*, 17–27.
6. Scott, K.P.; Thome, K.J.; Brownlee, M.R. Evaluation of Railroad Valley playa for use in vicarious calibration. *Proc. SPIE* **1996**, *2818*, 158–166.
7. Thome, K.; Schiller, S.; Conel, J.; Arai, K.; Tsuchida, S. Results of the 1996 Earth Observing System vicarious calibration joint campaign at Lunar Lake Playa, Nevada (USA). *Metrologia* **1998**, *35*, 631–638. [CrossRef]
8. Tsuchida, S.; Sato, I. Cross validation of calibration coefficients for AVNIR and OCTS sensors at the Rogers Dry Lake in USA. *J. Remote Sens. Soc. Jpn.* **1997**, *17*, 418–423.
9. Villa-Aleman, E.; Kurzeja, R.J.; Pendergast, M.M. Temporal, spatial, and spectral variability at the Ivanpah Playa vicarious calibration site. *Proc. SPIE* **2003**, *5093*, 320–330.
10. Rondeaux, G.; Steven, M.; Clark, J.; Mackay, G. La Crau: A European test site for remote sensing validation. *Int. J. Remote Sens.* **1998**, *19*, 2775–2788. [CrossRef]
11. Teillet, P.M.; Thome, K.J.; Fox, N.P.; Morisette, J.T. Earth observation sensor calibration using a global instrumented and automated network of test sites (GIANTS). *Proc. SPIE* **2001**, *4540*, 246–254.
12. Bhaskaran, S.; Phillip, N.; Rahman, A.; Mallick, J. Applications of satellite data for Aerosol Optical Depth (AOD) retrievals and validation with AERONET data. *Atmos. Clim. Sci.* **2011**, *1*, 61–67. [CrossRef]
13. Prata, A.J. Land surface temperature determination from satellites. *Adv. Space Res.* **1994**, *14*, 15–26. [CrossRef]
14. Barry, P.; Jarecke, P.J.; Pearlman, J.; Jupp, D.; Lovell, J.; Campbell, S. Radiometric calibration validation of the Hyperion instrument using ground truth at a site in Lake Frome, Australia. *Proc. SPIE* **2002**, *4480*, 242–246.
15. Wu, D.; Yin, Y.; Wang, Z.; Gu, X.F.; Verbrugghe, M.M.; Guyot, G. Radiometric characterisation of Dunhuang satellite calibration test site (China). In Proceedings of the 7th International Symposium on Physical Measurements and Signatures in Remote Sensing, Courchevel, France, 7–11 April 1997; Volume 4.
16. Wang, Z.M. Ground test sites for absolute radiometric calibration of China resources satellite. *Remote Sens. Land. Resour.* **1999**, *41*, 40–46.
17. Li, C.R.; Ma, L.L.; Tang, L.L.; Gao, C.X.; Qian, Y.G.; Wang, N.; Wang, X.H. A comprehensive calibration site for high resolution remote sensors dedicated to quantitative remote sensing and its applications. *Natl. Remote Sens. Bull.* **2021**, *25*, 198–219.
18. Thome, K.J.; Crowther, B.G.; Biggar, S.F. Reflectance- and irradiance-based calibration of Landsat-5 Thematic Mapper. *Can. J. Remote Sens.* **1997**, *23*, 309–317. [CrossRef]
19. Thome, K.J.; Markharn, B.L.; Barker, J.L.; Slater, P.N.; Biggar, S.F. Radiometric calibration of Landsat. *Photogramm. Eng. Remote Sens.* **1997**, *63*, 853–858.
20. Markham, B.L.; Barker, J.L.; Kaita, E.; Barsi, J.A.; Helder, D.L.; Palluconi, F.D.; Schott, J.R.; Thome, K.J.; Morfitt, R.; Scaramuzza, P. Landsat-7 ETM+ radiometric calibration: Two years on-orbit. In Proceedings of the Scanning the Present and Resolving the Future. Proceedings. IEEE 2001 International Geoscience and Remote Sensing Symposium, Sydney, NSW, Australia, 9–13 July 2001; Volume 1, pp. 518–520.

21. Thome, K.J. Absolute radiometric calibration of Landsat 7 ETM+ using the reflectance-based method. *Remote Sens. Environ.* **2001**, *78*, 27–38. [CrossRef]
22. Czapla-Myers, J.S.; Anderson, N.J.; Biggar, S.F. Early ground-based vicarious calibration results for Landsat 8 OLI. *Proc. SPIE* **2013**, *8866*, 205–214.
23. Czapla-Myers, J.; Mccorkel, J.; Anderson, N.; Thome, K.; Biggar, S.; Helder, D.; Aaron, D.; Leigh, L.; Mishra, N. The Ground-Based Absolute Radiometric Calibration of Landsat 8 OLI. *Remote Sens.* **2015**, *7*, 600–626. [CrossRef]
24. Liu, Y.K.; Ma, L.L.; Wang, N.; Qian, Y.G.; Qiu, S.; Li, C.R.; Tang, L.L. Vicarious radiometric calibration/validation of Landsat-8 operational land imager using a ground reflected radiance-based approach with Baotou site in China. *J. Appl. Remote Sens.* **2017**, *11*, 1. [CrossRef]
25. Biggar, S.F.; Dinguirard, M.C.; Gellman, D.I.; Henry, P.J.; Jackson, R.D.; Moran, M.S.; Slater, P.N. Radiometric calibration of SPOT 2 HRV: A comparison of three methods. *Proc. SPIE* **1991**, *1493*, 155–162.
26. Thome, K.J.; Whittington, E.E.; Smith, N.; Nandy, P.; Zalewski, E.F. Ground-reference techniques for the absolute radiometric calibration of MODIS. *Proc. SPIE* **2000**, *4135*, 51–59.
27. Thome, K.J.; Biggar, S.F.; Choi, H.J. Vicarious calibration of Terra Aster, Misr, and Modis. *Proc. SPIE* **2004**, *5542*, 290–299.
28. Zhang, X.W.; Fu, Q.Y.; Han, Q.J.; Pan, Z.Q.; Yang, L. The field radiometric calibration and validation of ZY-3 multispectral sensor. *Spectrosc. Spect. Anal.* **2014**, *34*, 2476–2480.
29. Gao, H.L.; Gu, X.F.; Yu, T.; Gong, H.; Li, J.G.; Li, X.Y. HJ-1A HSI on-orbit radiometric calibration and validation research. *Sci. China Technol. Sci.* **2010**, *53*, 3119–3128. [CrossRef]
30. Brook, A.; Ben-Dor, E. Supervised vicarious calibration (SVC) of multi-source hyperspectral remote-sensing data. *Remote Sens.* **2015**, *7*, 6196–6223. [CrossRef]
31. Li, H.W.; Zhang, H.; Zhang, B.; Chen, Z.C.; Yang, M.H.; Zhang, Y.Q. A Method Suitable for Vicarious Calibration of a UAV Hyperspectral Remote Sensor. *IEEE J. Sel. Top. Appl. Earth Obs. Remote Sens.* **2015**, *8*, 3209–3223. [CrossRef]
32. Brook, A.; Polinova, M.; Ben-Dor, E. Fine tuning of the SVC method for airborne hyperspectral sensors: The BRDF correction of the calibration nets targets. *Remote Sens. Environ.* **2018**, *204*, 861–871. [CrossRef]
33. Zhang, H.; Zhang, B.; Chen, Z.C.; Huang, Z.H. Vicarious Radiometric Calibration of the Hyperspectral Imaging Microsatellites SPARK-01 and -02 over Dunhuang, China. *Remote Sens.* **2018**, *10*, 120. [CrossRef]
34. United Nations. *Transforming Our World: The 2030 Agenda for Sustainable Development*; United Nations: New York, NY, USA, 2015.
35. He, J.T.; Lu, Y.H. The Measurement and Evaluation of Bidirectional Reflectance Characteristics of Dunhuang Radiative Calibration Site. *Natl. Remote Sens. Bull.* **1997**, *1*, 246–251.
36. Hu, X.; Liu, J.; Sun, L.; Rong, Z.; Li, Y.; Zhang, Y.; Zheng, Z.; Wu, R.; Zhang, L.; Gu, X. Characterization of CRCS Dunhuang test site and vicarious calibration utilization for Fengyun (FY) series sensors. *Can. J. Remote Sens.* **2010**, *36*, 566–582. [CrossRef]
37. Hu, X.Q.; Zhang, Y.X.; Qiu, K.M. In-flight Radiometric Calibration for VIR Channels of FY-1C Satellite Sensor by Using Irradiance-based Method. *Natl. Remote Sens. Bull.* **2003**, *7*, 458–464.
38. Fu, Q.Y.; Min, X.J.; Li, X.C.; Sha, C.M.; Li, X.Y.; Ma, G.Q.; Pan, Z.Q.; Guo, Y.; Li, Q.M.; Liu, G.D. In-flight Absolute Calibration of the CBERS-02 CCD Sensor at the Dunhuang Test Site. *Natl. Remote Sens. Bull.* **2006**, *10*, 433–439.
39. Zhang, H.; Zhang, B.; Chen, D.M.; Li, J.S.; Zhao, G.N. Influence of Filter Band Function on Retrieval of Aerosol Optical Depth from Sunphotometer Data. *J. Atmos. Ocean. Technol.* **2013**, *30*, 929–941. [CrossRef]
40. Zhang, H.; Chen, Z.C.; Zhang, B.; Chen, C.Y. Improved Algorithm for Water Vapor Retrieval and Field Calibration in the Channell 940-nm of Sun-photometer. In Proceedings of the 23rd CALCON Technical Conference, Logan, UT, USA, 11–14 August 2014; pp. 11–14.
41. Zhang, B.; Zhang, H.; Chen, Z.C. *Radiometric Calibration of Optical Remote Sensing*; Science Press: Beijing, China, 2020; p. 27.
42. Gao, H.L. Study on the Calibration and Validation of HSI Sensor on HJ-1ASatellite. Ph.D. Thesis, Institute of Remote Sensing Applications, Chinese Academy of Sciences, Beijing, China, 2010.
43. Gou, Z.Y.; Yan, L.; Chen, W.; Jing, X.; Yin, Z.Y.; Duan, Y.N. In-flight Absolute Radiometric Calibration of UAV Hyperspectral Camera and Its Validation Analysis. *Spectrosc. Spect. Anal.* **2012**, *32*, 430–434.
44. Eck, T.F.; Holben, B.N.; Reid, J.S.; Dubovik, O.; Smirnov, A.; O'Neill, N.T.; Slutsker, I.; Kinne, S. Wavelength dependence of the optical depth of biomass burning, urban, and desert dust aerosols. *J. Geophys. Res.* **1999**, *104*, 31333–31349. [CrossRef]
45. Zhang, Y.X.; Zhang, G.S.; Qiu, K.M.; Wang, Z.Q.; Zhang, L.J. Vicarious Radiometric Calibration of the Remote Sensing Data from Airborne Moderate Resolution Imaging Spectrometer. *J. Appl. Meteor. Sci.* **1998**, *9*, 106–112.
46. Thome, K.J.; Gellman, D.I.; Parada, R.J.; Biggar, S.F.; Slater, P.N.; Moran, M.S. In-flight radiometric calibration of Landsat-5 Thematic Mapper from 1984 to present. *Proc. SPIE* **1993**, *1938*, 126–130.
47. Pang, W.W.; Zheng, X.B.; Lu, J.H.; Li, J.J. Reflectance Characteristics of Dunhuang Radiometric Calibration Test Site. *J. Atmos. Environ. Optics.* **2015**, *10*, 472–481.
48. Thome, K.J. *Validation Plan for MODIS Level 1 at-Sensor Radiance*; Remote Sensing Group of the Optical Sciences Center, University of Arizona: Tucson, AZ, USA, 1999; pp. 23–28.
49. Biggar, S.F.; Slater, P.N.; Gellman, D.I. Uncertainties in the in-flight calibration of sensors with reference to measured ground sites in the 0.4–1.1 μm range. *Remote Sens. Environ.* **1994**, *48*, 245–252. [CrossRef]

50. Liu, X.; Bhartia, P.K.; Chance, K.; Froidevaux, L.; Spurr, R.J.D.; Kurosu, T.P. Validation of Ozone Monitoring Instrument (OMI) ozone profiles and stratospheric ozone columns with Microwave Limb Sounder (MLS) measurements. *Atmos. Chem. Phys.* **2010**, *10*, 2539–2549. [CrossRef]
51. Slater, P.N.; Biggar, S.F.; Thome, K.J.; Gellman, D.I.; Spyak, P.R. Vicarious radiometric calibrations of EOS sensors. *J. Atmos. Ocean. Technol.* **1996**, *13*, 349–359. [CrossRef]
52. Berk, A.; Anderson, G.P.; Acharya, P.K.; Shettle, E.P. *MODTRAN®5.2.1 User's Manual*; Spectral Science Inc.: Burlington, MA, USA, 2011.
53. Bao, S.W.; Chen, H.Y.; Zhang, L.M.; Wang, X.H. Absolute Radiometric Calibration of Optical Remote Sensor Based on Gray-Scale Targets Method and Reflectance Inversion Validation. *Acta Opt. Sin.* **2021**, *41*, 204–212.

remote sensing

MDPI

Article

High Accuracy Solar Diffuser BRDF Measurement for On-Board Calibration in the Solar Reflective Band

Zhiyuan Zhang [1,2], Hongyao Chen [2,*], Wenxin Huang [2], Xiaobing Zheng [2] and Liming Zhang [2]

[1] Science Island Branch of Graduate School, University of Science and Technology of China, Hefei 230026, China
[2] Key Laboratory of Optical Calibration and Characterization, Hefei Institutes of Physical Science, Chinese Academy of Sciences, Hefei 230031, China
[*] Correspondence: hychen@aiofm.ac.cn

Abstract: In the solar reflective band, an on-board calibration method based on a solar diffuser (SD) can realize full aperture, full field of view, and end-to-end absolute radiometric calibration of optical remote sensors. The SD's bidirectional reflectance distribution function (BRDF) is a key parameter that affects the accuracy of the on-board calibration. High-accuracy measurement of the SD BRDF is required in the laboratory before launch. Due to the uncertainty of the goniometer system, polarization effects, and other factors, the measurement uncertainty of the SD BRDF at large incident angles is much higher than that at a 0° incident zenith angle and 45° reflection zenith angle. In this paper, an absolute BRDF measurement facility is reported. The goniometric system consists of a high-brightness integrating sphere as a radiation source, a six-axis robot arm, and a large rotation stage. The measurement wavelength range was from 350 nm to 2400 nm. An improved data processing method based on the reciprocity theorem was proposed to reduce the measurement uncertainty of the SD BRDF at large incident angles. At an incident zenith angle of 75°, the improved data processing method reduced the measurement uncertainty of the SD BRDF by 52% at 410 nm to 480 nm, by 70% at 480 nm to 1000 nm, and by 20% at other bands compared to the absolute measurement method. The influence of the radiation source, goniometer system, detection system, and other factors on the measurement uncertainty are analyzed in this paper. The results show that the measurement uncertainty (coverage factor k = 2) of the SD BRDF was better than 1.04% at 350 nm to 410 nm, 0.60% at 410 nm to 480 nm, 0.43% at 480 nm to 1000 nm, and 0.86% at 1000 nm to 2400 nm.

Keywords: remote sensing; on-board calibration; BRDF measurement; reciprocity theorem; uncertainty budget

1. Introduction

An optical remote sensor requires absolute radiometric calibration to quantify the sensor's response to a known radiometric input [1]. Although a remote sensor has been calibrated with high accuracy before launch, its performance inevitably changes during its transportation, launch, and on-orbit operation. Therefore, on-orbit calibration is necessary [2,3]. Among the various on-orbit calibration methods, an on-board calibration method based on a solar diffuser (SD) has the advantages of high frequency, full aperture, a full field of view, and end-to-end characteristics. This is currently the dominant on-board calibration method in the solar reflective band. In this calibration method, the solar radiance reflected from an SD is used as a high radiance source. The radiance at the sensor aperture when viewing the SD is directly influenced by the SD BRDF [4–9]. Therefore, the SD BRDF is a key parameter that determines the uncertainty of on-board calibration.

The BRDF can accurately describe the spatial and spectral characteristics of an object. It is defined as the ratio of irradiance falling on the object's surface from a given direction, and its contribution, to the radiance that is reflected in another direction. The initial SD

BRDF for an on-board calibration can only be measured in a laboratory before launch [10]. After entering orbit, the stability of the BRDF can be monitored through an SD stability monitor or a backup SD [11,12].

In recent years, many institutions have conducted research on absolute BRDF measurement. The Robotic Optical Scattering Instrument (ROSI) developed by the National Institute of Standards and Technology (NIST) realized a high-accuracy BRDF measurement in the range of 250 nm to 2400 nm. In the range of 250 nm to 1050 nm, the uncertainty (coverage factor k = 2) of a BRDF measurement at standard in-plane geometry (0°/45°), i.e., with an incident zenith angle of 0° and a reflected zenith angle of 45°, was less than 0.82% and increased to 1.92% for 1050 nm to 2400 nm [13,14]. The Physikalisch-Technische Bundesanstalt (PTB) developed a gonioreflectomter based on a five-axis robot. The wavelength range covered 250 nm to 1700 nm. The uncertainty (k = 2) of the BRDF measurement in standard geometry was less than 0.5% at 400 nm to 1700 nm and 1.5% at 250 nm to 400 nm [15–17]. The National Research Council (NRC, Canada) gonioreflectometer applied an array spectroradiometer as the radiation measurement device to achieve high-accuracy BRDF measurement at 250 nm to 1700 nm. The k = 2 uncertainty in the standard geometry was less than 0.5% in the range of 400 nm to 900 nm [18]. The Consejo Superior de Investigaciones Científicas (CSIC) Research Institute developed a measurement facility named Gonio-EspectroFótometro Español (GEFE), which could measure the BRDF and bidirectional transmittance distribution function (BTDF) of samples. The wavelength range covered 380 nm to 780 nm. The overall BRDF measurement uncertainty was less than 1% [17,19]. All uncertainties above are the measurement uncertainties for a BRDF of a nearly perfect diffuser.

The measurement facilities of the above national metrology laboratories are mainly used for standard transfer, rather than SD BRDF measurement for on-board calibration. At present, most SD BRDF measurement facilities achieve measurements that are traceable to a reflectance standard, i.e., a reference sample. The reflectance of the reference is known for the specified geometry and wavelength range. The facility is first calibrated with reference to the standard, and then the sample is substituted for the standard. For example, two scatterometers at NASA GSFC's Diffuser Calibration Lab (DCL) were used to effect NIST traceable BRDF scale transfers and enable the prelaunch testing of large-area solar diffuser flight unit panels in their on-orbit configurations [20]. The relative measurement method is simple and convenient. However, the disadvantages are obvious. The standard transfer is needed in this method. The major uncertainty then lies in the BRDF of the standard. The reference sample can easily be contaminated, which affects the measurement results [10].

To further reduce measurement uncertainty, some laboratories have developed absolute BRDF measurement facilities, such as the Polarization and Scatter Characterization Analysis of Lambertian materials (PASCAL) at Raytheon [21]. An absolute measurement means that the BRDF is obtained by measuring the incident and reflected radiation according to the definition of the BRDF. Additionally, no reference standard is introduced into the absolute measurement. Compared to relative measurement, an absolute measurement system needs a linear receiver over a sufficiently wide dynamic range [10].

In the on-board calibration process, the relative geometric relationship with respect to the SD between the incident direction of the sunlight and the observed direction of a remote sensor determines the incident and reflected angles of the SD BRDF measurement in the laboratory. Due to the limitations of satellite orbits, on-board calibration times, and other factors, it is necessary to measure SD BRDF with high accuracy at in- and out-of-plane and large angle geometries. In this case, the BRDF measurement's uncertainty increases compared to measuring in the standard geometry. There are many reasons for this problem, such as angle uncertainty and polarization [22].

An SD is often made of polytetrafluoroethylene (PTFE), sand-blasted aluminum, etc. [23]. According to the spatial and spectral characteristics of the SD, an improved data processing method could be designed based on the reciprocity theorem. This method could reduce the dependence of the measurement uncertainty on the angular accuracy of the goniometer system. An integrating sphere radiator was used as a source to ensure high-irradiance uniform illumination while reducing the influence of polarization on the measurement results.

This article is organized as follows. In Section 2, the principle of absolute BRDF measurement is introduced. In Section 3, an absolute BRDF measurement facility is introduced, including the configuration, calibration methods, and results of each part. The workflow of the facility is also presented. In Section 4, the determination of the SD BRDF measurement angle and the causes of large-angle incidence and detection in on-board calibration are discussed. An improved data processing method based on the reciprocity theorem is also introduced. The uncertainty of the SD BRDF measurement is analyzed based on the system calibration results from the overall facility in Section 5. The measurement results of the SD BRDF for on-board calibration are provided in Section 6. Finally, conclusions are given in Section 7.

2. Absolute BRDF Measurement Principle

The BRDF can be defined as [24]

$$f_r(\theta_i, \varphi_i, \theta_r, \varphi_r, \lambda) = \frac{dL_r(\theta_i, \varphi_i, \theta_r, \varphi_r, \lambda)}{dE_i(\theta_i, \varphi_i, \lambda)}, \quad (1)$$

where dL_r represents the radiance reflected in a given direction by an infinitesimal element; dE_i represents the irradiance from the incident direction that falls on this infinitesimal surface element; θ_i, φ_i represent the zenith angle and azimuth angle in the incident direction; θ_r, φ_r represent the zenith angle and azimuth angle in the reflected direction; and λ stands for the wavelength. The geometric relationship of $\theta_i, \varphi_i, \theta_r, \varphi_r$ is shown in Figure 1.

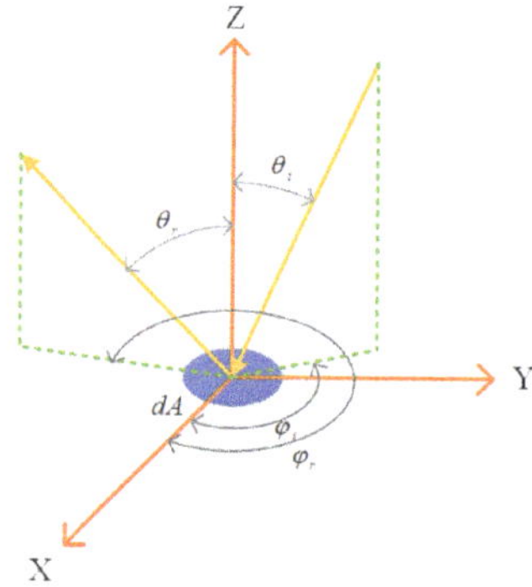

Figure 1. Geometry of incident and reflected beams.

According to the definition of radiance and irradiance,

$$dE_i(\theta_i, \varphi_i, \lambda) = \cos\theta_i dE_\perp(\lambda) = \cos\theta_i L_\perp(\lambda) d\Omega, \quad (2)$$

where $dE_\perp(\lambda)$ and $dL_\perp(\lambda)$ represent the radiance and irradiance, respectively, when the direction of incidence is perpendicular to the sample, and Ω represents the illumination solid angle.

Equation (2) can be substituted into Equation (1), integrating both sides at the same time:

$$\int f(\theta_i, \varphi_i; \theta_r, \varphi_r; \lambda) \cos\theta_i L_\perp(\lambda) d\Omega = \int dL_r(\theta_i, \varphi_i; \theta_r, \varphi_r; \lambda). \quad (3)$$

These measurements can be made at narrow solid angles and on small surfaces; the output of the radiation source is uniform (in spatial direction). The SD for on-board calibration is a nearly perfect diffuser. Therefore, $f(\theta_i, \varphi_i; \theta_r, \varphi_r; \lambda)$ and $L_\perp(\lambda)$ can be considered constants. Then

$$f(\theta_i, \varphi_i; \theta_r, \varphi_r; \lambda) = \frac{\int dL_r(\theta_i, \varphi_i; \theta_r, \varphi_r; \lambda)}{\cos\theta_i L_\perp(\lambda) \int d\Omega} = \frac{L_r(\theta_i, \varphi_i; \theta_r, \varphi_r; \lambda)}{L_\perp(\lambda)} * \frac{R^2}{A\cos\theta_i}, \quad (4)$$

where R represents the distance from the radiation source opening to the center of the sample, and A represents the area of the radiation source opening. Equation (4) shows that the required SD BRDF could be obtained by measuring the incident radiance, the illumination solid angle, and the reflected radiance.

The incident and reflected radiance are measured using the same detection system, and the final BRDF calculation requires only the ratio of the incident radiance to reflected radiance; therefore, it is not necessary to know the absolute value of the measured radiance. Thus, Equation (4) can be expressed as

$$f(\theta_i, \varphi_i; \theta_r, \varphi_r; \lambda) = \frac{DN_r(\theta_i, \varphi_i; \theta_r, \varphi_r; \lambda)}{DN_\perp(\lambda)} * \frac{R^2}{A\cos\theta_i}, \quad (5)$$

where $DN_r(\theta_i, \varphi_i; \theta_r, \varphi_r; \lambda)$ and $DN_\perp(\lambda)$ represent the digital number (DN) of the detector for the reflected and incident radiance, respectively. To simplify the expression, DN was subtracted from the dark currents in this paper.

3. Description and Calibration of the Facility

An absolute BRDF measurement facility is shown in Figure 2. It consists of three parts: an irradiation system, a goniometer system, and a detection system. All devices are connected to the same industrial control computer. The required BRDF of a sample can be automatically measured by the control software.

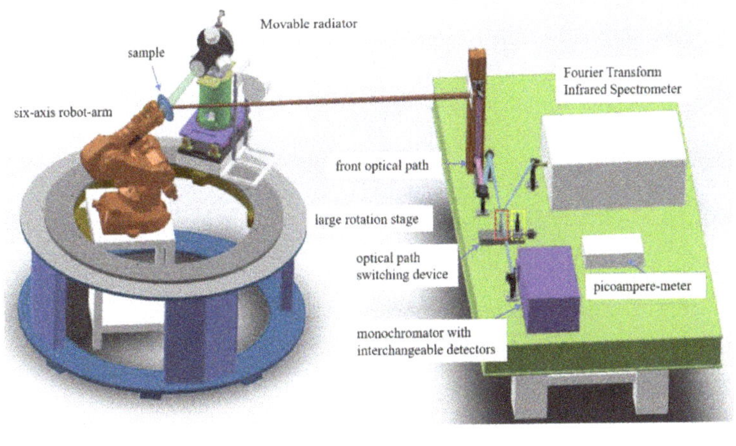

Figure 2. Absolute BRDF measurement facility.

3.1. Irradiation System

An integrating sphere was chosen as the radiation source to ensure that the brightness and uniformity requirements of the irradiation system were met. It also reduced the effect of polarization on the measurement results.

The inner coating of the sphere radiator is shown in Figure 3. It was made of pressed PTFE with a reflectivity above 99%. The inner diameter of the sphere was 150 mm, and three 400 W halogen lamps were built in. The sphere was equipped with hydrocooling and air-cooling systems. The power supply was set to 1000 W (approximately 80% of the rated power) when the sphere radiator operated. A stray light hood was added in front of the sphere radiator opening to reduce the influence of stray light during measurement.

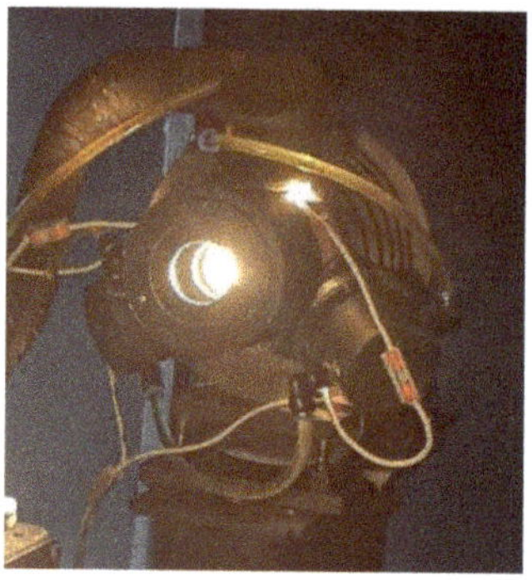

Figure 3. Integrating sphere radiator.

To verify the homogeneity of the sphere radiator, a special homogeneity measurement facility was built. Imaging optics for radiation at a 1:1 reproduction scale were chosen and mounted on a two-dimensional xy-translation stage. The step interval was set to 2 mm, and the scanning area covered the whole opening of the sphere radiator. The results show that the nonuniformity of the opening was less than 0.5%.

The stability of the sphere radiator could be detected by the facility's detection system. The facility was adjusted to the position for measuring incident radiation. The output power of the high-precision current source was set to 1000 W, and the measurement wavelength was 350 nm. After 30 min of preheating, the measurement lasted 80 min with a sampling interval of 20 ms. The results are shown in Figure 4. Stability was expressed as relative standard deviation, and the instability of the light source was less than 0.06% (80 min).

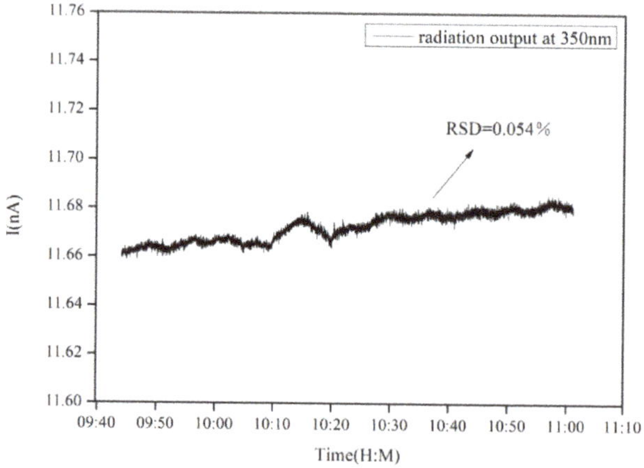

Figure 4. Stability test results of the sphere radiator.

3.2. Goniometer System

A six-axis robot arm and a large rotation stage from the goniometer system of the facility are shown in Figure 5. The robot arm held the sample for three-dimensional rotation, and the rotation stage loaded the irradiation system for one-dimensional rotation. The detection system was fixed to ensure high-accuracy radiation measurement.

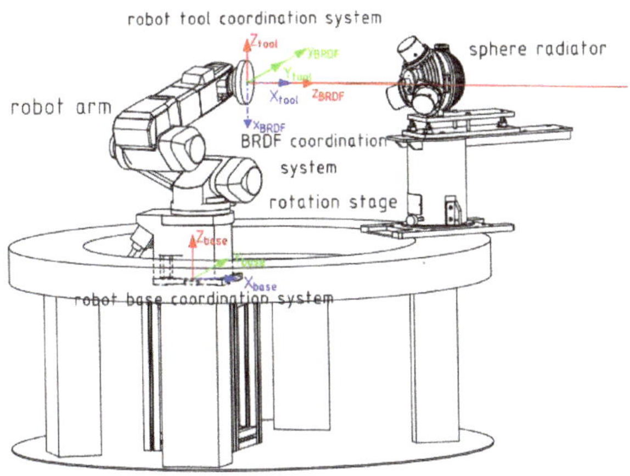

Figure 5. Goniometer system.

The angle ranges for BRDF measurement at the facility are shown in Table 1.

Table 1. Angle ranges for BRDF measurement.

Angle Parameter	Range (°)	Notes
Incident zenith angle (θ_i)	0–75	Due to obstruction of the radiation source, the angle between the incident and reflected rays is greater than 10°
Incident azimuth angle (φ_i)	0–360	
Reflection zenith angle (θ_r)	0–75	
Reflection azimuth angle (φ_r)	0–360	

Two coordinate systems were defined on the surface of the SD: the robot tool coordinate system XYZ and the BRDF coordinate system xyz. They both shared a common coordinate origin, which was located at the center of the sample. This point was also the intersection point of the rotation axis during the rotation stage and the centerline of the opening of the sphere radiator. The relative positions of the two coordinate systems are shown in Figure 5.

The angle of rotation of the robot arm around the XYZ axis of the robot tool coordinate system could be defined as α, β, γ, and the rotation angle of the rotation stage was δ. According to the different expression relationships between the incident vector, the reflected vector and the sample normal in the two coordinate systems, the relationship between $\theta_i, \varphi_i, \theta_r, \varphi_r$ and $\alpha, \beta, \gamma, \delta$ could be established as follows:

$$\cos\delta = \cos\theta_i \cos\theta_r + \sin\theta_i \sin\theta_r \cos(\varphi_r - \varphi_i) \tag{6}$$

$$\tan\alpha = \frac{\cos\theta_i \cos\delta - \cos\theta_r}{\cos\theta_i \sin\delta} \tag{7}$$

$$\sin\beta = \frac{\sin(\varphi_r - \varphi_i)\sin\theta_i \sin\theta_r}{\sin\delta} \tag{8}$$

$$\tan\gamma = \frac{\sin\theta_i \cos\theta_r \cos\varphi_i - \sin\theta_r \cos\theta_i \cos\varphi_r}{\sin\theta_i \cos\theta_r \sin\varphi_i - \sin\theta_r \cos\theta_i \sin\varphi_r}. \tag{9}$$

For a given BRDF angle $(\theta_i, \varphi_i, \theta_r, \varphi_r)$, the rotational angle of the robot arm and rotation stage $(\alpha, \beta, \gamma, \delta)$ could be calculated using Equations (6)–(9). The goniometer system was rotated based on the calculated $\alpha, \beta, \gamma, \delta$ to achieve the geometric relationships required for the measurement.

3.3. Detection System

An SD can reflect the incident radiation into the hemisphere so that only a small portion of the incident radiation is collected during measurement. The detection solid angle of the facility was very small, approximately 0.003 sr, and depending on the different incident angles of the radiation source, there were approximately four orders of magnitude of difference between the incident radiation and the reflected radiation. For different measurement wavelengths, the difference could even exceed four orders of magnitude. In this large dynamic range, non-linearity in the detection system needed to be determined and corrected.

The detection system had two configurations. One was a combination of a monochromator, detectors, and a picoamperemeter. The other was a Fourier transform infrared (FTIR) spectrometer. These two detection devices shared the same front optical path systems. Depending on the wavelength required, a combination of a plane reflector and an electric translation platform was used to switch the optical path.

The front optical path is shown in Figure 6. It was composed of five plane mirrors, three concave mirrors and one electric translation platform. The radiation reflected from the sample was parallel to the optical platform plane after being reflected by plane mirror 1 and plane mirror 2 and then converging on concave mirror 1. It was then reflected by plane mirror 3 to the optical path-switching device, which was composed of an electric translation platform and a plane mirror. After passing through the optical path-switching device, the concave mirror converged the light to the optical inlet of the monochromator or FTIR spectrometer.

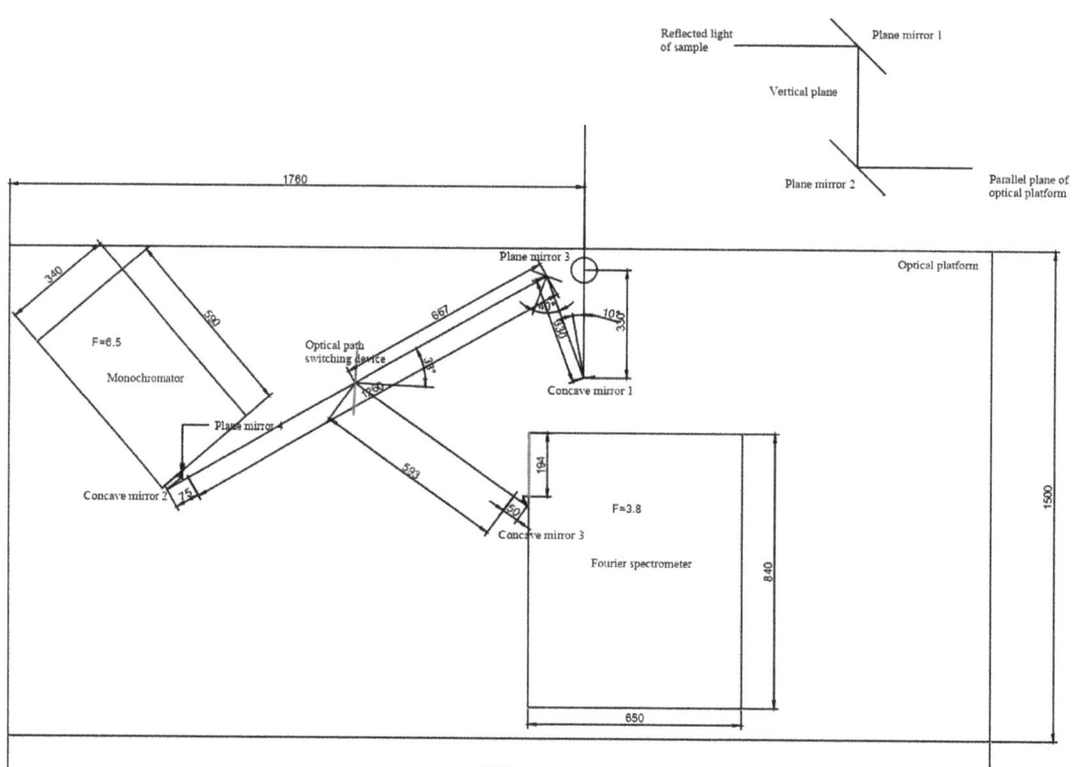

Figure 6. Optical path diagram of the detection system.

The electric translation platform controlled the back-and-forth movement of the plane mirror. From Figure 2, when the plane mirror was in the red dashed box position on the optical path-switching device, it reflected the radiation to the optical inlet of the FTIR spectrometer. Additionally, when it was completely removed (in the yellow dashed box position), the radiation entered the optical inlet of the monochromator.

The performance of the detection devices in these two configurations was as follows:

1. Monochromator, detector and picoamperemeter: The monochromator had a spectral resolution of approximately 3 nm at 350 nm to 600 nm and 6 nm at 600 nm to 1700 nm. Depending on the different measurement wavelengths, two types of detectors were equipped: Si (350 nm to 1000 nm) and InGaAs (1000 nm to 1700 nm, two-stage TE cooled, −40 °C) detectors. All photoelectric signals were detected by the picoamperemeter, and the measurement results were transmitted to a computer for data storage and calculation.
2. FTIR spectrometer: The FTIR spectrometer was a VERTEX80 [25] from the BRUKER company. In the measurement facility, the InGaAs detector was selected as the detector for the VERTEX80. The measurement spectral range was set at 11,764 cm^{-1} (850 nm) to 4148 cm^{-1} (2410 nm), and the spectral resolution was set at 64 cm^{-1} (4.6 nm to 36.6 nm). This resolution was selected because of the spectral characteristics of the SD's reflectivity (spectral flatness) and the improvement in the signal-to-noise (SNR) ratio when measuring the reflected radiation. In the process of inverse Fourier transform, the Blackman Harris 4-Term was used as the apodization function; the zero filling factor was 2; the power spectrum mode was selected for phase correction; and the absolute largest value was selected for the peak position retrieval mode.

When measuring the reflected radiation, the circuit amplification coefficient used by the FTIR spectrometer was different from that used for the incident radiation to ensure a sufficiently high SNR. The calculation method for scaling factor η was as follows:

$$\eta = \frac{DN_{rFTIR}(0°;45°;\lambda)/DN_{iFTIR}(\lambda)}{DN_{rInGaAs}(0°;45°;\lambda)/DN_{iInGaAs}(\lambda)}. \tag{10}$$

where $DN_{rFTIR}(0°;45°;\lambda)$ and $DN_{iFTIR}(\lambda)$ represent the DN of the FTIR spectrometer for the reflected radiation at 0°/45° and incident radiation, respectively, and $DN_{rInGaAs}(0°;45°;\lambda)$ and $DN_{iInGaAs}(\lambda)$ represent the DN of the InGaAs detector for reflected radiation at 0°/45° and incident radiation, respectively.

The DN ratio of the incident radiation and reflected radiation from the InGaAs detector at 0°/45° was compared with the DN ratio from the FTIR spectrometer under the same conditions. Data at the 1200 nm and 1300 nm bands were selected for calculation. The average scaling factor was 99.10.

3.4. System Calibration

3.4.1. Geometric Calibration

The geometric calibration of the facility was divided into two main aspects: determining the initial position (the rotation center of the robot arm and the origin of the rotation stage) of the BRDF measurement and verifying the uncertainty of the rotation angle.

The rotation center of the robot arm was located at the intersection point of the rotation axis of the rotation stage and the normal to the center of the sphere radiator opening. Before measuring, the initial position of the sample needed to be determined.

With the laser beam emitted from the sphere radiator opening as the indicator and the robot arm as the measuring tool (the measuring principle is shown in Figure 7a), the space linear equation of the rotation stage was obtained by measuring the beam emitted from the sphere radiator opening at different angles (the measuring schematic is shown in Figure 7b), and the transformation relationship between the robot's base coordinate system and the laboratory coordinate system was preliminarily established. Then, the laser beam was adjusted perpendicular to the rotation axis of the rotation stage, and the position of the laboratory coordinate origin and the changing relationship between the two coordinate systems were determined according to the position of the incident laser beam on the Y-axis of the tool coordinate system and the axis equation. A calibration model of the instrument was established. The position and attitude of the rotation center in space were calculated, and the initial position of the sample was determined. This calibration method is described in great detail in the literature [26].

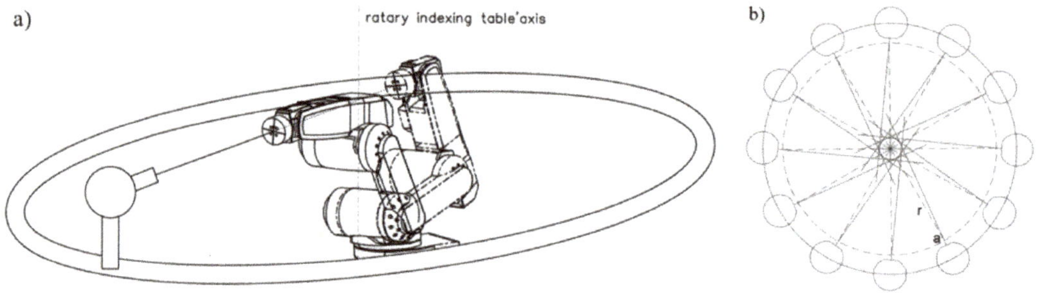

Figure 7. (**a**) Principle of measuring the beam. (**b**) Principle of determining the rotation axis by measuring the beam.

The origin of the rotation stage (red points marked in Figure 8) was corrected using a coordination measurement with the overall system. The radiation source was placed at the origin before correction, and the sample held by the robot arm was placed at the initial position determined and described in the previous paragraph. The radiation source was rotated counterclockwise by 30° to a starting point. Then, the reflected radiation was collected at 5° intervals. The measuring schematic is shown in Figure 8a. After acquisition was complete, the radiation source was returned to the origin, and the sample was rotated 180° around the z-axis by the robot arm. The radiation source was rotated clockwise 30° to the starting point. Then, the reflected radiation was collected again at 5° intervals. The measuring schematic is shown in Figure 8b. A schematic diagram of the whole correction operation follows:

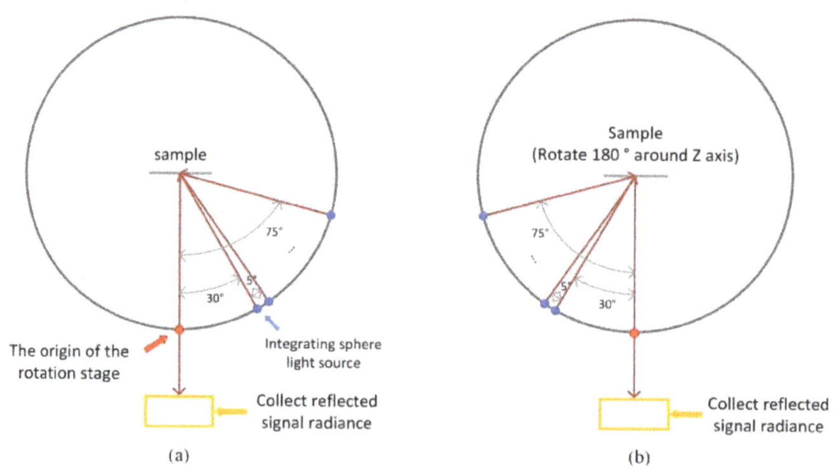

Figure 8. (**a**) Rotating the radiation source counterclockwise and collecting the reflected radiation. (**b**) Rotating the radiation source clockwise and collecting the reflected radiation.

If the angle of counterclockwise rotation was $\theta_{a1}, \theta_{a2} \cdots \theta_{a10}$, the angle of clockwise rotation was $\theta_{b1}, \theta_{b2} \cdots \theta_{b10}$, and the origin angle before correction was $\Delta\theta$, then,

$$\frac{DN(\theta_{ai})}{\cos(\theta_{ai} - \Delta\theta)} = \frac{DN(\theta_{bi})}{\cos(\theta_{bi} + \Delta\theta)}. \tag{11}$$

The correction value at the origin of the rotation stage could be calculated using Equation (11). After correction, the reflected radiation is shown in Figure 9. The relative deviation of $\frac{DN(\theta_{ai})}{\cos(\theta_{ai})}$ and $\frac{DN(\theta_{bi})}{\cos(\theta_{bi})}$ was less than 0.2%.

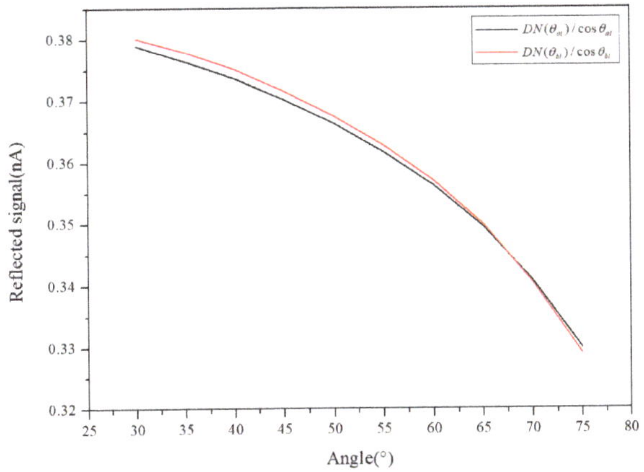

Figure 9. Reflected radiation at different incident angles of the incident after correction.

A detection method based on the principle of specular reflection was used to verify the angular uncertainty of the goniometer system. The origin of the coordinate paper was placed at the center spot of the incident light beam after rotating the radiation source 180° counterclockwise. Assuming that the counterclockwise rotation angle of the radiation source was $\Delta\delta$, the robot was adjusted so that the reflector rotated $\Delta\delta/2$ about the Y-axis, and the position of the center spot of the reflected light beam was recorded at this time. The experimental diagram is shown in Figure 10.

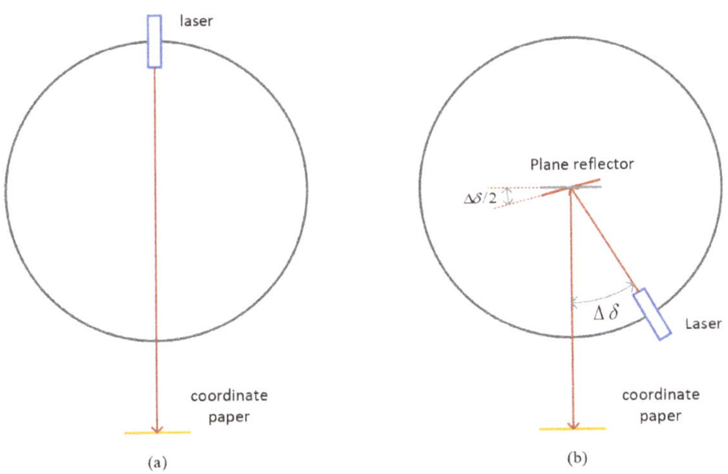

Figure 10. (**a**) Determining the origin of the coordinate paper. (**b**) Specular reflection test.

The range of $\Delta\delta$ was set to 6°–75°, and measurements were taken every 3°. The distance between the coordinate paper and the rotation center was 1900 mm. The experimental results are shown in Figure 11, and the incident angle when measuring the spot is marked beside each point.

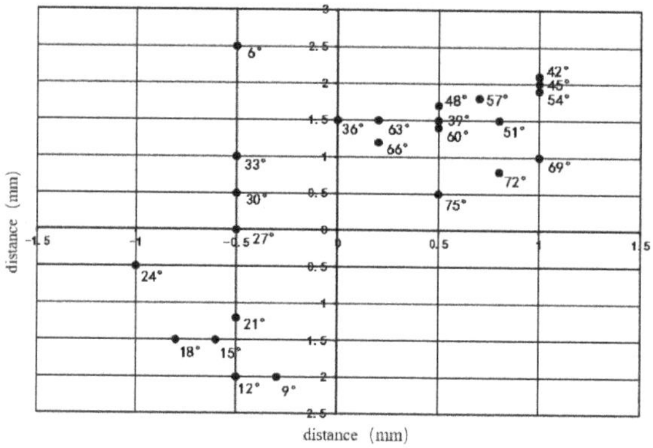

Figure 11. Measurement of reflected spot position data based on mirror reflection (incident angles are marked beside the points).

The angle error $\Delta\theta$ could be expressed as

$$\Delta\theta = \arctan\frac{\sqrt{{x_m}^2 + {y_m}^2}}{l}, \tag{12}$$

where x_m, y_m represent the horizontal and vertical coordinates, respectively, of the reflected light spot on the coordinate paper at the current angle; l represents the distance between the coordinate paper and the rotation center. Calculating the results based on Figure 11 and using Equation (12), the maximum deviation angle was 0.0769°; this means that the angular accuracy of the goniometer system was less than 0.1°.

3.4.2. Nonlinear Calibration of the Detection System

- 350 nm to 1000 nm: A method based on the flux addition of two light-emitting diodes (LEDs) in an integrating sphere was used to detect nonlinearity in the detector system [27]. The measurement range was from 12 pA to 364 nA, which matched the dynamic range of the facility. The measurement results are shown in Figure 12.

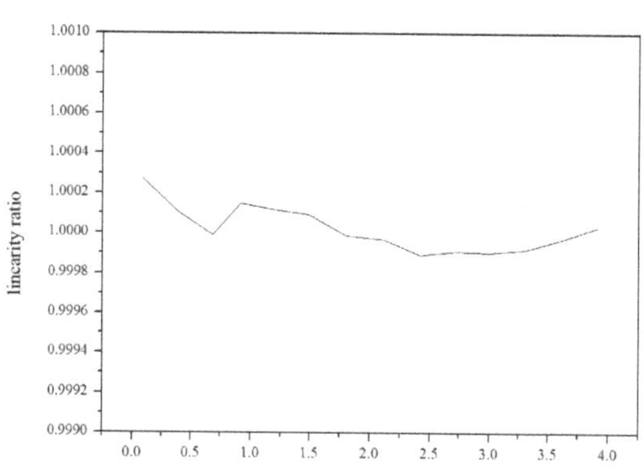

Figure 12. Measurement results for detector nonlinearity.

- 1000 nm to 2400 nm:

The linear uncertainty of the FTIR spectrometer as measured by the manufacturer was 0.02%.

3.5. Absolute Measurement Workflow for the Facility

The measurement steps for the facility are outlined as follows.

First, the robot held the sample and moved it to the avoidance position, and the radiation source was rotated 180° counterclockwise from the origin of the rotation stage. The incident radiation at the required wavelength was measured. Then, the software controlled the radiation source and the sample to the position required for measurement and measured the reflected radiation from the sample. After the reflected radiation measurements were complete, the facility returned to measure the incident radiation again. The software could control the acquisition frequency of the incident radiation by setting the number of reflected radiation acquisitions. Generally, the number of reflected radiation acquisitions was set to 10 to ensure that the incident radiation measurement interval did not exceed 15 min. Photos of the measurement process are shown in Figure 13a,b.

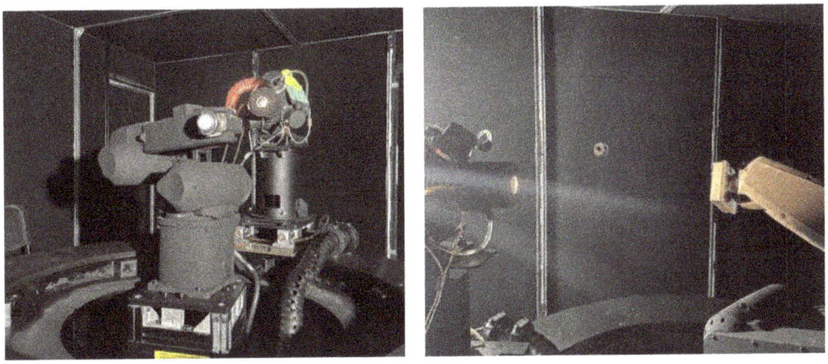

(a) (b)

Figure 13. (**a**) Measurement of the radiation of the radiation source. (**b**) Measurement of the reflected radiation of the sample.

4. BRDF Measurement by the SD for On-Board Calibration

4.1. Determination of the SD BRDF Measurement Angle

On-board calibration is generally conducted after satellite lighting and before subsatellite point lighting (Figure 14). Based on the analysis of observations from hyperspectral instruments on international satellites, on-board calibration is susceptible to scattered and reflected stray light from the Earth and the atmosphere. The appropriate Sun–Earth–satellite (SES) angle at on-board calibration times is selected to reduce this influence. For example, the SES angle for on-board calibration was just above 105° on the Greenhouse Gases Observing Satellite (GOSAT) [28].

After determining the calibration time on the satellite and the satellite orbital parameters, the elevation angle and azimuth angle of the incident beam were also determined. According to the design requirements of an on-board calibration facility based on an SD, the direction of the central axis of the payload (the observation direction) could also be determined at the calibration time.

The attitude of the SD was determined from the vectors of the incident and observation directions. It was required that the zenith angles of the incident and observation directions be as small as possible, and mirror positions of the incident and observation azimuth angles should be avoided.

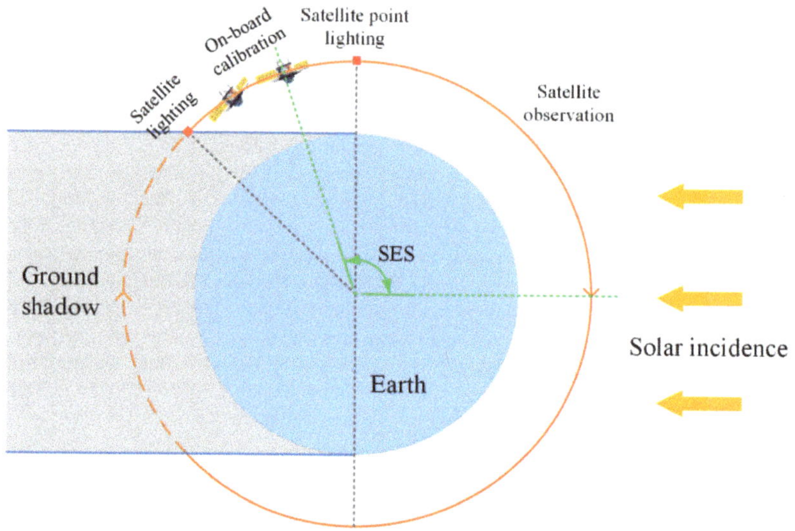

Figure 14. Schematic diagram of the on-board calibration time.

However, if the angle between the incident and observation vectors was large, the zenith angles of the incidence and observation directions could not be reduced, no matter how the SD was placed. If the zenith angle in the observation direction was too large, the size of the optical surface of the SD would increase in order to ensure that the whole detector could be covered. An oversized SD takes up a large space, which affects the design of other systems in the satellite. Therefore, it is common to prepare a design for a large observation zenith angle and a small reflection zenith angle.

4.2. Improved Data Processing Method Based on Reciprocity Theorem for Measuring Large Angle BRDF

According to Equation (5), the uncertainty of an absolute BRDF measurement could be related to the uncertainty of the incident angle, and the relative uncertainty [29] component associated with the incident angle (u_{θ_i}) was

$$\begin{aligned} u_{\theta_i} &= \frac{\frac{\partial f_r}{\partial \theta_i} * u(\theta_i)}{f_r} \\ &= \left(\frac{\partial DN_r}{\partial \theta_i}\frac{1}{DN_r} + \tan\theta_i\right) * u(\theta_i). \end{aligned} \tag{13}$$

Assuming that $u(\theta_i) = 0.1°$, the trend of $\tan\theta_i * u(\theta_i)$ and $\frac{\partial DN_r}{\partial \theta_i}\frac{1}{DN_r} * u(\theta_i)$ with θ_i could be calculated as shown in Figure 15. Note that the results here are for pressed or sintered PTFE. $DN_r(\theta_i)$ was obtained by fitting data from a 10° to 75° (5° interval) incident zenith angle and 0° reflected zenith angle. It was obvious that the greater the incident angle, the greater $\tan\theta_i * u(\theta_i)$ and $\frac{\partial DN_r}{\partial \theta_i}\frac{1}{DN_r} * u(\theta_i)$ were. At an 80° incident angle, $u_{\theta_i} > 1\%$ when $u(\theta_i) = 0.1°$, which caused a significant impact on BRDF measurement.

For pressed PTFE, the angle had a small effect on $DN_r(\theta_i, \varphi_i; \theta_r, \varphi_r; \lambda)$. In the extreme case of large incident and reflected angles, the relative uncertainty of the reflected radiation component associated with the angle was less than 0.15% when $u(\theta) = 0.1°$. At large incident angles, the effect of the angle's uncertainty on $DN_r(\theta_i, \varphi_i; \theta_r, \varphi_r; \lambda)$ was much smaller than that on $\cos\theta_i$. Therefore, an improved data processing method for measuring the BRDF at a large angle was designed. This method was based on the reciprocity theorem. The principle of this method was that, with vertical detection as a reference, the BRDF of a tilted observation could be obtained by comparing tilted observation and vertical observation at the same illumination. This improved method could reduce the dependence of the measurement uncertainty on the angular accuracy of the goniometer system.

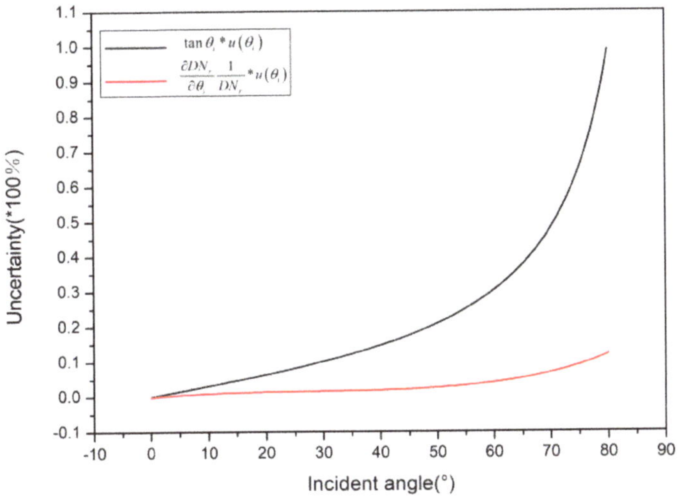

Figure 15. Relative uncertainty component associated with the incident angle.

The reciprocity theorem describes the reversibility of an optical path. It refers to the concept that the energy emitted by a light beam after arbitrary reflection or refraction on a surface or in a passive medium is equal to the energy emitted by the beam along the opposite path. When applied to the BRDF, BRDF remained unchanged when the incident angle and reflection angle were interchanged:

$$f(\theta_i, \varphi_i; \theta_r, \varphi_r; \lambda) = f(\theta_r, \varphi_r; \theta_i, \varphi_i; \lambda). \tag{14}$$

Compared with tilted illumination ($\theta_i \neq 0°$), the influence of u_{θ_i} on Equation (12) was the least when vertically illuminated ($\theta_i = 0°$). Therefore, the BRDF measured at $0°/45°$ was used as a reference to obtain the BRDF for other angles at the same irradiance. Then

$$f(0°; 45°; \lambda) = \frac{DN_r(0°; 45°; \lambda)}{DN_i(\lambda)} \frac{R^2}{A}, \tag{15}$$

$$f(0°; \theta_r; \lambda) = \frac{DN_r(0; \theta_r; \lambda)}{DN_r(0°; 45°; \lambda)} f(0°; 45°; \lambda), \tag{16}$$

where DN_i represents the DN of the incident radiance.

From (14),

$$f(0; \theta_r, \varphi_r; \lambda) = f(\theta_r, \varphi_r; 0; \lambda). \tag{17}$$

For other nonvertical irradiance conditions,

$$\begin{aligned} f(\theta_i, \varphi_i; \theta_r, \varphi_r; \lambda) &= \frac{DN_r(\theta_i, \varphi_i; \theta_r, \varphi_r; \lambda)}{DN_r(\theta_i, \varphi_i; 0; \lambda)} f(\theta_i, \varphi_i; 0; \lambda) \\ &= \frac{DN_r(\theta_i, \varphi_i; \theta_r, \varphi_r; \lambda)}{DN_r(\theta_i, \varphi_i; 0; \lambda)} f(0; \theta_i, \varphi_i; \lambda) \\ &= \frac{DN_r(\theta_i, \varphi_i; \theta_r, \varphi_r; \lambda)}{DN_r(\theta_i, \varphi_i; 0; \lambda)} \frac{DN_r(0; \theta_i, \varphi_i; \lambda)}{DN_r(0°; 45°; \lambda)} f(0°; 45°; \lambda) \end{aligned} \tag{18}$$

The measurement process for the improved data processing method is shown in Figure 16.

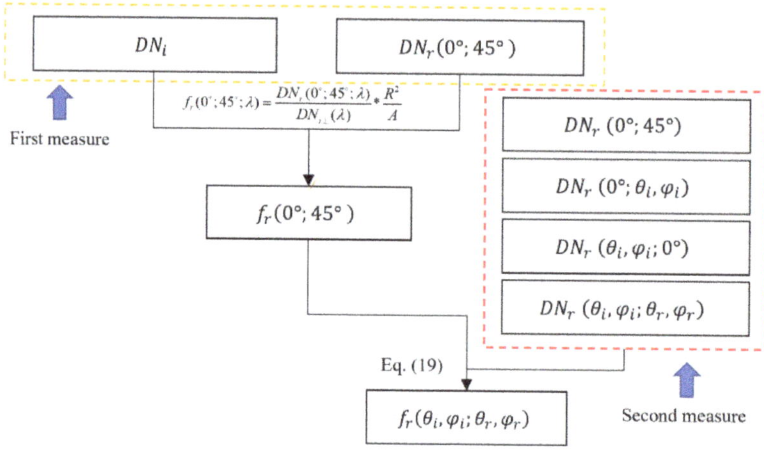

Figure 16. Measurement process of improved data processing method.

5. Uncertainty Budget

5.1. Relative Uncertainty Budget for Absolute Measurement of the SD BRDF at 0°/45°

According to Equation (15), the relative uncertainty of the BRDF at 0°/45° could be given by

$$u_{f(0°;45°;\lambda)} = \sqrt{u_R{}^2 + u_A{}^2 + u_{L_i}{}^2 + u_l{}^2 + u_\lambda{}^2 + u_\theta{}^2 + u_{DN_i}{}^2 + u_{DN_r}{}^2 + u_\eta{}^2 + u_p{}^2 + u_s{}^2}, \tag{19}$$

where u_R and u_A represent the relative uncertainty components associated with the measurement geometric factor; u_{L_i} represents the relative uncertainty component associated with the stability of the light source; u_l, u_{DN_i}, u_{DN_r}, u_η represent the relative uncertainty components associated with the detection system, which correspond to the linearity and repeatability of the incident and reflected radiation detection and solution of the amplification factor; u_λ represents the relative uncertainty component associated with the wavelength; u_θ represents the relative uncertainty component associated with the incident angle; u_p represents the relative uncertainty component associated with the polarization; and u_S represents the relative uncertainty component associated with stray light.

5.1.1. Measurement Geometric Factors

Distance R was measured using a 1000 mm Vernier caliper, a dial gauge, and a connecting rod. Area A of the opening was measured using an inner diameter ruler. The R and A were obtained from the average of multiple measurements. The measurement accuracy was 0.01 mm. The measurement process is shown in Figure 17.

The measurement uncertainties of the distance and area were described by the standard deviation of multiple measurements. The relative uncertainty components associated with R and A, respectively, were as follows:

$$u_R = \left(\frac{\partial f_r}{\partial R}/f_r\right) * u(R) = 0.083\%, \tag{20}$$

$$u_D = \left(\frac{\partial f_r}{\partial D}/f_r\right) * u(D) = 0.048\%. \tag{21}$$

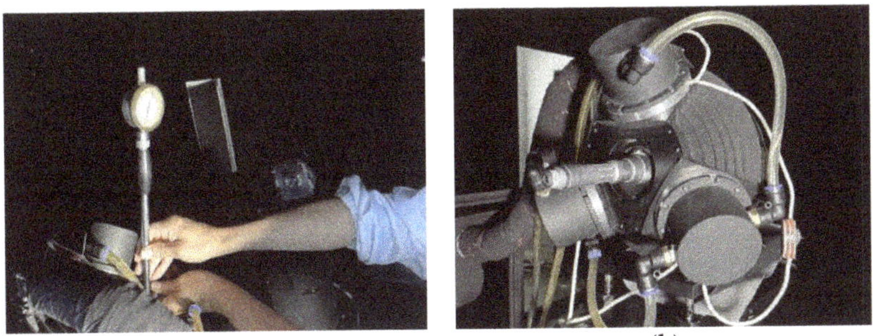

(a) (b)

Figure 17. (**a**) Measuring the distance from the integrating sphere opening to the sample surface. (**b**) Measuring the area of the light opening.

5.1.2. Stability of the Light Source

According to the description in Section 3.1, the measurement interval of the incident radiation did not exceed 15 min. According to Figure 3, $u_{L_i} \leq 0.015\%$.

5.1.3. Linearity

The linear uncertainty could be expressed as

$$\varepsilon = \frac{|1 - C_n|}{C_n}, \tag{22}$$

where C_n represents the nth order corresponding to the nonlinear correction factor and ε represents the nonlinearity uncertainty. The calculation steps for C_n are described in detail in [30].

According to the results in Section 3.4.2, the relative uncertainty component associated with the Si detector could be calculated to be $u_{l(SI)} = 0.033\%$.

From the factory test report of the FTIR spectrometer, the linearity of the spectrometer was better than 0.02%, that is, $u_{l(FTIR)} = 0.02\%$.

5.1.4. Wavelength

From the characteristics of PTFE, the reflectance of the SD was flat along the spectrum (excluding the absorption band near 2200 nm); therefore, $(1/f_r) \cdot (\partial f_r/\partial\lambda) * u(\lambda) \approx 0$.

5.1.5. Angle

According to Figure 11, the angle uncertainty (0.1°) had little impact on the reflected signal at 0°/45°. Therefore, $(1/f_r) \cdot (\partial f_r/\partial\theta) \cdot u(\theta) \approx 0$.

5.1.6. Repeatability of Incident and Reflected Signal Detection

The DN at the detection system was obtained from an average of 36 repeated measurements. The uncertainty calculation method for n repeated measurements of the DN was as follows:

$$u_{DN} = \sqrt{\frac{\sum_{i=1}^{n}(x_i - \overline{x})^2}{n(n-1)(\overline{x})^2}} * 100\%, \tag{23}$$

where x_i represents the DN for the i-th measurement and $\overline{x}$ represents the average of n measurements.

The relative uncertainty component associated with the DN of the picoamperemeter is shown in Tables 2–4, and the relative uncertainty was the maximum value within the band range.

Table 2. Uncertainty (k = 1) of the incident radiation of the detectors.

Incident Angle	Detector			
	Si			InGaAs
	350–410 nm	410–480 nm	480–1000 nm	1000–1600 nm
0°	<0.01%	<0.01%	<0.01%	<0.01%

Table 3. Uncertainty (k = 1) of the SD reflected radiation of the detectors at 0° reflection.

Incident Zenith Angle	Detector			
	Si			InGaAs
	350–410 nm	410–480 nm	480–1000 nm	1000–1600 nm
15°	0.10%	0.03%	0.01%	0.04%
30°	0.12%	0.04%	0.02%	0.05%
45°	0.16%	0.05%	0.02%	0.05%
60°	0.22%	0.07%	0.02%	0.06%
75°	0.31%	0.15%	0.04%	0.16%

Table 4. Uncertainty (k = 1) of the SD reflected radiation of the detectors at 0° incident.

Incident Zenith Angle	Detector			
	Si			InGaAs
	350–410 nm	410–480 nm	480–1000 nm	1000–1600 nm
15°	0.10%	0.02%	0.01%	0.02%
30°	0.11%	0.03%	0.01%	0.03%
45°	0.11%	0.03%	0.01%	0.03%
60°	0.11%	0.04%	0.01%	0.04%
75°	0.12%	0.04%	0.01%	0.04%

The relative uncertainty component associated with the DN data collected by the FTIR spectrometer is shown in Figures 18–20.

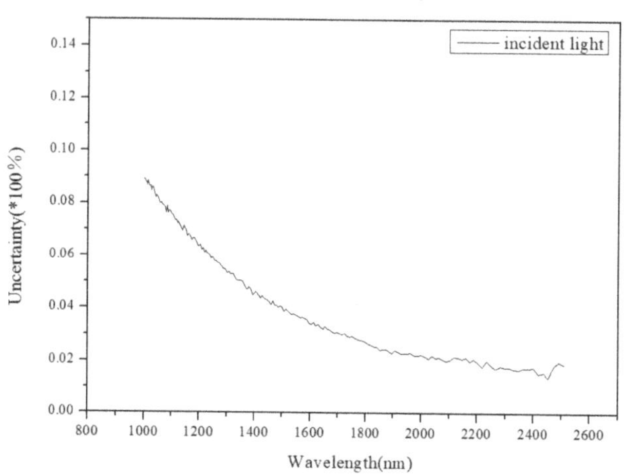

Figure 18. Uncertainty (k = 1) of the incident radiation of the FTIR spectrometer.

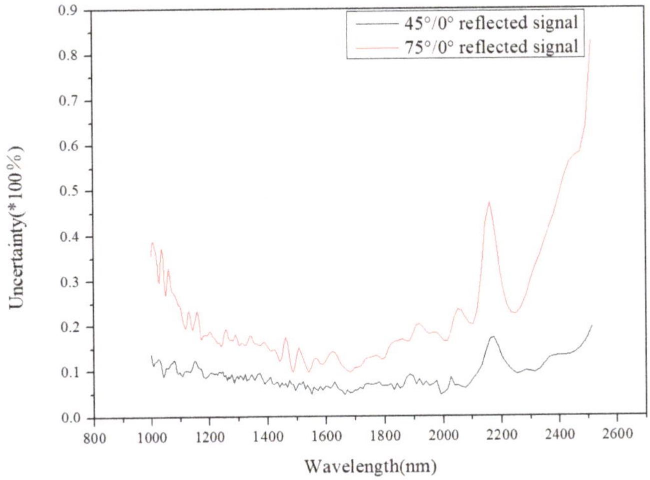

Figure 19. Uncertainty (k = 1) of the reflected radiation of the FTIR spectrometer at 0° reflection.

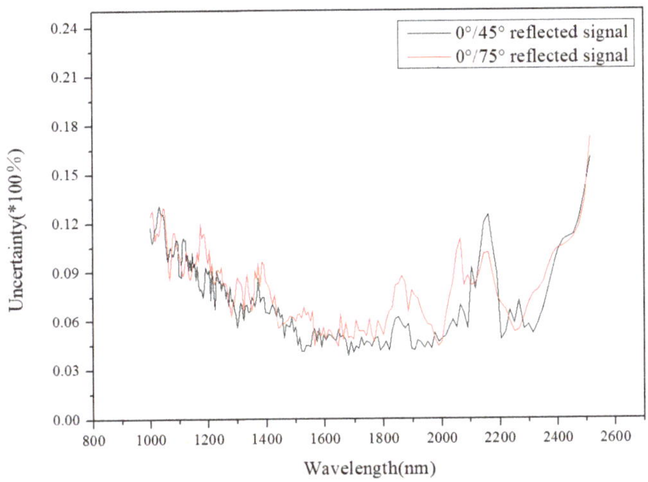

Figure 20. Uncertainty (k = 1) of the reflected radiation of the FTIR spectrometer at 0° incidence.

5.1.7. Scaling Factor

According to (5), $u_\eta = \sqrt{u^2{}_{DN_{rInGaAs}(0°;45°;1200nm)} + u^2{}_{DN_{iInGaAs}(1200nm)}} = 0.028\%$.

5.1.8. Stray Light

The transfer coefficient could be calculated as

$$\frac{\Delta f_r(\lambda)}{f_r(\lambda)} = \frac{q_r(\lambda) - q_i(\lambda)}{1 + q_i(\lambda)}, \tag{24}$$

where $q_i(\lambda)$ represents the signal ratio of the stray light to the incident flux and $q_r(\lambda)$ represents the signal ratio of the stray light to the reflected flux [31]. The relative uncertainty component associated with the stray light was estimated to be 0.1% based on an actual condition.

The uncertainty (k = 2) of the absolute measurement of the SD BRDF at 0°/45° was calculated as shown in Table 5 [1].

Table 5. Absolute measurement uncertainty (k = 2) of the SD BRDF at 0°/45°.

Component of Uncertainty	Type	Relative Uncertainty (%)
Distance	A	0.166%
Aperture area	A	0.096%
Source stability	A	0.03%
Detector linearity	B	0.066%(350–1000 nm) 0.04%(1000–2500 nm)
Wavelength	B	<0.01%
Rotation angle	B	<0.01%
Incident radiation	A	<0.01%
Reflected radiation	A	0.22% (350–410 nm) 0.06% (410–480 nm) 0.02% (480–1000 nm) 0.24% (1000–2500 nm)
Scaling factor	B	0.056% (1000–2500 nm)
Stray light	B	0.2%
Total		0.36% (350–410 nm) 0.29% (410–480 nm) 0.29% (480–1000 nm) 0.37% (1000–2500 nm)

5.2. *Relative Uncertainty Budget for SD BRDF Measurement at Large Angles*

According to Equation (18), the relative uncertainty budget for SD BRDF measurement at a large angle was

$$u_f = \sqrt{u^2{}_{DN_r(\theta_i,\varphi_i;\theta_r,\varphi_r;\lambda)} + u^2{}_{DN_r(0°;\theta_i,\varphi_i;\lambda)} + u^2{}_{DN_r(\theta_i,\varphi_i;0°;\lambda)} + u^2{}_{DN_r(0°;45°;\lambda)} + u^2{}_{f(0°;45°;\lambda)} + u^2{}_{\theta}}, \tag{25}$$

where $u^2{}_{DN_r(\theta_i,\varphi_i;\theta_r,\varphi_r;\lambda)}$ represents the relative uncertainty component associated with the repeatability of the reflected radiation at $\theta_i, \varphi_i; \theta_r, \varphi_r$; $u^2{}_{DN_r(0°;\theta_i,\varphi_i;\lambda)}$ represents the relative uncertainty component associated with the repeatability of the reflected radiation at $0°; \theta_i, \varphi_i$; $u^2{}_{DN_r(\theta_i,\varphi_i;0°;\lambda)}$ represents the relative uncertainty component associated with the repeatability of the reflected radiation at $\theta_i, \varphi_i; 0°$; $u^2{}_{DN_r(0°;45°;\lambda)}$ represents the relative uncertainty component associated with the repeatability of the reflected radiation at 0°/45°; $u^2{}_{f(0°;45°;\lambda)}$ represents the relative uncertainty of the absolute measurement of the SD BRDF at 0°/45°; $u^2{}_{\theta}$ represents the relative uncertainty component associated with the angle.

From the figure and table in Section 5.1.6, the relative uncertainty of the radiation at a 0° incident angle with different reflection angles was basically the same as that of the radiation at 0°/45°. At a large incident angle, the relative uncertainty of the radiation with different reflection angles was also the same as that of the signal with a 0° reflection angle. That is,

$$u_{DN_r(0°;\theta_i,\varphi_i;\lambda)} = u_{DN_r(0°;45°;\lambda)}, \tag{26}$$

$$u_{DN_r(\theta_i,\varphi_i;\theta_r,\varphi_r;\lambda)} = u_{DN_r(\theta_i,\varphi_i;0°;\lambda)}. \tag{27}$$

When $u(\theta) = 0.1°$, the relative uncertainty component of the reflected radiation at a 0° incident angle or 0° reflection angles ($DN_r(0°; \theta_r, \varphi_r; \lambda)$, $DN_r(\theta_i, \varphi_i; 0°; \lambda)$) was small and could be ignored. In the extreme case of large angles of incidence and reflection, this relative uncertainty of reflected radiation ($DN_r(\theta_i, \varphi_i; \theta_r, \varphi_r; \lambda)$) was less than 0.15% with the same angular uncertainty.

According to Equation (25), the uncertainty of the improved method based on the reciprocity theorem is shown in Table 6.

Table 6. Uncertainty (k = 2) of measurement of the SD BRDF using the improved method.

Component of Uncertainty	Relative Uncertainty (k = 2) (×100%)			
	350–410 nm	410–480 nm	480–1000 nm	1000–2500 nm
$DN_r(\theta_i, \varphi_i; \theta_r, \varphi_r; \lambda)$	0.62	0.30	0.08	0.45
$DN_r(\theta_i, \varphi_i; 0; \lambda)$	0.62	0.30	0.08	0.45
$DN_r(0°; \theta_i, \varphi_i; \lambda)$	0.22	0.06	0.02	0.24
$DN_r(0°; 45°; \lambda)$	0.22	0.06	0.02	0.24
$f(0°; 45°; \lambda)$	0.36	0.29	0.29	0.37
θ_i	0.30	0.30	0.30	0.30
Total	1.04	0.60	0.43	0.86

5.3. Comparison of the Measurement Uncertainty at Large Angles between Improved Method and Absolute Measurement Method

The uncertainty of the SD BRDF measured by the two methods in Sections 2 and 4.2 for a large incident angle (75°) is shown in Figure 21. The uncertainty budget for the absolute measurement method was similar to that given in Section 5.1. It differed only in Section 5.1.5. The absolute BRDF measurement of an SD at a 75° incident angle (incident zenith angle is 75°) was

$$u_{\theta_i} = \left(\frac{\partial DN_r}{\partial \theta_i} \frac{1}{DN_r} + \tan\theta_i \right) * u(\theta_i) = 0.8\%. \tag{28}$$

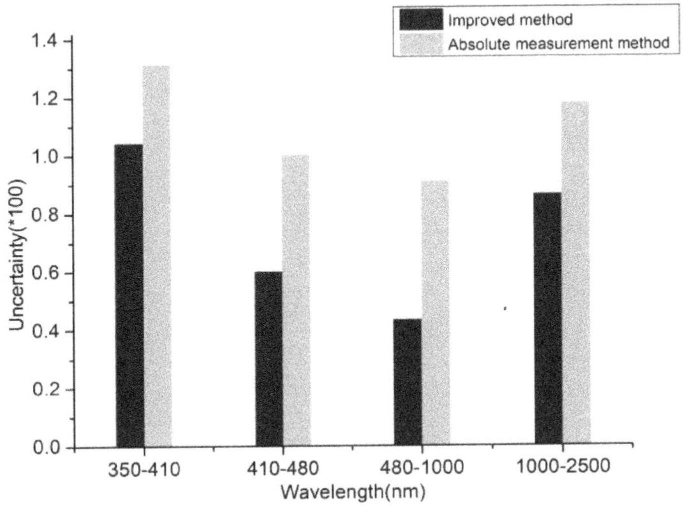

Figure 21. Uncertainty (k = 2) of the two methods at a 75° incident angle.

The uncertainty of the absolute BRDF measurement method could be obtained according to Equation (20).

In the range of 410 nm to 1000 nm, the improved method based on the reciprocity theorem reduced the measurement uncertainty by more than 50%. The measurement accuracy was significantly improved because the dependence of the measurement uncertainty on the angular accuracy of the goniometer system was reduced by the comparison measurement. However, in the 350 nm to 410 nm and 1000 nm to 2500 nm ranges, improvement in the measurement accuracy was limited. The reason for this was that the SNR of the reflected radiation in the current band was low.

6. Results

The SD BRDF at 0°/45° was measured in the range of 350 nm to 2400 nm. The results are shown in Figure 22.

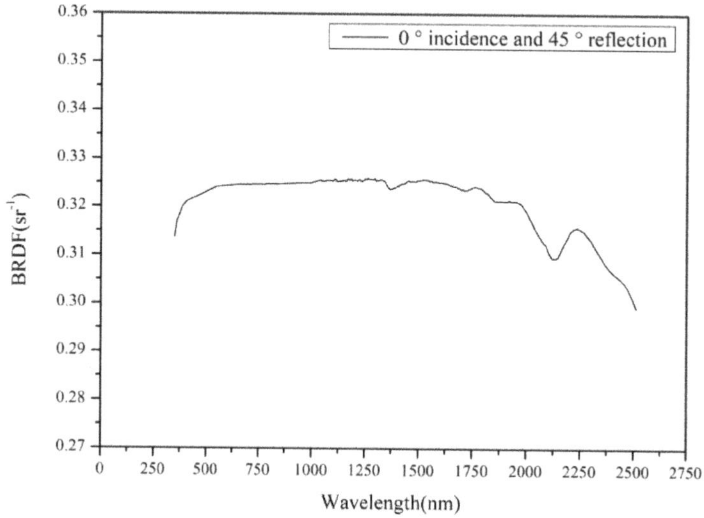

Figure 22. SD BRDF at 0°/45°.

With the angle of incidence or reflection fixed at 0°, the SD BRDF (at 650 nm) was measured, and the result is shown in Figure 23.

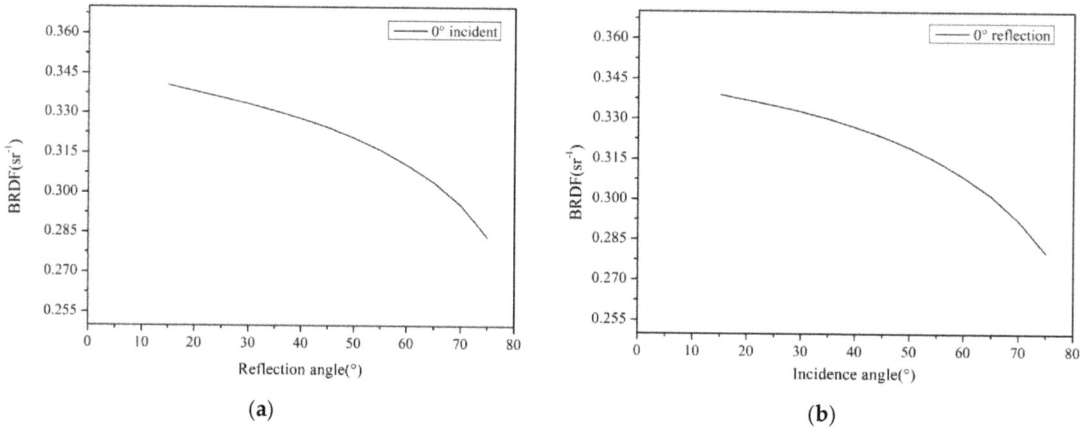

Figure 23. (**a**) SD BRDF for different reflected angles at 0° incidence. (**b**) SD BRDF for different incident angles at 0° reflection.

Some measurement results of the SD BRDF that were obtained using improved data processing based on the reciprocity theorem are shown in Figure 24. The angles of incidence and reflection come from the on-board calibration system in HY-1C.

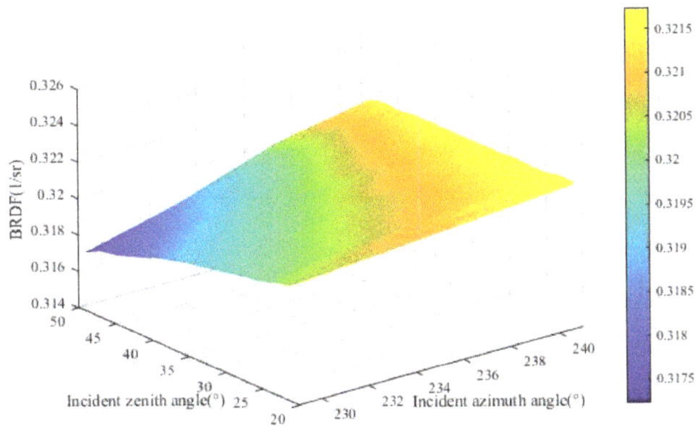

Figure 24. SD BRDF for different incident azimuth angles and incident zenith angles when the reflection zenith angle was 41.4° and azimuth angle was 117° at 565 nm.

7. Conclusions

Measurement of the SD BRDF at in- and out-of-plane and large angle geometries for on-board calibration is inevitable under the restrictions of the satellite orbit, the on-board calibration time, and other factors. An improved data processing method based on the reciprocity theorem to realize the SD BRDF at in- and out-of-plane and large angle geometries for on-board calibration in the solar reflective band was suggested. The results show that the measurement uncertainty (k = 2) was better than 1.04% at 350 nm to 410 nm, 0.60% at 410 nm to 480 nm, 0.43% at 480 nm to 1000 nm, and 0.86% at 1000 nm to 2400 nm.

Compared with the absolute measurement method, at 410 nm to 1000 nm, the measurement accuracy of the improved method was significantly improved, and the measurement uncertainty was reduced by 1/2. In the ultraviolet and near-infrared bands, improvement was limited because the improved method transferred the dependence of the uncertainty of the absolute measurement method on the angle to a dependence on the radiation detection accuracy. Similarly, this was the reason for the high uncertainty of measurement in the ultraviolet and near-infrared bands. Different lamps could be used to improve the incident radiation. We aim to explore this to improve the measurement accuracy of the reflected radiation in future work.

The literature [19] suggests that the effect of polarization on the measurement results may be negligible if the radiation source is an unpolarized source. However, some devices in the detection system, such as the grating monochromator, could be polarization-sensitive in the near infrared band, which may affect the measurement results. It is hoped that further research will be carried out in this area.

Author Contributions: Z.Z. was responsible for data analysis and writing the manuscript. H.C. led the experiment and participated in data processing. L.Z. contributed to the main research ideas. W.H. and X.Z. reviewed the manuscript and provided valuable suggestions. All authors have read and agreed to the published version of the manuscript.

Funding: Supported by the Instrument Developing Project of the Chinese Academy of Sciences, Grant No. YZ201125.

Data Availability Statement: Not applicable.

Acknowledgments: The authors thank Alfred Schirmacher for the comments and suggestions on this manuscript.

Conflicts of Interest: The authors declare no conflict of interest.

References

1. Tansock, J.; Bancroft, D.; Butler, J.; Cao, C.; Datla, R.; Hansen, S.; Helder, D.; Kacker, R.; Latvakoski, H.; Mylnczak, M.; et al. Guidelines for Radiometric Calibration of Electro-Optical Instruments for Remote Sensing. 2015. Available online: https://nvlpubs.nist.gov/nistpubs/hb/2015/NIST.HB.157.pdf (accessed on 15 May 2023).
2. Dinguirard, M.; Slater, P.N. Calibration of Space-Multispectral Imaging Sensors: A Review. *Remote Sens. Environ.* **1999**, *68*, 194–205. [CrossRef]
3. Gu, X.F.; Tian, G.L.; Yu, T.; Li, X.; Gao, H.; Xie, Y. *Principle and Method of Radiometric Calibration for Aerospace Optical Remote Sensor*; Science Press: Beijing, China, 2013.
4. Guenther, B.; Xiong, X.; Salomonson, V.; Barnes, W.; Young, J. On-orbit performance of the Earth Observing System Moderate Resolution Imaging Spectroradiometer; first year of data. *Remote Sens. Environ.* **2002**, *83*, 16–30. [CrossRef]
5. Xiong, X.; Barnes, W. An overview of MODIS radiometric calibration and characterization. *Adv. Atmos. Sci.* **2006**, *23*, 69–79. [CrossRef]
6. Xiong, X.; Sun, J.; Barnes, W.; Salomonson, V.; Esposito, J.; Erives, H.; Guenther, B. Multiyear On-Orbit Calibration and Performance of Terra MODIS Reflective Solar Bands. *IEEE Trans. Geosci. Remote Sens.* **2007**, *45*, 879–889. [CrossRef]
7. Olij, C.; Schaarsberg, J.G.; Werij, H.G.; Zoutman, E.; Baudin, G.; Chommeloux, B.; Bezy, J.-L.; Gourmelon, G. Spectralon diffuser calibration for MERIS. In Proceedings of the SPIE 3221, Sensors, Systems, and Next-Generation Satellites, London, UK, 31 December 1997; pp. 63–74. [CrossRef]
8. Sun, J.; Wang, M. VIIRS Reflective Solar Bands Calibration Progress and Its Impact on Ocean Color Products. *Remote Sens.* **2016**, *8*, 194. [CrossRef]
9. Haas, E.; Moyer, D.; De Luccia, F.; Rausch, K.; Fulbright, J. VIIRS solar diffuser bidirectional reflectance distribution function (BRDF) degradation factor operational trending and update. In Proceedings of the Earth Observing Systems XVII, San Diego, CA, USA, 15 October 2012; Volume 8510, pp. 443–450.
10. Renotte, E.; Novi, A.; Labate, D.; Melozzi, M.; Giunti, C.; Meynart, R. Solar diffuser preflight calibration set-up. In Proceedings of the Advanced and Next-Generation Satellites II, Taormina, Italy, 27 January 1997; Volume 2957, pp. 355–372.
11. Xiong, X.J.; Angal, A.; Sun, J.; Choi, T.J.; Johnson, E. On-orbit performance of MODIS solar diffuser stability monitor. *J. Appl. Remote Sens.* **2014**, *8*, 083514. [CrossRef]
12. Huang, W.X.; Zhang, L.M.; Si, X.L.; Xu, W.W.; Zou, P.; Wang, J.X.; Zhu, X.M. Design of solar diffuser degeneration monitor and its performance evaluation. *Opt. Precis. Eng.* **2016**, *24*, 732. [CrossRef]
13. Patrick, H.J.; Zarobila, C.J.; Germer, T.A. The NIST Robotic Optical Scatter Instrument (ROSI) and its application to BRDF measurements of diffuse reflectance standards for remote sensing. In Proceedings of the Earth Observing Systems XVIII, San Diego, CA, USA, 23 September 2013; Volume 8866, pp. 364–375.
14. Patrick, H.J.; Cooksey, C.C.; Germer, T.A.; Nadal, M.E.; Zarobila, C.J. Bidirectional reflectance capabilities of the NIST Robotic Optical Scattering Instrument. *Appl. Opt.* **2021**, *60*, 8774–8786. [CrossRef] [PubMed]
15. Hünerhoff, D.; Grusemann, U.; Höpe, A. New robot-based gonioreflectometer for measuring spectral diffuse reflection. *Metrologia* **2006**, *43*, S11. [CrossRef]
16. Holopainen, S.; Manoocheri, F.; Ikonen, E.; Hauer, K.-O.; Höpe, A. Comparison measurements of 0:45 radiance fac-tor and goniometrically determined diffuse reflectance. *Appl. Opt.* **2009**, *48*, 2946–2956. [CrossRef] [PubMed]
17. Bernad, B.; Ferrero, A.; Strothkämper, C.; Campos, J.; Pons, A.; Quast, T.; Hauer, K.-O.; Schirmacher, A. Deviation of white diffuse reflectance standards from perfect reflecting diffuser at visible and near-infrared spectral ranges. *Metrologia* **2019**, *56*, 055005. [CrossRef]
18. Baribeau, R.; Neil, W.S.; Côté, E. Development of a robot-based gonioreflectometer for spectral BRDF measurement. *J. Mod. Opt.* **2009**, *56*, 1497–1503. [CrossRef]
19. Rabal, A.M.; Ferrero, A.; Campos, J.; Fontecha, J.L.; Pons, A.; Rubiño, A.M.; Corróns, A. Automatic gonio-spectrophotometer for the absolute measurement of the spectral BRDF at in- out-of-plane and retroreflection geometries. *Metrologia* **2012**, *49*, 213–223. [CrossRef]
20. Zeng, J.; Butler, J.J.; Xiong, X. Comparison of BRDF results from two different scatterometers for instrument validation in support of satellite instrumentation. In Proceedings of the Earth Observing Systems XXVII, San Diego, CA, USA, 30 September 2022; Volume 12232, pp. 9–18.
21. Murgai, V.; Johnson, L.; Moskun, E.M. BRDF Characterization of Solar Diffuser for JPSS J1 using PASCAL. In Proceedings of the Earth Observing Systems XIX, San Diego, CA, USA, 2 October 2014; Volume 9218, pp. 346–355.
22. Georgiev, G.T.; Butler, J.J. The effect of incident light polarization on Spectralon BRDF measurements. In Proceedings of the Sensors, Systems, and Next-Generation Satellites VIII, Maspalomas, CI, USA, 4 November 2004; Volume 5570, pp. 492–502.
23. Guzman, C.T.; Palmer, J.M.; Slater, P.N.; Bruegge, C.J.; Miller, E.A. Requirements of a solar diffuser and measurements of some candidate materials. In Proceedings of the Calibration of Passive Remote Observing Optical and Microwave Instrumentation, Orlando, FL, USA, 1 August 1991; Volume 1493, pp. 120–131.
24. Nicodemus, F.E.; Richmond, J.C.; Hsia, J.J.; Ginsberg, I.W.; Limperis, T. *Geometrical Considerations and Nomenclature for Reflectance*; Final Report National Bureau of Standards 1977; US Department of Commerce, National Bureau of Standards: Washington, DC, USA, 1977. [CrossRef]

25. Bruker. The Flyer of VERTEX80. Available online: https://www.bruker.com/en/products-and-solutions/infrared-and-raman/ft-ir-research-spectrometers/vertex-research-ft-ir-spectrometer/vertex-80-80v-ft-ir-spectrometer/_jcr_content/root/sections/more_information/sectionpar/search.download-asset.pdf/7670882f-08ac-4fa2-9fb4-86ffa5764a69/VERTEX80_Flyer_EN.pdf (accessed on 16 May 2023).
26. Guzman, C.T.; Palmer, J.M.; Slater, P.N.; Bruegge, C.J.; Miller, E.A. Calibration method of BRDF absolute measurement device based on six axis robot. *Chin. J. Quantum Electron.* **2022**, *39*, 707–719.
27. Shin, D.-J.; Park, S.; Jeong, K.-L.; Park, S.-N.; Lee, D.-H. High-accuracy measurement of linearity of optical detectors based on flux addition of LEDs in an integrating sphere. *Metrologia* **2013**, *51*, 25–32. [CrossRef]
28. Yoshida, Y.; Kikuchi, N.; Yokota, T. On-orbit radiometric calibration of SWIR bands of TANSO-FTS onboard GOSAT. *Atmos. Meas. Tech.* **2012**, *5*, 2515–2523. [CrossRef]
29. Taylor, B.N.; Kuyatt, C.E. Guidelines for Evaluating and Expressing the Uncertainty of NIST Measurement Results. Rev. Diane Publishing. 2009. Available online: https://www.nist.gov/pml/nist-technical-note-1297 (accessed on 15 February 2023).
30. Liu, H.; Chen, H.; Si, X.; Zhang, L. Linear Calibration of Wide Dynamic Range Response of Silicon Detectors Based on Double LEDs Light Source Integrating Sphere. *Acta Opt. Sin.* **2017**, *37*, 0904001. [CrossRef]
31. Barnes, P.Y.; Early, E.A.; Parr, A.C. Spectral reflectance. *NIST Spec. Publ.* **1998**. [CrossRef]

MDPI AG
Grosspeteranlage 5
4052 Basel
Switzerland
Tel.: +41 61 683 77 34

Remote Sensing Editorial Office
E-mail: remotesensing@mdpi.com
www.mdpi.com/journal/remotesensing

www.ingramcontent.com/pod-product-compliance
Lightning Source LLC
Chambersburg PA
CBHW040814250926
51850CB00007BA/1170
9783725868605